Natural Products and Therapeutics

Medicine & Health
New York

Natural Products and Therapeutics

Medicinal Plants and COVID-19
Asita Elengoe, PhD (Editor)
2023. ISBN: 979-8-88697-958-9 (Hardcover)
2023. ISBN: 979-8-88697-987-9 (eBook)

Herbal Prescriptions for Signs and Symptoms: GMP Herbal Formulations for Treating Clinical Manifestations of Diseases
Sun-Chong Wang
2023. ISBN: 979-8-88697-835-3 (Hardcover)
2023. ISBN: 979-8-88697-901-5 (eBook)

Management of Diabetes Mellitus Based on Natural Products
Yachana Mishra, PhD
and Vijay Mishra, PhD (Editors)
2023. ISBN: 979-8-88697-853-7 (Hardcover)
2023. ISBN: 979-8-88697-970-1 (eBook)

Health Benefits of Phenolic Antioxidants
Aman B. Upaganlawar, PhD
and Vipin V. Dhote, PhD (Editors)
M. K. Mohan Maruga Raja, PhD (Editor)
2022. ISBN: 979-8-88697-058-6 (Hardcover)
2022. ISBN: 979-8-88697-120-0 (eBook)

Ginger and its Health Benefits
Isla Kermode (Editor)
2022. ISBN: 978-1-68507-695-5 (Softcover)
2022. ISBN: 978-1-68507-699-3 (eBook)

More information about this series can be found at
https://novapublishers.com/product-category/series/natural-products-and-therapeutics/

Asita Elengoe
Editor

Medicinal Plants and COVID-19

nova
Medicine & Health
New York

DOI: https://doi.org/10.52305/QQUM5776

NOTICE TO THE READER

Library of Congress Cataloging-in-Publication Data

Names: Elengoe, Asita, editor.
Title: Medicinal plants and COVID-19 / editor, Asita Elengoe, PhD, Associate Professor, Biotechnology, Lincoln University College, Selangor, Malaysia.
Description: New York : Nova Science Publishers, [2023] | Series: Natural products and therapeutics | Includes bibliographical references and index. |
Identifiers: LCCN 2023027563 (print) | LCCN 2023027564 (ebook) | ISBN 9798886979589 (hardcover) | ISBN 9798886979879 (adobe pdf)
Subjects: LCSH: Materia medica, Vegetable. | COVID-19 (Disease)--Treatment.
Classification: LCC RS164 .M3525 2023 (print) | LCC RS164 (ebook) | DDC 615.3/21--dc23/eng/20230703
LC record available at https://lccn.loc.gov/2023027563
LC ebook record available at https://lccn.loc.gov/2023027564

Published by Nova Science Publishers, Inc. † New York

Specially for my beloved mother, Thavamani
My lovely sisters, Suguna and Vaani
and
My wonderful brother, Tevanraj

Contents

Preface

In December 2019, a novel coronavirus was discovered in Wuhan city, China. People had symptoms of "pneumonia of undetermined cause," which resembled severe acute respiratory syndrome coronavirus (SARS) in 2003. This viral infection was connected to the Huanan seafood market in Southern China. Scientists took sample from patients' lower respiratory tract and sequenced its genome. They found that the virus had 70% homology with the SARS-CoV and 96.3% genetic similarity with the Yunnan bat coronavirus RaTG13. The World Health Organization (WHO) declared that a novel coronavirus found in 2019 (2019- nCoV) or SARS-CoV-2 was the source of this epidemic outbreak on January 12, 2020. The name of the disease is coronavirus disease 2019 (COVID-19). SARS-CoV-2 belongs to beta coronavirus family. It is single-stranded, positive and spherical RNA. It consists of four structural components such as spike (S), envelope (E), membrane (M) and nucleocapsid (N). It is a deadly virus. It is highly transmitted from human to human. Furthermore, there are different types of variants such as variant being monitored (VBM), variant of interest (VOI), variant of concern (VOC) and variant of high consequence (VOHC). Currently, Omicron is considered as VOC because of the reduction in vaccine effectiveness, a disproportionately high number of infections in vaccinated persons, very low vaccine-induced protection against severe disease, reduction of susceptibility to approved therapeutics, failure of diagnostic tests, the severity of the disease, a high number of hospitalization cases, etc.

Recently, medicinal plants and their phytochemicals become one of the main focuses of interest to hunt for effective, safe to humans and affordable drugs to cope with the current necessities. Indigenous traditional herbal therapy has a long history of treating a wide range of chronic and infectious diseases such as hypertension, diabetes mellitus, obesity, lung, heart and kidney-related diseases, etc. Researchers have demonstrated that medicinal plants that possess anti-viral properties such as *Andrographis paniculata*, *Lindera chunii*, *Dioscorea bulbifera*, *Wistaria floribunda*, *Xanthoceras sorbifoli* and *Aegle marmelos* showed significant results for anti-HIV activity. A variety of bioactive compounds can inhibit viral entry, viral attachment to the host receptor, viral replication, viral transcription, viral translation, etc.

This book presents 17 chapters. Most of the contributors wrote on plant-derived anti-SARS-CoV-2 viral compounds; metabolomics study of phytochemicals and their therapeutic prospects in SARS-CoV-2 and the role of bioinformatics in the development of sensitive, specific and cost-effective antiviral agents from medicinal plants. These novel approaches can be a feasible strategy that can combat COVID-19 and prevent its comorbidities. Mechanisms of the phytochemicals against SARS-CoV-2 are explained through a variety of new pathways. It is vital to design a plant-based drug for COVID-19 patients. To date, there is no specific treatment for COVID-19. We have also attempted to revisit the pathogenesis of coronavirus through the host, clinical features and diagnostic approach on COVID-19, counteracting SARS-CoV-2 and managing COVID-19 impacts using natural products and medicinal plants and traditional plant-based therapies for SARS-CoV-2 infection.

Acknowledgments

Foremost, I would like to express my sincere gratitude to GOD for His blessings because I could not finish my book on time without Him.

This book acknowledges wholeheartedly and gratefully all my colleagues for their valuable contributions. I cannot resist my temptation to applaud the enthusiasm of these experts which catalyzes my academic passion and catapults me beyond my limitations.

I highly appreciate our contributors for accepting my invitation and sharing their knowledge and research. I produced an excellent quality book with a good teamwork effort. I would like to thank Nova Science Publishers for their kind co-operation at every stage of the book production.

My sincere gratitude also goes to my dear friends who supported me during the completion of this book. Last but not the least, I would like to thank my mother, Thavamani who supported me spiritually throughout my life.

Chapter 1

Introduction: Medicinal Plants and COVID-19

Asita Elengoe[*], PhD

Department of Biotechnology, Faculty of Applied Science, Lincoln University College, Selangor, Malaysia

Abstract

Early in 2020, a serious health problem emerged. An outbreak of the novel SARS-CoV-2 (severe acute respiratory syndrome coronavirus 2) epidemic took place in Wuhan, China, in December 2019. It quickly spread around the world. On February 11th, 2020, the World Health Organization (WHO) designated the disease as coronavirus disease (COVID-19). There are four classes of SARS-CoV-2 variants such as variant being monitored (VBM), variant of interest (VOI), variant of concern (VOC) and variant of high consequence (VOHC). However, the Omicron variant is a variant of concern at present because it can develop a more severe disease; reduce the neutralization of antibodies; reduce the effectiveness of treatments or vaccines, etc. Thus, scientists are required to find a new or modified vaccine or treatment for the novel variants of SARS-CoV-2. Phytoconstituents from medicinal plants have been proven to treat a variety of viral infections by inhibiting the pathogenesis of the virus (e.g., viral entry, attachment, replication, etc.). In this chapter, we will discuss the timeline of the COVID-19 pandemic development, SARS-CoV-2 and its variants, and medicinal plants and their biological activities.

Keywords: COVID-19, medicinal plants, SARS-CoV-2, variants

---

[*] Corresponding Author's Email: asitaelengoe@yahoo.com.

In: Medicinal Plants and COVID-19
Editor: Asita Elengoe
ISBN: 979-8-88697-958-9

Introduction

Coronaviruses are a vast family of zoonotic viruses that cause a wide range of illnesses, from severe respiratory infections to the common cold. They can be spread from animals to humans (Elengoe and Selvam, 2022; Elengoe et al., 2022). Certain coronaviruses are found in a population of animals. However, they are not harmful to humans because they are non-infectious diseases. SARS-CoV-2 is the most recent virus that infects humans. It is highly pathogenic to humans. SARS-CoV-2 infection is spread from person to person through droplets developed by infected people's respiratory systems, most commonly while coughing or sneezing. It takes between 2 to 14 days for exposure to onset of symptoms, according to current reports, with a five-day average. Two other recent outbreaks of coronavirus were reported. Middle East Respiratory Syndrome (MERS-CoV) was found to be transmitted to humans from dromedary camels in 2012 (Shereen et al., 2019). The transmission from civet cats to humans was found to be severe acute respiratory syndrome (SARS-CoV) in 2002 (Liu et al., 2020).

In 1965, the history of human coronaviruses started when Tyrrell and Bynoel discovered that a virus known as B814 can pass through them (Kahn and McIntosh, 2005). In human embryonic tracheal organ cultures collected from the respiratory system, it was discovered. The virus developed a common cold symptom. Later that decade, researchers discovered a collection of related human and animal viruses. They named them based on their crown-like appearance. They found that 7 different types of coronaviruses can infect humans. SARS was first discovered in 2002 in southern China. It rapidly spread to 28 other countries. More than 8,000 persons have been infected by July 2003, with 774 of them died. In 2004, there were just four more occurrences in a brief outbreak. The coronavirus causes fever, headache, as well as respiratory problems including cough and breathing shortness. MERS began in 2012 in Saudi Arabia (Rossi et al., 2020). Approximately all of the nearly 2,500 cases had resulted in individuals living in or traveling to the Middle East. This coronavirus is less infectious, but more lethal than its SARS-CoV because it caused the death of 858 individuals. It has the same respiratory symptoms as the flu, but it can also lead to kidney failure (Rossi et al., 2020).

SARS-CoV-2 infection, which would be the fifth pandemic after the 1918 influenza epidemic, has currently infected people all across the world. For the time being, we can follow the first warning and subsequent breakout of a cluster of unique instances of human pneumonia in Wuhan, China, which began in late December 2019 (Elengoe, 2020). 1st December 2019 was the earliest date of symptom initiation. The patients had symptoms such as fever, malaise, dry cough, and dyspnea. These symptoms were identified as viral pneumonia. Initially, due to the area and pneumonia symptoms, the disease was called Wuhan pneumonia by the press. Results of whole-genome sequencing indicated that a novel coronavirus is a causal factor (Singhal, 2020). As a result, this virus is the 7th member of the coronavirus family to infect humans (Kumar et al., 2020). The virus can cause pneumonia, respiratory failure, respiratory difficulties, liver problems, septic shock, and death (Kumar et al., 2020).

The World Health Organization (WHO) temporarily termed the new 2019 coronavirus virus novel (2019-nCoV) on 12th January 2020, then on 12th February 2020, it was formally named coronavirus disease 2019 (COVID-19) (Guo et al., 2020; Elengoe, 2020). Since COVID-19 first appeared in China three years ago, it has grown and expanded rapidly to other countries around the world, posing a global threat. Following the 1918 Spanish flu (H1N1), 1957 Asian flu (H2N2), 1968 Hong Kong flu (H3N2), and 2009 Pandemic flu (H1N1), which killed an estimated 50 million, 1.5 million, 1 million, and 300,000 people, the WHO eventually determined that COVID-19 was a pandemic (Liu et al., 2020). COVID-19 killed 6,730,560 million people (Worldometer, 2022).

Timeline of COVID-19 Pandemic Development

On 31 December 2019, Wuhan Municipal Health Commission notified the World Health Organization about the pneumonia cases in Wuhan City, Hubei Province, China, with an unidentified cause. A novel coronavirus was then eventually discovered. The unknown illness was initially termed as 2019-nCoV and then was referred to as COVID-19. On 1st January 2020, the WHO set up an IMST (Incident Management Support Team) across the three levels of the organization: headquarters, regional headquarters and country level, placing the organization on the immediate ground to deal with the outbreaks. Authorities shut down the Huanan seafood market, which was speculated to be the origin of the mysterious illness. This is because from the first 27 recorded patients, most cases were epidemiologically related to the Huanan Seafood Wholesale Market, a wet market located in the city of Wuhan, which sells not only seafood and live animals, including poultry and wildlife, where some of the patients infected with the pneumonia-like disease were the market dealers and vendors. According to a retrospective report, the first reported case dates back to 8th December 2019 (Jenny and Sara, 2020; World Health Organization (WHO), 2020).

At the beginning of January 2020, China announced a total of 44 alleged mystery disease patients. WHO posted on social media that there was a cluster of cases of pneumonia -with no deaths- in Wuhan, Hubei province on 4th January 2020 (Jenny and Sara, 2020; World Health Organization (WHO), 2020). The isolated virus from bronchoalveolar lavage fluid samples from patients with extreme pneumonia, Chinese researchers have detected beta coronavirus as the pathogenic factor of this new infectious illness. The novel coronavirus is then reported by China as the source of the epidemic. Coronaviruses are a broad group of viruses that can develop different degrees of illness, from mild colds to serious lethal diseases. Typically present in animals, some can infect humans and spread among humans. Both SARS and MERS are attributable to coronavirus (Hu et al., 2020; World Health Organization (WHO), 2020).

China records the first death linked to the new coronavirus on 9th January 2020. A 61-year-old man was admitted to Wuhan hospital with many serious health issues. Other than respiratory failure and extreme pneumonia, the man also had abdominal tumors and chronic liver disease (Jenny and Sara, 2020). The first genetic sequence of the novel coronavirus was released on the virological website on 10th January and on 12th January 2020. China shares the genetic sequence of the novel coronavirus publicly, allowing countries to test and track any potentially infected individuals (Hu et al., 2020).

On 13 January 2020, authorities announced a case of COVID-19 in Thailand, the first case reported outside of China. The patient is a 61-year-old female from Wuhan with no record of visiting the Huanan Seafood Market. On 16th January 2020, Japan announced another imported case of COVID-19. The Chinese national, a male in his 30s, also did not report a connection to the Huanan seafood market. The first case of 2019-nCoV was confirmed by South Korea on 20th January 2020. The patient did not register a visit to the markets in Wuhan (Jenny and Sara, 2020).

Subsequently, many patients with no history of Huanan Seafood Wholesale Market exposure were found. Several familial clusters of infection were reported. Infections also occurred within healthcare facilities. All of these events presented ample proof of human-to-human transmission of the new virus (Hu et al., 2020).

Since this epidemic clashed with the lunar New Year, movement between towns and villages before the festival promoted the spread of viruses in China.

This new coronavirus pneumonia eventually came to other areas in the region of Hubei as well as other parts of China. Within 1 month, the virus had dispersed to all 34 provinces of China. The number of reported cases unexpectedly increased, with thousands of new cases confirmed daily at the end of 15th January 2020 (Hu et al., 2020; World Health Organization (WHO), 2020).

On 25th January 2020, 2019-nCoV entered more nations. Australia announced the first case of 2019-nCoV on the continent, with three more cases identified by health officials later in the day. France reported a new case, the first in Europe. Malaysia also recorded the first four cases, while Canada announced the first case. On 30th January 2020, the WHO announced the outbreak of coronavirus a global health crisis of international concern. On 11th February 2020, the International Committee on Virus Taxonomy called the novel coronavirus as 'SARS-CoV-2' and the WHO named the disease 'COVID-19.' The outbreak of COVID-19 in China peaked in February. According to the National Health Commission of China, the cumulative number of cases kept growing sharply at the beginning of February at an average rate of more than 3,000 newly reported cases each day (Hu et al., 2020; Jenny and Sara, 2020). China has introduced extraordinarily stringent public health initiatives to regulate COVID-19. The city of Wuhan was closed on 23rd January 2020, and all travel and transportation connections to the city were stopped. In the next few weeks, all outdoor events and meetings were prohibited, and public facilities were closed both in most towns and in the countryside. As a result of these steps, the regular number of cases in China has begun to decline gradually. However, amid the gradual decrease in China, the worldwide transmission of COVID-19 has escalated since late February (Hu et al., 2020).

At the beginning of March 2020, the number of COVID-19 cases exceeded 100,000 and more than 100 countries registered COVID-19 cases (Jenny and Sara, 2020). The higher frequency transmission of SARS-CoV-2 and the accessibility of international flights made it possible for COVID-19 to spread rapidly throughout the world. On 11th March 2020, WHO Director-General Tedros Adhanom Ghebreyesus called the worldwide outbreak of COVID-19 a pandemic. Since March, while COVID-19 has been effectively regulated in China, the number of cases in Europe, the USA, India and other regions has jumped dramatically (WHO,2020). As of 16th January 2023, 671,375,204 positive cases; 6,730,560 death cases and 642,681,552 recovered cases have been reported throughout the world (Worldometer, 2022).

SARS-CoV-2

The SARS-CoV-2 is a beta coronavirus, which has a genome of roughly 30 kb and is large, spherical, encapsulated, positive-sense, single-stranded RNA (Richman et al., 2016). Spike glycoprotein (S), membrane (M), envelope (E), and nucleocapsid (N) proteins make up its four primary structural components (Fehr and Perlman, 2015). SARS-CoV-2 uses its spike to block the activity of neutralizing antibodies because neutralizing antibodies are playing an important role in blocking viral particles from binding with the host cell for infection to occur. Spike protein contains S1 and S2 domains. The S1 domain interacts with the host cell receptor known as Angiotensin Converting Enzyme 2 (ACE-2) to trigger a conformational shift in the S protein. The virus penetrates the host cell by mediating membrane fusion with the host cell membrane through the S2 domain.

Since the start of the COVID-19 pandemic, genetic lineages of SARS-CoV-2 have been emerging and spreading globally. There are four classes of SARS-CoV-2 variants such as variants being monitored (VBM), variant of interest (VOI), variant of concern (VOC) and variant of high consequence (VOHC). Alpha (B.1.1.7 and Q lineages), Beta (B.1.351 and descendent lineages), Gamma (P.1 and descendent lineages), Delta (B.1.617.2 and AY lineages), Epsilon (B.1.427 and B.1.429), Eta (B.1.525), Iota (B.1.526), Kappa (B.1.617.1), 1.617.3, Mu (B.1.621, B.1.621.1) and Zeta (P.2) are the variants being monitored (Table 1). According to the World Health Organization (WHO) data, there are no variants of interest as of 12th October 2022. However, Omicron (B.1.1.529, BA.1, BA.1.1, BA.2, BA.3, BA.4 and BA.5 lineages) are the variants of concern (Table 2). Currently, there are no variants of high consequence have been identified in the United States of America. Omicron was identified in South Africa. Its first outbreak occurred in November 2021. It has the potential to increase transmissibility; cause more severe disease (e.g., increased hospitalizations, morbidity and mortality rates); reduce neutralization by some emergency use authorization (EUA) monoclonal antibody treatments; and reduction in neutralization by post-vaccination sera. Moreover, it has the potential to reduce the effectiveness of treatments or vaccines, or diagnostic detection failures. Therefore, there is an urgency to develop a new or modified vaccine or effective treatments for new variants of SARS-CoV-2.

Table 1. List of SARS-CoV-2 variants being monitored

WHO label	PANGO lineage	Nextstrain clade	First outbreak	Earliest sample	Designated VOC	Designated VOI	Designated VBM	Current circulation	Mutations
Alpha	B.1.1.7 and Q lineages	20I (V1)	United Kingdom	20th Sept 2020	December 29, 2020	-	September 21, 2021	No	69-70del, N501Y, P681H
Beta	B.1.351 and descendent lineages	20H (V2)	South Africa	May 2020	December 29, 2020	-	September 21, 2021	No	K417N, E484K, N501Y
Gamma	P. 1 and descendent lineages	20J (V3)	Brazil	Nov 2020	December 29, 2020	-	September 21, 2021	No	K417T, E484K, N501Y
Delta	B.1.617.2 and AY lineages	21A	India	Oct 2020	June 15, 2021	-	April 14, 2022	No	L452R, T478K, P681R
Epsilon	B.1.427 B.1.429	21C	United States	July 2020	March 19, 2021	February 26, 2021 June 29, 2021	September 21, 2021	No	I4205V, D1183Y, S13I, W152C, L452R
Eta	B.1.525	21D	United Kingdom	December 2020	-	February 26, 2021	September 21, 2021	No	E484K, ΔH69/ΔV70 deletion, N439K, Y453F
Iota	B.1.526	21F	United States	November 2020	-	February 26, 2021	September 21, 2021	No	E484K, S477N, L5F, T95I, D253G, E484K, D614G and A701V

Table 1. (Continued)

WHO label	PANGO lineage	Nextstrain clade	First outbreak	Earliest sample	Designated VOC	Designated VOI	Designated VBM	Current circulation	Mutations
Kappa	B.1.617.1	21B	India	December 2020	-	May 7, 2021	September 21, 2021	No	L452R, E484Q, P681R
N/A	B.1.617.3	19B	India	February 2021	-	May 7, 2021	September 21, 2021	No	L452R, E484Q, D614G, P681R
Zeta	P.2	21K	Brazil	January 2021	-	February 26, 2021	September 21, 2021	No	E484K, D614G
Mu	B.1.621 B.1.621.1	21H	Colombia	January 2021	-	-	VBM: September 21, 2021	No	R346K, E484K, N501Y, D614G, P681H

Table 2. List of SARS-CoV-2 variants concern lineages under monitoring (WHO) (as of 13th January 2023)

WHO Label	Pango lineage	Nextstrain clade	First outbreak	Relation to circulating VOCs	Earliest sample	Mutations
Omicron	BA.2.75	22D	India	BA.2 sublineage	31.12.2021	BA.2.75: BA.2 + S:K147E, S:W152R, S:F157L, S:I210V, S:G257S, S:D339H, S:G446S, S:N460K, S:Q493R reversion BA.2.75.2: BA.2.75 + S:R346T, S:F486S, S:D1199N
Omicron	BF.7	22B	China	BA.5 sublineage	24.01.2022	BA.5 + S:R346T
Omicron	BQ.1	22E	United Kingdom	BA.5 sublineage	07.02.2022	BQ.1 and BQ.1.1: BA.5 + S:R346T, S:K444T, S:N460K
Omicron	XBB	22F	United States	Recombinant of BA.2.10.1 and BA.2.75 sublineages, i.e., BJ1 and BM.1.1.1, with a breakpoint in S1	13.08.2022	BA.2+ S:V83A, S:Y144-, S:H146Q, S:Q183E, S:V213E, S:G252V, S:G339H, S:R346T, S:L368I, S:V445P, S:G446S, S:N460K, S:F486S, S:F490S XBB.1.5: XBB + S:F486P

Table 3. List of medicinal plants and their mechanism

No.	Plant species	Bioactive compound	Virus	Mechanism	Reference
1.	*Andrographis paniculata*	Andrographolide	SARS-CoV-2	– Inhibits viral replication	Shi et al., 2020
2.	*Justicia adhatoda* L.	Anisotine	SARS-CoV-2, Human influenza virus	– Blocks RBD of the SARS-CoV-2 spike protein and prevents host attachment through ACE2 receptor	Ghosh et al., 2021
3.	*Allium sativum* L.	Ajoene, allicin, alliin, allyl methyl thiosulfinate, methyl allyl thiosulfinat, allitridin, diallyl sulfide, garlicin,lectin	Human influenza virus, SARS-CoV-2	– Inhibits viral replication	Rouf et al., 2020
4.	*Mangifera indica*	Mangiferin	Human influenza virus	– Inhibits viral adsorption	Al-Rawi et al., 2019
5.	*Aloe vera* (L.) Burm.f.	Feralolide, 9-dihydroxyl-2-O-(z)-cinnamoyl-7-methoxy-aloesin, aloeresin, quercetin, catechin hydrate, and kaempferol	SARS-CoV-2	– Inhibits viral replication	Mpiana et al., 2020
6.	*Bombax ceiba* L.	Kaempferol-3-O-(6″-O-Ep-coumaroyl)-β-D-glucopyranoside	SARS-CoV-2	– Inhibits the ORF 3a protein of SARS-CoV-2	Zhang et al., 2015
7.	*Cyperus rotundus* L.	Humulene epoxide, and caryophyllene oxide	SARS-CoV-2	– Inhibits the activity of spike glycoprotein, papain-like protease (PLpro), 3-chymotrypsin-like protease (3CLpro) and RNA-dependent RNA polymerase (RdRp)	Amparo et al., 2021
8.	*Azadirachta indica* A. Juss.	Gedunin, pongamol, and azadirachtin	SARS-CoV-2	– Inhibits viral replication	Nesari et al., 2021
9.	*Piper nigrum* L.	Guaiol	SARS-CoV-2	– Inhibits viral replication	Pandey et al., 2021

Table 3. (Continued)

No.	Plant species	Bioactive compound	Virus	Mechanism	Reference
10.	*Aegle marmelos* (L.)	Seselin	SARS-CoV-2	– Blocks the activity of main protease (Mpro)	Nivetha et al., 2021
11.	*Citrus sinensis* (L.)	Hesperidin, luteolin, and vitamin C	SARS-CoV-2	– Inhibits spike protein and Mpro activity	Bellavite and Donzelli, 2020
12.	*Zingiber officinale*	6-gingerol, and gingeronone A	SARS-CoV-2	– Inhibits SARS CoV-2 by interacting viral proteases, RNA binding protein, and spike protein – Inhibits main protease and ORF8 of the SARS-CoV-2	Pandey et al., 2021
13.	*Curcuma longa* L.	Curcumin	SARS-CoV-2	– Interacts directly with viral membrane proteins, disrupts viral envelope, inhibits viral protease and induces host antiviral response by boosting immunity	Thimmulappa et al., 2021
14.	*Salvia rosmarinus*	Carnosic acid, carnosol or rosmanol, genkwanin, cirsimaritin or homoplantaginin, ursolic acid	SARS-CoV-2	– Blocks SARS-CoV-2 infection and reduce inflammation	Satoh et al., 2020
15.	*Ocimum tenuiflorum*	Eugenol	SARS-CoV-2	– Inhibits the interaction between spike S1 and ACE2 – Reduces SARS-CoV-2 spike S1-induced activation of NF-κB and the expression of IL-6, IL-1β and TNFα	Paidi et al., 2021

Medicinal Plants

Since ancient times, people use traditional herbs to treat or cure a variety of chronic diseases such as hypertension, diabetes mellitus, lung, kidney and heart-related diseases, etc. Indians use turmeric in their cooking as a spice. They also use neem leaves for chickenpox treatment. Phytoconstituents are a class of promising antiviral agents. They have been intensively to treat different types of viruses such as influenza virus, chicken pox virus, hepatitis B virus, human papillomavirus, rhinovirus, dengue virus type-2, herpes simplex virus-1, poliovirus, adenovirus, Epstein-Barr virus, respiratory syncytial virus, hepatitis C virus, human immunodeficiency virus (HIV), etc. Furthermore, many research studies have been demonstrated that traditional medicine gives better and more effective results than commercial drugs for viral infection. It is cost-effective, has low or no side effects and has a productive therapeutic effect. Ethanopharmacology helps to find novel biologically active compounds from plant sources. Bioactive compounds from medicinal plants have been proven to be effective against viral infections by inhibiting various steps in the pathogenesis of the virus such as viral entry, viral attachment to the host receptor, viral replication, endocytosis, transcription of the viral genome, protein translation, cleavage of the polypeptide, protein synthesis, etc. The uses of medicinal plants in viral infection were collected from the previous research findings (Table 3).

Conclusion

SARS-CoV-2 is a highly pathogenic virus. It can spread rapidly from one person to another person. There are different types of variants. Currently, World Health Organization (WHO) declared Omicron variants as a variant of concern (VOC). In conclusion, natural compounds from medicinal plants may be used as potential drugs for COVID-19 treatment. They can increase the effectiveness of the treatment against Omicron variants. The discovery of novel bioactive substances aids in developing new, safe and effective drugs.

References

Al Rawi AAS, Al Dulaimi, HSH, Al Rawi MAA. Antiviral activity of mangifera extract on influenza virus cultivated in different cell cultures. *Journal of Pure and Applied Microbiology* (2019) 13:455-458.

Amparo TR, Seibert JB, Silveira BM, Costa FSF, Almeida TC, Braga SFP. Brazilian essential oils as source for the discovery of new anti-COVID-19 drug: a review guided by *in silico* study. *Phytochemistry Reviews* (2021):1-20.

Bellavite P, Donzelli A. Hesperidin and SARS-CoV-2: New light on the healthy function of citrus fruits. *Antioxidants (Basel)* (2020) 9:742.

Elengoe A. COVID-19 outbreak in Malaysia. *Osong Public Health and Research Perspectives* (2020) 11:93.

Elengoe A. Indian spices boost the immune system against COVID-19. *The Annals of the University Dunarea de Jos of Galati. Fascicle VI-Food Technology* (2020) 44(2):189-206.

Elengoe A, Selvam S. Overview on origin, transmission, treatment and clinical researches of SARS-CoV-2. *Journal of Health Sciences & Surveillance System* (2022) 10:2-12.

Elengoe A, Selvam S, Selvarajah M, Manikam H, Vijaya R. Sequence analysis and structure prediction of Malaysia SARS-CoV-2 strain's structural and accessory proteins. *Biointerface Research in Applied Chemistry* (2022) 12:3259-3304.

Fehr AR, Perlman S. Coronaviruses: An overview of their replication and pathogenesis. In *Coronaviruses*. New York (NY): Humana Press, 2015. pp.1-23.

Ghosh R, Chakraborty A, Biswas A, Chowdhuri, S. Identification of alkaloids from *Justicia Adhatoda* as potent SARS CoV-2 main protease inhibitors: An *in silico* perspective. *Journal of Molecular Structure* (2020) 1229:129489.

Guo YR, Cao QD, Hong ZS, Tan YY, Chen SD, Jin HJ, Yan Y. The origin, transmission and clinical therapies on coronavirus disease 2019 (COVID-19) outbreak–an update on the status. *Military Medical Research* (2020) 7:1-10.

Hu B, Guo H, Zhou P. Characteristics of SARS-CoV-2 and COVID-19. *Nature Reviews Microbiology* (2020).

Jenny LR, Sara J. (2020). *COVID-19 - a timeline of the coronavirus outbreak.* Accessed 12 December 2022. https://www.devex.com/news/covid-19-a-timeline-of-the-corona virus-outbreak-96396.

Kahn JS, McIntosh K. History and recent advances in coronavirus discovery. *The Pediatric Infectious Disease Journal* (2005) 24:S223-S227.

Kumar S, Nyodu R, Maurya VK, Saxena SK. Morphology, genome organization, replication, and pathogenesis of severe acute respiratory syndrome coronavirus 2 (SARS-CoV-2). In *coronavirus disease 2019* (COVID-19), Springer, Singapore, 2020, pp. 23-31.

Liu YC, Kuo RL, Shih SR. COVID-19: The first documented coronavirus pandemic in history. *Biomedical Journal* (2020) 43:328-333.

Mpiana PT, Ngbolua KT, Tshibangu DST, Kilembe JT, Gbolo BZ, Mwanangombo DT. Identification of potential inhibitors of SARS-CoV-2 main protease from *aloe vera* compounds: A molecular docking study. *Chemical Physics Letters* (2020) 754:137751.

Nesari TM, Bhardwaj A, Shrikrishna R, Ruknuddin G, Ghildiyal S, Das A. Neem (*Azadirachta Indica A. Juss*) Capsules for prophylaxis of COVID-19 Infection: A pilot, double-blind, randomized controlled trial. *Alternative Therapies In Health And Medicine* (2021) 23:1-8.

Nivetha R, Bhuvaragavan S, Janarthanan, S. Inhibition of multiple SARS-CoV-2 proteins by an antiviral biomolecule, seselin from aeglemarmelos deciphered using molecular docking analysis. *Research Square* (2021).

Paidi RK, Jana M, Raha S, McKay M, Sheinin M, Mishra RK, Pahan K. Eugenol, a component of holy basil (tulsi) and common spice clove, inhibits the interaction between SARS-CoV-2 spike S1 and ACE2 to induce therapeutic responses. *Journal of Neuroimmune Pharmacology* (2021) 16(4):743-755.

Pandey P, Singhal D, Khan F, Arif M. An *in silico* screening on *Piper nigrum*, *Syzygium aromaticum* and *Zingiber officinale* roscoe derived compounds against SARS-COV-2: A drug repurposing approach. *Biointerface Research in Applied Chemistry* (2021) 11.

Richman DD, Whitley RJ, Hayden FG. *Clinical Virology*, 4th ed. Washington (WA): ASM Press, 2016.

Rossi GA, Sacco O, Mancino E, Cristiani L, Midulla F. Differences and similarities between SARS-CoV and SARS-CoV-2: spike receptor-binding domain recognition and host cell infection with support of cellular serine proteases. *Infection* (2020) 48:665-669.

Rouf R, Uddin SJ, Sarker DK, Islam MT, Ali ES, Shilpi JA. Antiviral potential of garlic (*allium sativum*) and its organosulfur compounds: A systematic update of pre-clinical and clinical data. *Trends Food Science Technology* (2020) 104:219-234.

Satoh T, Trudler D, Oh CK, Lipton SA. Potential therapeutic use of the rosemary diterpene carnosic acid for Alzheimer's disease, Parkinson's disease, and long-COVID through NRF2 activation to counteract the NLRP3 inflammasome. *Antioxidants (Basel)* (2022) 11(1):124.

Shereen MA, Khan S, Kazmi A, Bashir N, Siddique R. COVID-19 infection: Origin, transmission, and characteristics of human coronaviruses. *Journal of Advanced Research* (2020) 24:91.

Shi TH, Huang YL, Chen CC, Pi WC, Hsu YL, Lo LC, Chen WY, Fu SL, Lin CH. Andrographolide and its fluorescent derivative inhibit the main proteases of 2019-nCoV and SARS-CoV through covalent linkage. *Biochemical and Biophysical Research Communications* (2020) 533:467-473.

Singhal T. A review of coronavirus disease-2019 (COVID-19). *The Indian Journal of Pediatrics* (2020) 87:281-286.

Thimmulappa RK, Mudnakudu-Nagaraju KK, Shivamallu C, Subramaniam KJT, Radhakrishnan A, Bhojraj S. Antiviral and immunomodulatory activity of curcumin: A case for prophylactic therapy for COVID-19. *Heliyon* (2021) 7:e06350.

World Health Organization (WHO). (2020). *Timeline -COVID-19.* Accessed 27 April 2020. https://www.who.int/news/item/27-04-2020-who-timeline---covid-19.

Worldometer. (2022). Accessed 15 January 2023. https://www.worldometers.info/corona virus/.

Zhang YB, Wu P, Zhang XL, Xia C, Li GQ, Ye WC. Phenolic compounds from the flowers of *Bombax malabaricum* and their antioxidant and antiviral activities. *Molecules* (2015) 20:19947-19957.

Chapter 2

The Pathogenesis of Coronavirus through the Host

Hemapriyaa Vijayan*, MSc
Rathimalar Ayakannu, PhD
Erlina Abdullah, PhD
and Chun Hoe Tan, PhD
Department of Biotechnology, Faculty of Applied Science,
Lincoln University College, Selangor, Malaysia

Abstract

The presence of the COVID-19 pandemic had brought a major impact on the healthcare and socioeconomic sectors. The emergence of SARS-CoV-2 had caused a high mortality rate across the globe. The virus was first identified in China in the year 2020 and further spread from the country across the world. Human-to-human transmission via respiratory droplets was reported as the major route of transmission. Transmission via fomite and airborne transmission in closed and poorly ventilated settings was also confirmed. Bats were identified as the reservoir for SARS-CoV-2 as bats host many viruses without presenting the clinical symptoms thus SARS-CoV-2 has been indicated as a zoonotic origin. The clinical spectrum of the disease varies from mild, moderate to severe. There are a few steps that involve the entry of SARS-CoV-2 into the human host: attachment, penetration, biosynthesis, maturation and release. Most asymptomatic patients (people without symptoms) can transmit the infection to people who are in close contact or proximity. Some patients have a mild influenza-like illness that cannot be

* Corresponding Author's Email: hemapriyaa@lincoln.edu.my.

In: Medicinal Plants and COVID-19
Editor: Asita Elengoe
ISBN: 979-8-88697-958-9

differentiated from a simple upper respiratory tract infection. Hospitalization is required for moderate and severe cases. Studies have shown that mainly elderly and immunocompromised patients are the individuals who progress to the severe stages of the disease. Lessons from the understanding of the nature of the virus and its pathogenesis would help in new approaches and improving human health.

Keywords: pathogenesis, reservoir, SARS-CoV-2, transmission

Introduction

In December 2020, SARS-CoV-2 emerged when a worker from the market which sells live animals and exotic animals in the Hubei province of China presented with fever, dizziness, and cough. After approximately 4 weeks, there were nearly 10000 cases reported across countries and within a short period, the infection had succumbed across the globe (Yuki et al., 2020). Globally, the pandemic has claimed the lives of more than 6.8 million, according to John Hopkins University. It caused substantial morbidity among patients who survived their illnesses. Researchers worked substantially and identified the causative agent of coronavirus disease-2019 (COVID-19) to be severe acute respiratory syndrome-associated coronavirus-2 (SARS-CoV-2) (Li et al., 2020). The virus is primarily transmitted between people through respiratory droplets and contact routes. The disease continues to spread at an alarming rate worldwide. It was declared as a pandemic by the World Health Organization (WHO) (Zheng et al., 2020).

SARS-CoV-2 has been reported to be more infectious compared to SARS-CoV as the reproductive number is higher and this facilitates the rapid spread (Table 1). SARS-CoV-2 has several features which make it more infectious. SARS-CoV-2 binds strongly to the ACE2 receptors due to its differences in the surface protein and this facilitates greater efficiency in the host cell invasion (Wrapp et al., 2020). In addition to that, SARS-CoV-2 also has a greater affinity for the mucosa of the conjunctiva and upper respiratory tract thus leads to a higher infectious rate of SARS-CoV-2 (Wolfel et al., 2020; Hui et al., 2020).

Table 1. The recent outbreak of respiratory viruses and their severity

Types of viruses	Severity of disease caused by the virus	References
SARS-CoV-2	Highly contagious with severe respiratory signs	Meyerowitz et al., 2021 CDC, 2020
MERS	Severe respiratory signs, High mortality rate	Yang et., 2015 Aly et al., 2017
SARS	Contagious and severe respiratory signs	Woodhead et al., 2003

Primary Reservoir and Host of Coronavirus

The origination and transmission of the coronavirus (SARS-CoV-2) are essential to be determined as a preventive strategy to contain the virus. Since the emergence of SARS-CoV-2, potential animal species have been thoroughly studied to identify the possible reservoirs, genomic evolution of the virus and the intermediate hosts (Ji et al., 2020) (Bassetti et al., 2020). Bats are known to harbor a large diversity of coronavirus as there is the presence of anti-SARS-CoV-2 antibodies found in the samples isolated from the *Rhinolopus* sp. which further confirms the viral replication and as a primary reservoir. In the current scenario, researchers have been focusing on raccoon dogs and palm civets (Xiao et al., 2008) as the major reservoir of the infection; however, only the sample obtained from China food market showed positive for coronavirus and this suggests that the palm civet as the secondary host. Snakes were also suggested as a possible host however after genomic similarity findings of a novel coronavirus, bat coronavirus demonstrated approximately 88% similarities, this supports that bats could be the key reservoir of coronavirus (Markotter et al., 2020). Several genomes with sequences similarities have been reported between bat's coronavirus and SARS-CoV and this shows a high genetic similarity between both viruses (Hu et al., 2017).

Researchers suggest that in between the bats and humans, there is an intermediate host which contributes to the role of production of recombinant viruses. They have been extensively studying samples from bats and have concluded that bats can be a potential source of zoonotic diseases due to their ability to harbor many viruses without presenting any clinical symptoms for several months (Misra, 2020). Samples obtained from bats that are infected with CoV can propagate the virus without immunological interference compared to other infected mammals and this makes them an ideal reservoir for viruses (Zhou et al., 2020). Viral outbreaks from several spillovers from bats reservoir have been suggested and this has been highlighted in the SARS-CoV emergence (Roes, 2020). There are a few factors that contribute to the spillover of viruses, particularly the coronavirus from bats such as wildlife trade and exotic animal marketing, disruption of the natural ecosystem and increased urbanization pressure from humans (Latinne et al., 2020).

Mode of Transmission and Dynamics

The transmission of coronavirus has been indicated as a zoonotic origin as the virus was believed to be transmitted from live animals to humans. Many factors have been reported to contribute to the rate of infection and transmission such as host infectiousness, environmental factors and contact patterns. The transmission of SARS-CoV-2 occurs via the infected respiratory droplets. This can be transmitted through direct or indirect contact with nasal and oral mucosa. The receptors which are the site for binding for SARS-CoV-2 are located mainly in the upper respiratory tract which facilitates the entry of the virus. Transmission via conjunctiva and gastrointestinal mode has also been proven as the mode of transmission of SARS-CoV-2 (Cevik et al., 2021). The most prevalent transmission occurs through close contact which is approximately 15 minutes of exposure in a close range within 2 meters and spreads rapidly in large gatherings. Besides that, getting in contact and sleeping in the same room, dining in close proximity, and food sharing with the infected individual increases the risk of infection. Transmission via aerosol can occur if a person is exposed in a prolonged timeframe and poorly ventilated indoor settings (European Centre for Disease Control and Prevention, 2020). Most super-spreading incidents mainly happened in indoor settings with no proper social distancing.

The spread of an infectious agent through the dissemination of droplet nuclei (aerosol) which is viable and remains suspended in air over a long period is known as airborne transmission. SARS-CoV-2 can be transmitted through this mode during medical procedures which generate aerosols such as manual ventilation, intubation and sputum induction. Airborne transmission has been demonstrated to be possible in poorly ventilated areas (Klompas et al., 2020).

SARS-CoV-2 thrive in low temperature and humid conditions and remains viable for an extended period on surfaces such as wood, glass, plastic and stainless steel and further transfers the infection from the contaminated surfaces to eyes, nasal and mouth mucosa when the person comes in contact with the surfaces. In many communal areas, these modes of transmission have contributed to the majority of cases and it is difficult to control without proper social distancing.

Mechanism of SARS-CoV-2 Invasion into Host Cell

SARS-CoV-2 is transmitted via aerosol and respiratory droplets from person to person (Cascella et al., 2020). The cycle of the virus involves a few steps: attachment, penetration, biosynthesis, maturation and release. Coronavirus is approximately 80 to 120 nm in size and is enveloped with a pleiomorphic appearance. The coronavirus virion has a positive-strand RNA with a large RNA genome and it is known as one of the largest RNA viruses (Lu et al., 2020). The presence of nucleocapsid (N), in the form of the viral helical capsid, is found within the viral membrane. The membrane of coronavirus has been confirmed for the presence of three viral proteins, in which the spike (S) and the type 1 glycoprotein forms peplomers on the surface of the virion which gives rise to the crown-like appearance or known as the corona. The membrane protein (M) has a short N-terminal ectodomain and a cytoplasmic tail and a small membrane protein (E) which is highly hydrophobic. The S (spike protein) which is responsible for attachment and penetration has a structure that protrudes and it can be found abundantly on the surface of the virion (Bosch et al., 2003). This protein has two subunits (S1 and S2) which the S1 is responsible for the binding to host cell receptors and S2 for the fusion of viral and host cellular membrane. The presence of Angiotensin-converting enzyme 2 (ACE2) functional receptors, has been identified and it is highly expressed in the pulmonary epithelial cells through this receptor, the S protein binds to the host cell and starts the invasion. ACE2 receptors are also highly expressed in the bladder, lung, heart, ileum and kidney (Chen et al., 2020; Zou et al., 2020).

After the successful binding of the virus SARS-CoV-2 and ACE2, this process triggers a conformational change in the spike protein and activation occurs via two-step protease cleavage: Firstly, for priming at the S1/S2 cleavage site and secondly for the cleavage for activation at a position adjacent to a fusion peptide within the S2 subunit (Walls et al., 2020). This facilitates the entry of the virus into the pulmonary alveolar epithelial cell and the viral contents are released once the entry is successful. Upon entry of the virus into the host cell, the virus undergoes replication and the formation of a negative RNA strand by the positive single-stranded RNA through the transcription process facilitated by RNA polymerase (Wan et al., 2020). The formation of negative RNA strands will serve as a template for the production of new positive RNAs which will then lead to the synthesis of new proteins in the cell cytoplasm by the translation process (Figure 1). The coronavirus spike is very unique among other viruses as a range of proteases can cleave and activate it. The existence of furin cleavage site (RPPA sequence) at the S1 and S2 site is a unique characteristic present in SARS-CoV-2 which makes this virus the ability of rapid infectivity and transmissibility potential (Ou et al., 2020).

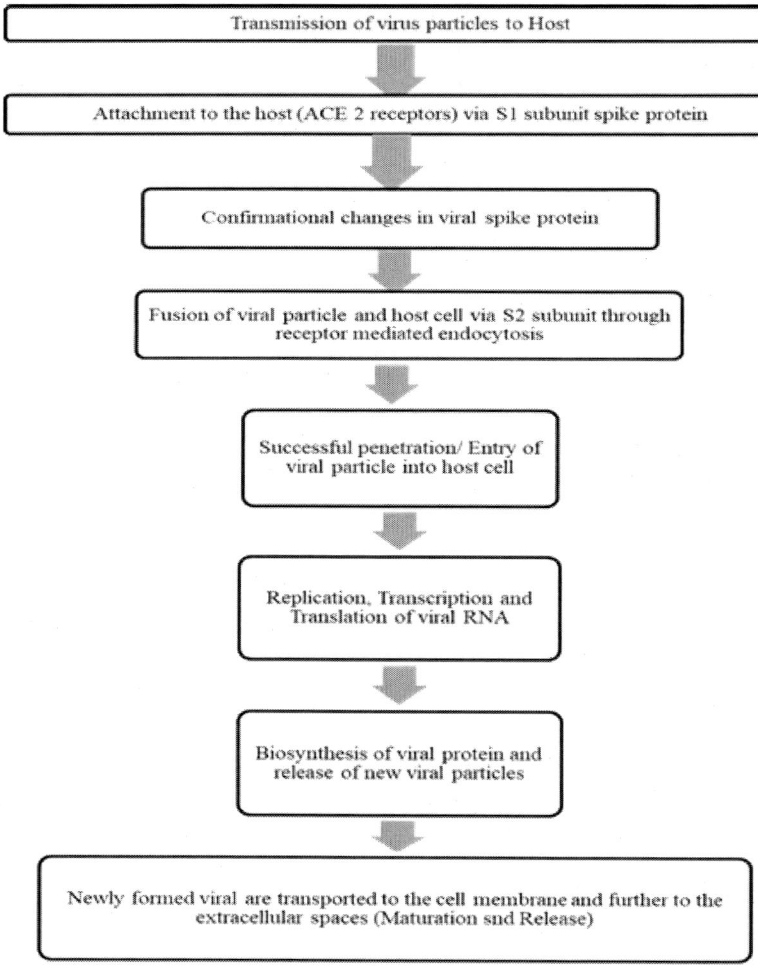

Figure 1. Severe Acute Respiratory Syndrome (SARS-CoV-2) life cycle.

Host and Disease Spectrum

Symptoms that are often seen in patients affected with SARS CoV-2 range from mild to severe with respiratory collapse and multiple damages of organs. The formation of pulmonary ground-glass opacity can be observed via CT scan even for asymptomatic patients. The virus can destroy the lung alveolar epithelial cells due to the high expression of ACE 2 (Zou et al., 2020). Dendritic cells, alveolar macrophages and epithelial cells are the major components of innate immunity in the airways. The dendritic cells are present underneath the epithelium and the macrophage reside at the apical side of the epithelium. This component (dendritic cells and macrophage) serves as innate immunity until adaptive immunity starts its defense mechanism (Guan et al., 2020).

Upon entry of the virus, the T cells are directed to the site of infection. The active replication and release of the virus in the lung cell leads to symptoms such as fever, myalgia, headache and respiratory symptoms (Burke et al., 2020). In patients who develop severe disease, SARS-CoV-2 exhibits trigger the immune response of the host actively. The lung tissue of SARS-CoV-2 patients' histological evidence has indicated the inflammatory nature of the injury with bilateral diffuse alveolar damage, hyaline-membrane formation, interstitial mononuclear inflammatory infiltrates, and desquamation consistent with acute respiratory distress syndrome (ARDS) (Xu et al., 2020). These characteristics were also shown in severe MERS and SARS infection but a very distinctive feature of SARS-CoV-2 infection is the formation of fibrinous exudate and mucus plugs in the respiratory tract, which leads to the collapse of the respiratory system and this justifies the severity of the infection (Iba et al., 2020). Lung parenchymal damage happens due to the excessive production of pro-inflammatory cytokines which accumulate and cause congestion. Multiple organ dysfunction followed by septic shock which further leads to death is also observed in patients with severe infection. Highly sensitive troponin and natriuretic peptides confirm the involvement of the cardiovascular system in COVID-19. Moreover, the presence of intra-alveolar hemorrhage and platelet-fibrin thrombin in small and minute arteries is seen. The accumulation of cytokine in the organs of some patients may cause extensive tissue damage or cytokine storm resulting in capillary leak, thrombus formation and organ dysfunction (Gu et al., 2020).

Phases of COVID-19 Infection

Asymptomatic Phase

This is a phase where the SARS-CoV-2 enters the nasal passage and binds to the epithelial cells and starts the replication process. ACE2 has been highly expressed in epithelial cells. The virus undergoes local replication and propagation and it's limited to innate immunity (Wan et al., 2020). The virus load is low and can be detected at this stage via nasal swabs. At this particular stage, the individuals are infectious and those who come in contact with the person can be infected. This phase lasts from 1 to 2 days (Wu et al., 2020). Some infected individuals have presented with very mild symptoms but mostly remain asymptomatic.

Involvement and Invasion of the Upper Respiratory Tract

The virus triggers a very aggressive innate immunity that propagates and migrates along the respiratory tract (Sims et al., 2005). At this stage, the involvement of the upper airways results in the manifestation of the disease as the appearance of early symptoms such as dry cough, malaise and fever (Lu et al., 2020). During this phase, the immune response is aberrant with the release of CXC motif chemokine ligand (CXCL-10) and interferons mainly the alpha and lambda interferons (IFN-λ and IFN-β) which are from the virus-infected cell. CXCL serves as a useful marker for SARS (Rockx et al., 2009). Usually, the virus will be contained and almost 80% of patients will be subjected to treatment at this phase and does not progress further.

Progression to Acute Respiratory Distress Syndrome (ARDS)

This is a very crucial phase where approximately 15-20% of the patients who are infected would progress to the development of pulmonary infiltrates. At this phase, the virus reaches the gaseous exchange units of the lungs and the alveolar type II cells are infected. SARS- CoV-2 usually affects alveolar type II cells compared to type 1 cells. Once SARS-CoV-2 starts propagating within the type II cells, a large number of virus particles are released and eventually, the cells undergo apoptosis and die (Yee et al., 2017). There are accumulating toxins and viral particles that infect type II cells in the adjacent units. A cytokine storm is produced by the release of many cytokines and inflammatory markers such as (IL120, IL12, IL8, IL6 and IL1), tumor necrosis factor, Monocyte chemoattractant protein (MCP-1), Macrophage inflammatory protein (MIP1-α) and CXCL 10 (Xu et al., 2020). The formation of cytokine storm causes the attraction of helper T cell and Cytotoxic T cell which is responsible to fight the virus and at the same time causes lung cell inflammation and injury (Tang et al., 2005; Yoshikawa et al., 2009).

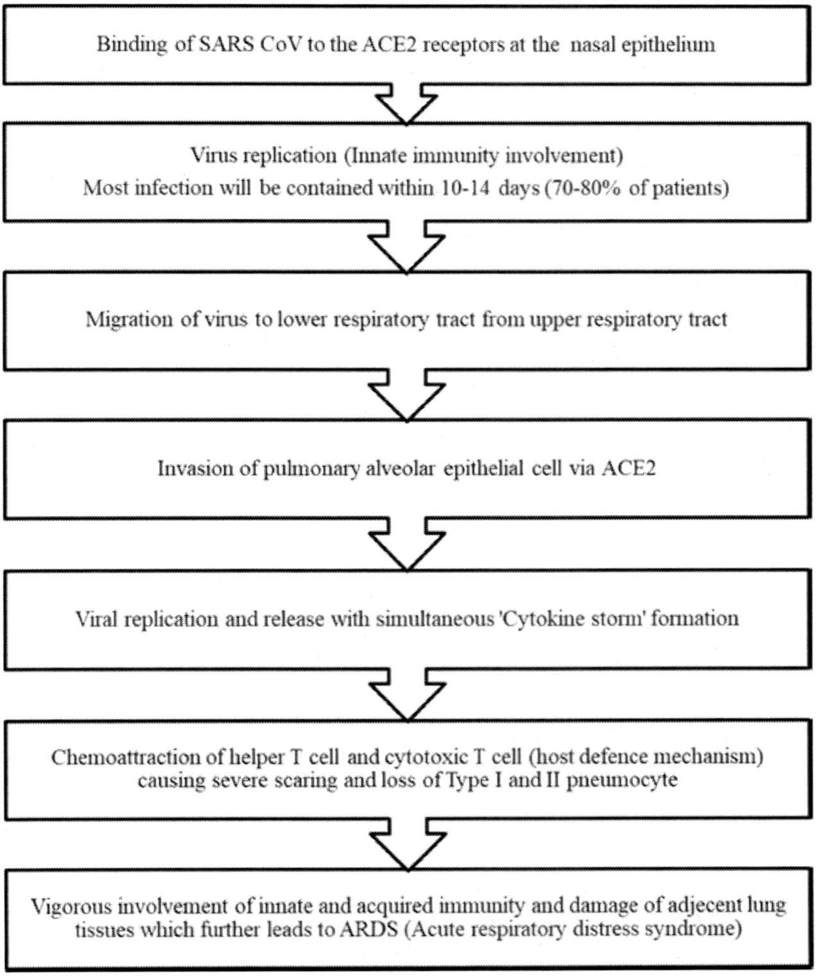

Figure 2. Pathophysiology of SARS-CoV-2.

Diffuse alveolar damage with a hyaline membrane rich in fibrin and the presence of a few multinucleated giant cells are found in the pathological result of patients with COVID-19. This is due to the inflammatory markers and cytokines which cause further inflammation by fighting the virus and the loss of Type I and Type II pneumocyte (Qian et al., 2013). The wound-healing process would lead to very severe scarring and fibrosis (Koutsogiannaki et al., 2019). This would cause a vigorous involvement of innate and acquired and immunity and epithelial regeneration. The highest risk group is the elderly patients due to the immunosuppression and the ability to repair damaged epithelium is impaired and they also have reduced mucociliary clearance (Saule et al., 2006). This leads to the extensive spread of the virus to the gaseous exchange units of the lungs. Thus, it causes acute respiratory distress syndrome (Figure 2).

Conclusion

The COVID-19 pandemic has affected people throughout the world. An understanding of pathogenesis and the etiological aspects of the virus would be very beneficial in the therapeutic formulation. An in-depth understanding of the nature of the virus, its reservoirs and the transmission dynamics are important for the early prediction, prevention and eradication of the virus.

References

Aly M, Elrobh M, Alzayer M, Aljuhani S, Balkhy H. Occurrence of the Middle East Respiratory Syndrome Coronavirus (MERS-CoV) across the Gulf Corporation Council countries: Four years update. *PLoS One* (2017) 12(10):e0183850.

Bassetti M, Vena A, Giacobbe D R. The novel Chinese coronavirus, (2019-nCoV) infections: Challenges for fighting the storm. *Eur. J. Clin. Invest.* (2020) 50:e13209.

Bosch B J, Van der Zee R, De Haan C A, Rottier P J. The coronavirus spike protein is a class I virus fusion protein: structural and functional characterization of the fusion core complex, *J. Virol.* (2003) 77:8801-8811.

Cascella M, Rajnik M, Aleem A, Dulebohn S C, Di Napoli R. *Features, evaluation and treatment coronavirus (COVID-19)*. Stat. Pearls Publishing (2020).

CDC (2020) *Novel Coronavirus (2019-nCoV) Situation Summary*.

Cevik M, Marcus J L, Buckee C, Smith T C. Severe Acute Respiratory Syndrome Coronavirus 2 (SARS-CoV-2) transmission dynamics should inform policy. *Clin. Infect. Dis.* (2021) 73:S170-S176.

ChenY, Guo Y, Pan Y, Zhao Z J. Structure analysis of the receptor binding of 2019-nCoV. *Biochem. Biophys. Res. Commun.* (2020) 525(1):135-140.

European Centre for Disease Prevention and Control. *Surveillance definitions for COVID-19.* (2020). Accessed 15 December 2022. https://www.ecdc.europa.eu/en/covid-19/surveillance/surveillance-definitions.

Gu J, Korteweg C. Pathology and pathogenesis of severe acute respiratory syndrome. *Am. J. Pathol.* (2007) 170:1136-1147.

Guan W J. Clinical characteristics of coronavirus disease 2019 in China. *N. Engl. J. Med.* (2020) 382:1708-1720.

He Y, Wang J, Li F, Shi Y. Main clinical features of COVID-19 and potential prognostic and therapeutic value of the microbiota in SARS-CoV-2 Infections. *Front. Microbiol.* (2020) 11:1302

Hu B. Discovery of a rich gene pool of bat SARS-related coronaviruses provides new insights into the origin of SARS coronavirus. *PLoS Pathog* (2017) 13:e1006698.

Hui K P, Cheung M C, Perera R A, Ng K C, Bui C H, Ho J C, Ng M M, Kuok D I, Shih K C, Tsao S W, Poon L L. Tropism, replication competence, and innate immune responses of the coronavirus SARS-CoV-2 in human respiratory tract and conjunctiva: an analysis in *ex-vivo* and *in-vitro* cultures. *Lancet Respir. Med.* (2020) 8:687-695.

Iba T, Levy J H, Levi M, Thachil J. Coagulopathy in COVID-19. *J. Thromb. Haemost.* (2020) 18:2103-2109.

Ji W, Wang W, Zhao X, Zai J, Li X. Cross-species transmission of the newly identifed coronavirus 2019-nCoV. *J. Med. Virol.* (2020); 92:433-440.

Klompas M, Baker M A, Rhee C. Airborne transmission of SARS-CoV-2: Theoretical considerations and available evidence. *JAMA* (2020) 324(5):441-442.

Koutsogiannaki S, Shimaoka M, Yuki K. The use of volatile anesthetics as sedatives for acute respiratory distress syndrome. *Transl. Perioper. Pain Med.* (2019) 6(2):27-38.

Latinne A, Hu B, Olival K J, Zhu G, Zhang L, Li H, Chmura A A, Field H E, Zambrana-Torrelio C, Epstein J H, Li B. Origin and cross-species transmission of bat coronaviruses in China. *Nat. Commun.* (2020) (11):4235.

Li Q, Guan X, Wu P, Wang X, Zhou L, Tong Y, Ren R, Leung K S, Lau E H, Wong J Y, Xing X. Early transmission dynamics in Wuhan, China, of novel coronavirus-infected pneumonia. *N. Engl. J. Med.* (2020) 382:1199-1207.

Lu R, Zhao X, Li J, Niu P, Yang B, Wu H, Wang W, Song H, Huang B, Zhu N, Bi Y. Genomic characterisation and epidemiology of 2019 novel coronavirus: implications for virus origins and receptor binding, *Lancet* (2020) 395:565-574.

Markotter W, Coertse J, de Vries L, Geldenhuys M, Mortlock M. Bat-borne viruses in Africa: A critical review. *J. Zool.* (2020) 311:77-98.

Meyerowitz E A, Richterman A, Bogoch I I, Low N, Cevik M. Towards an accurate and systematic characterisation of persistently asymptomatic infection with SARS-CoV-2. *Lancet Infect. Dis.* (2021) 6:e163-e169.

Misra V. Bats and viruses. *Lancet Infect. Dis.* (2020) 20(12):1380.

Ou X, Liu Y, Lei X, Li P, Mi D, Ren L, Guo L, Guo R, Chen T, Hu J, Xiang Z. Characterization of spike glycoprotein of SARS-CoV-2 on virus entry and its immune cross-reactivity with SARS-CoV. *Nat. Commun.* (2020) 11:1620.

Qian Z, Travanty E A, Oko L, Edeen K, Berglund A, Wang J, Ito Y, Holmes K V, Mason R J. Innate immune response of human alveolar type II cells infected with severe acute respiratory syndrome-coronavirus. *Am. J. Respir. Cell Mol. Biol.* (2013) 48:742-748.

Rockx B, Baas T, Zornetzer G A, Haagmans B, Sheahan T, Frieman M, Dyer M D, Teal T H, Proll S, Van Den Brand J, Baric R. Early upregulation of acute respiratory distress syndrome-associated cytokines promotes lethal disease in an aged-mouse model of severe acute respiratory syndrome coronavirus infection. *J. Virol.* (2009) 83:7062-7074.

Roes F L. On the evolution of virulent zoonotic viruses in bats. *Biol. Theory* (2020) 15(4):223-225.

Saule P, Trauet J, Dutriez V, Lekeux V, Dessaint J P, Labalette M. Accumulation of memory T cells from childhood to old age: central and effector memory cells in CD4($^+$) versus effector memory and terminally differentiated memory cells in CD8($^+$) compartment, *Mech. Ageing Dev.* (2006) 127(3):274-281.

Sims A C, Baric R S, Yount B, Burkett S E, Collins P L, Pickles R J. Severe acute respiratory syndrome coronavirus infection of human ciliated airway epithelia: role of ciliated cells in viral spread in the conducting airways of the lungs. *J. Virol.* (2005) 79:15511–15524.

Tang N L, Chan P K, Wong C K, To K F, Wu A K, Sung Y M, Hui D S, Sung J J, Lam C W. Early enhanced expression of interferon-inducible protein-10 (CXCL-10) and other chemokines predicts adverse outcome in severe acute respiratory syndrome. *Clin. Chem.* (2005) 51:2333-2340.

Walls A C, Park Y J, Tortorici M A, Wall A, McGuire A T, Veesler D. Structure, function, and antigenicity of the SARS-CoV-2 spike glycoprotein, *Cell* (2020) 181(2):281-292.

Wan Y, Shang J, Graham R, Baric R S, Li F. Receptor recognition by novel coronavirus from Wuhan: An analysis based on decade-long structural studies of SARS. *J. Virol.* (2020) 94: e00127-20.

Wolfel R, Corman V M, Guggemos W, Seilmaier M, Zange S, Müller M A, Niemeyer D, Jones T C, Vollmar P, Rothe C, Hoelscher M. Virological assessment of hospitalized patients with COVID-2019. *Nature* (2020) 581:465-469.

Woodhead M, Ewig S, Torres A. Severe acute respiratory syndrome (SARS). *Eur. Respir. J.* (2003) 21:739-740.

Wrapp D, Wang N, Corbett K S, Goldsmith J A, Hsieh C L, Abiona O, Graham BS, McLellan JS. Cryo-EM structure of the 2019-nCoV spike in the prefusion conformation. *Science* (2020) 367:1260-1263.

Wu Z, McGoogan J M. Characteristics of and important lessons from the coronavirus disease 2019 (COVID-19) outbreak in China: summary of a report of 72314 cases from the Chinese Center for Disease Control and Prevention. *JAMA* (2020).

Xiao Y. Pathological changes in masked palm civets experimentally infected by severe acute respiratory syndrome (SARS) coronavirus. *J. Comp. Pathol.* (2008) 138:171-179.

Xu Z, Shi L, Wang Y, Zhang J, Huang L, Zhang C, Liu S, Zhao P, Liu H, Zhu L, Tai Y. Pathological findings of COVID-19 associated with acute respiratory distress syndrome. *Lancet Respir. Med.* (2020) 8:420-422.

Yang Y, Liu C, Du L, Jiang S, Shi Z, Baric R S, Li F. Two mutations were critical for bat-to-human transmission of Middle East respiratory syndrome coronavirus. *J. Virol.* (2015) 89(17):9119-9123.

Yee M, Domm W, Gelein R, Bentley K L, Kottmann R M, Sime P J, Lawrence B P, O'Reilly M A. Alternative progenitor lineages regenerate the adult lung depleted of alveolar epithelial type 2 cells. *Am. J. Respir. Cell Mol. Biol.* (2017) 56:453-464.

Yoshikawa T, Hill T, Li K, Peters C J, Tseng C T. Severe acute respiratory syndrome (SARS) coronavirus-induced lung epithelial cytokines exacerbate SARS pathogenesis by modulating intrinsic functions of monocyte-derived macrophages and dendritic cells, *J. Virol.* (2009) 83(7):3039-3048.

Yuki K, Fujiogi M, Koutsogiannaki S. COVID-19 pathophysiology: a review. *Clin. Immunol.* (2020) 215:108427.

Zheng M, Gao Y, Wang G, Song G, Liu S, SunD, Xu Y, Tian Z. Functional exhaustion of antiviral lymphocytes in COVID-19 patients, *Cell Mol. Immunol.* (2020) 17(5):533-535.

Zhou, P. A pneumonia outbreak associated with a new coronavirus of probable bat origin. *Nature* (2020) 579:270–273.

Zou X, Chen K, Zou J, Han P, Hao J, Han Z. Single-cell RNA-seq data analysis on the receptor ACE2 expression reveals the potential risk of different human organs vulnerable to 2019-nCoV infection. *Front. Med.* (2020) 14(2):185-192.

Chapter 3

Clinical Symptoms and Diagnoses of COVID-19

Rathimalar Ayakannu*, PhD
Hemapriyaa Vijayan, MSc
Chun Hoe Tan, PhD
and Erlina Abdullah, PhD
Department of Biotechnology, Faculty of Applied Science,
Lincoln University College, Selangor, Malaysia

Abstract

The most common clinical manifestations of COVID-19 are fever, cough, flu-like symptoms, loss of smell and taste, dyspnea, and myalgia or fatigue. Lesser common, clinical manifestations included nausea or vomiting, dizziness, diarrhea, rashes and chills.

Many cases of COVID-19 have moderate or nonspecific symptoms, and some patients have severe respiratory problems, including shortness of breath and decreased vocal cord trill on palpation. Laboratory findings and radiological findings are needed for the clinical diagnosis of COVID-19. Based on recent evidence, we summarize the current clinical features and diagnostic approaches to manage the disease. As information from COVID-19 is evolving rapidly, this review will help the researcher to consider the possible clinical symptoms and potential diagnostic approaches based on up-to-date knowledge. Furthermore, this chapter summarizes the advantages and disadvantages of various diagnostic methods currently used for COVID-19. The accurate diagnosis and treatment of co-infection poses a significant challenge due to the

* Corresponding Author's Email: rathimalar@lincoln.edu.my.

In: Medicinal Plants and COVID-19
Editor: Asita Elengoe
ISBN: 979-8-88697-958-9

overlapping clinical and laboratory parameters. Therefore, confirmatory diagnostic tests are necessary for accurate and timely diagnosis and patient management.

Keywords: clinical features, COVID-19, diagnostic approach

Introduction

The coronavirus disease 2019 (COVID-19) is an enduring global pandemic of coronavirus illness impelled by severe acute respiratory syndrome coronavirus 2 (SARS-CoV-2) (Cui et al., 2019). Infected individuals are highly transmittable and can spread the sickness even if they are asymptomatic (Cascella et al., 2022).

Older people and those with underlying medical conditions such as cancer, diabetes, cardiovascular disease, or chronic respiratory disease are more likely to develop acute illnesses (Dhama et al., 2020). As demonstrated by the origins of the COVID-19 pandemic, children infected with severe acute respiratory syndrome coronavirus 2 (SARS-CoV-2) are mostly asymptomatic or have mild symptoms (Nikolopoulou and Maltezou, 2022). However, children may experience any of the same symptoms as adults (Pierce et al., 2022).

The average incubation period for SARS-CoV-2 is estimated to be 5.1 days, and the majority of patients will develop symptoms within 11.5 days of infection (Lauer et al., 2020). Clinical features are critical for accurate disease detection (Israfil et al., 2020). Symptoms of COVID-19 are variable, ranging from mild signs to severe infections (Grant et al., 2020). Common symptoms include fever, myalgia, fatigue, headache, loss of smell and taste, nasal congestion and runny nose, nausea, shortness of breath, cough, muscle pain, sore throat, diarrhea, rashes and dyspnoea (Larsen et al., 2020).

The key prevention includes screening for infection to promote its early detection, with prompt treatment of those found to be infected, even if they are asymptomatic (Vetter et al., 2020). Diagnostic tools are major players, not only to identify possible infections but also to further limit the spread and control of the extent of disease development (Kumar et al., 2020). Studies of clinical findings, epidemiological history, and assessments are crucial for the clinical diagnosis of COVID-19 (Ratre et al., 2021). This chapter aims to systematically summarize the clinical characteristics of COVID-19 patients and categorize their diagnostic methods to help doctors manage COVID-19 patients better.

Clinical Features of COVID-19

Asymptomatic carriers, if not detected and isolated early, can increase disease transmission to unmanageable levels (Yu et al., 2020). Hence, fast and precise screening of possible virus carriers along with the critical observation of individuals without any symptomatic signs is very much critical for monitoring the spread of COVID-19 (Cascella et al., 2022). Patients with typical clinical signs suspicious of COVID-19 such as fever, cough, sore throat, loss of taste or smell, malaise, and myalgia should be promptly tested for SARS-CoV-2 (Cascella et al., 2022).

Frequent Symptoms

Fever

A body temperature of 100.4°F (38°C) or higher is considered a fever (Ogoina, 2011). According to patient data collected by Larsen et.al, fever is often one of the early symptoms shown after a SARS-CoV-2 infection (Larsen et al., 2020). Persistent fever beyond 7 days after the onset of infection identifies patients who might be at risk for adverse outcomes from COVID-19 (Ng et al., 2020). Although not everyone who has been exposed to coronavirus will develop a fever, most people affected by COVID-19 will have a fever or a higher than usual temperature (Stokes et al., 2020).

Cough

Coughing is one of the most common indicators of COVID-19 (Esmaeilzadeh et al., 2022). Based on a cross-sectional study of Arab COVID-19 patients, the most recorded symptom was found to be cough (Mahmoud et al., 2023). Coronaviruses have been shown to cause respiratory infections that lead to prolonged coughing and wheezing (Esmaeilzadeh et al., 2022).

Myalgia and Fatigue

Myalgia is a rheumatic disorder characterized by diffuse soft tissue pain and tender points, sleep disruption, and fatigue (Bockman and Eckerson, 2007). Myalgia is a common indicator in patients with viral infections such as the 2019 novel coronavirus disease and influenza (Kucuk et al., 2020). COVID-19 can cause musculoskeletal pain and causes infection in the pulmonary system (Cure and Cumhur Cure, 2020). Myalgia and fatigue may last longer in patients with COVID-19 than in other viral infections and may not respond to conventional pain medications (Kucuk et al., 2020).

Loss of Smell and Taste

A large number of people infected with SARS-CoV-2 developed new loss of smell or taste (Printza et al., 2021). Anosmia is sudden onset, usually profound, and occurs during the course of the disease (Iravani et al., 2020). A quarter of patients develop anosmia before other COVID-19 symptoms (Printza et al., 2021). Odor assessment not only aids in the diagnosis of COVID-19 infection during pre-test screening, but also in direct post-test triage of patients at all levels (Yan et al., 2020). A large number of the patients (88%) recovered their sense of smell by 61 days (Printza et al., 2021). Anosmia is emerging as an important clinical factor in the diagnosis and likely outcome of COVID-19 (Yan et al., 2020).

Nasal Congestion and Runny Nose (Flu)

Influenza is another common infection that affects the respiratory tract (Arshad Ali et al., 2020). In general, high-risk groups are more possible to get severe cases of the flu, such as people 65 and older, people with underlying health conditions, and those who are immuno-compromised (Ghebrehewet et al., 2016). Distinguishing early COVID-19 from influenza based on symptoms alone is nearly impossible due to the prevalence of influenza-like illness (ILI) in COVID-19 patients (Liu et al., 2020). When a new infection arises, it is critical to quickly detect its hallmark signs and symptoms to help differentiate it from other diseases illnesses (Alemi et al., 2022).

Nausea

Of note, some patients reported nausea/vomiting as an early clinical symptom of COVID-19, which is often ignored (Zhang et al., 2021). Today it is revealed that not only the lungs but also the gastrointestinal tract (GI) can be infected by SARS-CoV-2 (Andrews et al., 2021) (Zhang et al., 2021). In general, more attention must be paid to perceptions of nausea as an early symptom of COVID-19 and its impact on the gastrointestinal tract of SARS-CoV-2 (Andrews et al., 2021).

Shortness of Breath / Dyspnoea

Abnormal respiration is often characterized by the feeling of suffocation or being short of breath (Feneley et al., 2017). Shortness of breath can make it hard to breathe, which can leave patients gasping for air (Jamil et al., 2021). SARS-CoV-2 can cause shortness of breath as a result of lung infection and damage (Jamil et al., 2021). Shortness of breath, also known as dyspnea, is a common and worrying symptom in people with COVID-19 (Hentsch et al., 2021). A study conducted in Northeast India, showed shortness of breath was the common complaint among COVID-19 patients (Jamil et al., 2021).

Sore Throat

Sore throat is also a common symptom of COVID-19 virus infections (Singhal, 2020). A sore throat can be caused by a viral or bacterial infection, and inflammation is what causes a sore throat, due to localized damage to the lining of the throat (Wise, 2022). Sore throat is one of the top five symptoms reported by COVID -19 patients, according to a study conducted in the UK (Iacobucci, 2022).

Diarrhea

Diarrhea is a common early symptom in a significant number of patients with SARS-CoV-2 infection (Megyeri et al., 2021). A study in Turkey shows diarrhea is more common among older people infected with the virus (Comoglu et al., 2021). This virus can infect and multiply in esophageal cells and intestinal cells, causing direct damage to intestinal epithelial cells (Megyeri et al., 2021). Le et al. conclude that SARS-CoV-2 invades the gut and may be the first symptom of COVID-19 (Li et al., 2020). Infection reduces the levels of the angiotensin-converting enzyme 2 receptor, thus altering the composition of the gut microbiota (Megyeri et al., 2021).

Rashes

Most prevalently, COVID-19 is related to respiratory and typical viral manifestations (Larsen et al., 2020). However, non-respiratory manifestations such as rashes are becoming more common (Gooch, 2021). Warning signs associated with COVID-19 rashes include skin discoloration, swelling and itching (Genovese et al., 2021). Reported dermatologic changes with COVID-19 include maculopapular and urticaria, chilblains and livedo reticularis, and less commonly, vesicular and ecchymotic rashes (Daneshgaran et al., 2020; Gooch, 2021).

Diagnostic Approaches for COVID-19

Patient screening for accurate diagnosis must be simple, affordable, fast and with the most consistent results (Ratre et al., 2021). Coronavirus exposure methods are largely based on the travel history of the person from the exaggerated areas along with the examination of their clinical signs as well as some supporting examinations (Kumar et al., 2020). Studies of epidemiological history, clinical verdicts, and testing are critical for the clinical diagnosis of COVID-19 (Ratre et al., 2021). This diagnostic method has advantages and disadvantages in the current environment (Table 1).

Generally, suspected individuals will endure an immediate chest X-ray or CT scan, depending on the severity of the condition (Shen et al., 2020). Chest images can provide insight into disease progression around the lungs (Pan et al., 2020). Consolidations, ground-glass opacities, vascular thickening and pleural effusions correlate with more severe disease (Sharif et al., 2022).

Notably, nucleic acid-based assays are important tools for the diagnosis of viral pathogens (Ratre et al., 2021). In specific, the real-time polymerase chain reaction (RT-PCR) is a "gold standard method" for virus detection due to rapid identification, high sensitivity and specificity (Ratre et al., 2021). However, nucleic acid methods have some disadvantages in their process, such as time-consuming, false-positive results, and a high risk of contamination (Ratre et al., 2021). Therefore, numerous biosensor-based technologies are being developed, and some have already been established for the diagnosis of COVID-19 pneumonia (Asif et al., 2020). Biosensor devices emerged as ideal replacement tools, which showed a fast response, high precision, enhanced sensitivity, and the possibility of early detection (Huang et al., 2018).

Next-generation sequencing, LAMP (Loop-mediated isothermal amplification) and immunological methods such as antigen-antibody response-based tests such as ELISA, have their own advantages and limitations in the present situation (Table 1).

Table 1. Diagnosis method available for COVID-19

Method	Procedure period	Working principle	Advantage	Disadvantage
CT scan	1 hour	Chest images	Show interstitial changes in the lung periphery, enhance the sensitivity of early detection	Difficulty in differentiating from other viral pneumonia
RT-PCR	3-4 hour	Specific primer-probe based detection	The established procedure, Rapid results, can be performed in a single step, has higher sensitivity, require a small amount of DNA	Time-consuming, Expensive due to the high cost of lab equipment and consumables
Next Generation Sequencing	1-2 day	Whole genome sequencing	Well established method, higher sensitivity, and rapid detection can detect novel strain	Expensive due to the high cost of lab equipment and consumables, technical expertise needed
LAMP (Loop-mediated isothermal amplification)	1 hour	Specific primers-pair-based detection	Single working temperature, highly accurate, low cost	Highly vulnerable to false positive results due to cross-contamination
Biosensors	30 seconds to 1 minute	Bioreceptor, transducer and signal detector, coupled with fluorescent/radioactive labeling	Cost-effective, early diagnosis, detect the small volume of the target analyte, highly sensitive, specific and rapid method	Combined detection methods might be needed for the accuracy of virus detection
POCT (Point of care testing)	15-30 minutes	Handheld POC device, Antigen/Antibodies IgG/ IgM detection	Easy, portable, user-friendly, cost-effective	Less sensitive, Prone to False positives due to longer time requirement (3 to 4 days) for the detectable antibodies after the infection
SHERLOCK assay (Specific High-Sensitivity Enzymatic Reporter UnLOCKing technique)	1 hour	CRISPR-mediated detection (Cas13 or Cas 12) via colorimetric readouts and fluorescence	High-sensitivity, multiplexed, single base pair specificity, portable, inexpensive	Highly rely on the fluorescent detection of a single target molecule in a single reaction
ELISA (Antigen-antibody detection)	4–6 hour	Antigen/Antibodies IgG/ IgM detection	Highly sensitive and specific	Prone to False positive due to the longer time requirement (4-7 days) for the detectable antibodies after the infection
Paper-based detection	50 minutes-1 hours	Wastewater (feces, urine)	Low-cost, user friendly	Complicated multi-step techniques

Conclusion

In this chapter, we evaluated the combination of clinical features and diagnostic approaches to COVID-19. We find references supporting the notion of the most common order of the different symptoms of COVID-19, which also differs from other prominent respiratory diseases. Global characterization of the clinical features of COVID-19 obtained from laboratory tests and diagnostic results will provide physicians with useful evidence to identify the disease and better management of the patients. There are many methods and devices for rapid analysis of COVID-19, in progress and various stages. These POCTs, biosensors and other alternative devices have the potential to become the technology of the future.

References

Alemi F, Vang J, Wojtusiak J, Guralnik E, Peterson R, Roess A A, & Jain P. Differential diagnosis of COVID-19 and influenza. *PLOS Glob. Public Health* (2022).

Andrews P L R, Cai W, Rudd J A, & Sanger G J. COVID-19, nausea, and vomiting. *J. Gastroenterol. Hepatol.* (2021) 36(3):646-656.

Arshad Ali S, Baloch M, Ahmed N, Arshad Ali A, & Iqbal A. The outbreak of Coronavirus Disease 2019 (COVID-19)-An emerging global health threat. *J. Infect. Public Health* (2019) 13(4):644-646.

Asif M, Ajmal M, Ashraf G, Muhammad N, Aziz A, Iftikhar T, Wang J, & Liu H. The role of biosensors in coronavirus disease-2019 outbreak. *Curr. Opin. Electrochem.* (2020) 23:174-184.

Bockman C S, & Eckerson J Myalgia, In S J Enna & D B Bylund (Eds.), *xPharm: The Comprehensive Pharmacology Reference* (pp. 1-7). Elsevier (2007).

Cascella M, Rajnik M, Aleem A, Dulebohn S C, & Di Napoli R. *Features, Evaluation, and Treatment of Coronavirus (COVID-19).* In StatPearls. StatPearls Publishing, Copyright ©, StatPearls Publishing LLC. (2022).

Comoglu Ş, Öztürk S, Kant A, Arslan M, Karakoc H N, & Yılmaz G. Evaluation of diarrhea in patients with COVID-19. *Dig. Dis.* (2021) 39(6):622-625.

Cui J, Li F & Shi Z L. Origin and evolution of pathogenic coronaviruses. *Nat. Rev. Microbiol.* (2019) 17(3): 181-192.

Cure E, & Cumhur Cure M Comment on "Organ-protective effect of angiotensin-converting enzyme 2 and its effect on the prognosis of COVID-19." *J. Med. Virol.* (2020) 92(9):1423-1424.

Daneshgaran G, Dubin D P, & Gould D J. Cutaneous manifestations of COVID-19: An evidence-based review. *Am. J. Clin. Dermatol.* (2020) 21(5):627-639.

Dhama K, Patel S K, Natesan S, Vora K S, Iqbal Yatoo M, Tiwari R, SS K, Singh K P, Singh R, & Malik Y S. COVID-19 in the elderly people and advances in vaccination approaches. *Hum. Vaccin. Immunother.* (2020) 16(12):2938-2943.

Esmaeilzadeh H, Sanaei Dashti A, Mortazavi N, Fatemian H, & Vali M. Persistent cough and asthma-like symptoms post COVID-19 hospitalization in children. *BMC Infect. Dis.* (2022) 22(1):244.

Feneley A, Clarke A, & Goldsack N. Patient with shortness of breath. *BMJ* (2017) 358, j1661.

Genovese G, Moltrasio C, Berti E, & Marzano A V. Skin manifestations associated with COVID-19: Current knowledge and future perspectives. *Dermatology* (2021) 237(1); 1-12.

Ghebrehewet S, MacPherson P, & Ho A. Influenza. *BMJ* (2016) *355*:6258.

Gooch M D. A New Rash Differential: CoVID-19. *Adv. Emerg. Nurs J.* (2021) 43(1):28-34.

Grant M C, Geoghegan L, Arbyn M, Mohammed Z, McGuinness L, Clarke E L, & Wade R G. The prevalence of symptoms in 24,410 adults infected by the novel coronavirus (SARS-CoV-2; COVID-19): A systematic review and meta-analysis of 148 studies from 9 countries. *PLoS One* (2020)15(6), e0234765.

Hentsch L, Cocetta S, Allali G, Santana I, Eason R, Adam E, & Janssens J P. Breathlessness and COVID-19: A Call for Research. *Respiration* (2021) 100(10):1016-1026.

Huang X, Xu D, Chen J, Liu J, Li Y, Song J, Ma X, & Guo J. Smartphone-based analytical biosensors. *Analyst* (2018) 143(22):5339-5351.

Iacobucci G. Covid-19: UK adds sore throat, headache, fatigue, and six other symptoms to official list. *BMJ* (2022) 377, o892.

Iravani B, Arshamian A, Ravia A, Mishor E, Snitz K, Shushan S, Roth Y, Perl O, Honigstein D, Weissgross R, Karagach S, Ernst G, Okamoto M, Mainen Z, Monteleone E, Dinnella C, Spinelli S, Mariño-Sánchez F, Ferdenzi C, Lundström J N. Relationship between odor intensity estimates and COVID-19 prevalence prediction in a Swedish population. *Chem. Senses* (2020)

Israfil S M H, Sarker M M R, Rashid P T, Talukder A A, Kawsar K A, Khan F, Akhter S, Poh C L, Mohamed I. N, & Ming L C. Clinical Characteristics and diagnostic challenges of COVID-19: An update from the global perspective. *Front. Public Health* (2020) 8:567395.

Jamil M, Bhattacharya P K, Barman B, Topno N, Barman H, Nongpiur V N, War G, Hynniewta Y, Saikia B, & Naku N. Clinical and demographic profile of COVID-19 patients: A tertiary level hospital-based study from northeast India. *Cureus* (2021) 13(10):e18881.

Kucuk A, Cumhur Cure M, & Cure E. Can COVID-19 cause myalgia with a completely different mechanism? A hypothesis. *Clin. Rheumatol.* (2020) 39(7):2103-2104.

Kumar R, Nagpal S, Kaushik S, & Mendiratta S. COVID-19 diagnostic approaches: different roads to the same destination. *Virusdisease* (2020) 31(2):97-105.

Larsen J R, Martin M R, Martin J D, Kuhn P, & Hicks J B. Modeling the onset of symptoms of COVID-19. (2020) *Front. Public Health*, *8*:473.

Li X-Y, Dai W J, Wu S N, Yang X Z, & Wang H G. The occurrence of diarrhea in COVID-19 patients. *Clinics and Research in Hepatology and Gastroenterology* (2020) 44(3):284-285.

Liu M, Deng L, Wang D & Jiang T. Influenza activity during the outbreak of coronavirus disease 2019 in Chinese mainland. *Biosaf. Health* (2020) 2(4):206-209.

Mahmoud N, Radwan N, Alkattan A, Hassanien M, Elkajam E, Alqahtani S, Haji A, Alfaifi A, Alfaleh A, & Alabdulkareem K. Post-COVID-19 syndrome: nature of symptoms and associated factors. *Journal of Public Health* (2023).

Megyeri K, Dernovics Á, Al-Luhaibi Z I I, & Rosztóczy A. COVID-19-associated diarrhea. *World J. Gastroenterol.* (2021) 27(23):3208-3222.

Ng D H L, Choy C Y, Chan Y H, Young B E, Fong S W, Ng L F P, Renia L, Lye D C, & Chia P Y. Fever patterns, cytokine profiles, and outcomes in COVID-19. *Open Forum Infect. Dis.* (2020) 7(9), ofaa375.

Nikolopoulou G B, & Maltezou H C. COVID-19 in Children: Where do we Stand? *Arch. Med. Res.* (2022) 53(1), 1-8.

Ogoina D. Fever, fever patterns and diseases called 'fever' – A review. *Journal of Infection and Public Health* (2011) 4(3):108-124.

Pan Y, Guan H, Zhou S, Wang Y, Li Q, Zhu T, Hu Q, & Xia L. Initial CT findings and temporal changes in patients with the novel coronavirus pneumonia (2019-nCoV): a study of 63 patients in Wuhan, China. *European Radiology* (2020) 30(6):3306-3309.

Pierce C A, Herold K C, Herold B C, Chou J, Randolph A, Kane B, McFarland S, Gurdasani D, Pagel C, Hotez P, Cobey S, & Hensley S E. COVID-19 and children. *Science* (2022) 377(6611): 1144-1149.

Printza A, Katotomichelakis M, Valsamidis K, Metallidis S, Panagopoulos P, Panopoulou M, Petrakis V, & Constantinidis J. Smell and taste loss recovery time in COVID-19 patients and disease severity. *J. Clin. Med.* (2021) 10(5).

Ratre Y K, Kahar N, Bhaskar L, Bhattacharya A, & Verma H K. Molecular mechanism, diagnosis, and potential treatment for novel coronavirus (COVID-19): a current literature review and perspective. *3 Biotech.* (2021) 11(2):94.

Sharif P M, Nematizadeh M, Saghazadeh M, Saghazadeh A, & Rezaei N. Computed tomography scan in COVID-19: a systematic review and meta-analysis. *Pol. J. Radiol.* (2022) 87:e1-e23.

Shen K, Yang Y, Wang T, Zhao D, Jiang Y, Jin R, Zheng Y, Xu B, Xie Z, Lin L, Shang Y, Lu X, Shu S, Bai Y, Deng J, Lu M, Ye L, Wang X, Wang Y. Global pediatric pulmonology, a diagnosis, treatment, and prevention of 2019 novel coronavirus infection in children: experts' consensus statement. *World Journal of Pediatrics* (2020) 16(3):223-231.

Singhal T. A Review of Coronavirus Disease-2019 (COVID-19). *Indian J. Pediatr.* (2020) 87(4):281-286.

Stokes E K, Zambrano L D, Anderson K N, Marder E P, Raz K M, El Burai Felix S, Tie Y, & Fullerton K E. Coronavirus Disease 2019 Case Surveillance - United States, January 22-May 30, 2020. *MMWR Morb. Mortal Wkly Rep.* (2020) 69(24):759-765.

Vetter P, Vu D L, L'Huillier A G, Schibler M, Kaiser L, & Jacquerioz F. Clinical features of COVID-19. *BMJ*, (2020) 369: m1470.

Wise J. Covid-19: Symptomatic infection with omicron variant is milder and shorter than with delta, study reports. *BMJ* (2022) 377:o922.

Yan C H, Faraji F, Prajapati D P, Ostrander B T, & DeConde A S. Self-reported olfactory loss associates with outpatient clinical course in COVID-19. *Int. Forum Allergy Rhinol.* (2020) 10(7): 821-831.

Yu C, Zhou M, Liu Y, Guo T, Ou C, Yang L, Li Y, Li D, Hu X, Shuai L, Wang B, & Zou Z. Characteristics of asymptomatic COVID-19 infection and progression: A multicenter, retrospective study. *Virulence*, (2020) 11(1):1006-1014.

Zhang T, Liu D, Tian D, & Xia L. The roles of nausea and vomiting in COVID-19: did we miss something? *J. Microbiol. Immunol. Infect.* (2021) 54(4):541-546.

Chapter 4

Counteracting SARS-CoV-2 and Managing the Impact of COVID-19 Using Natural Products and Medicinal Plants

Idris Adewale Ahmed*, PhD

Department of Biotechnology, Faculty of Applied Science, Lincoln University College, Petaling Jaya, Selangor, Malaysia

Abstract

Historically, different plant varieties have been explored for their therapeutic benefits against various illnesses, including viral diseases such as bronchitis, dengue virus and influenza virus. Therefore, this narrative review focuses on evaluating the healing potential of common medicinal plants and natural products against the novel coronavirus disease of 2019 (COVID-19) from the severe acute respiratory syndrome coronavirus-2 (SARS-CoV-2). Considerations were given to most studies involving natural products and medicinal plants as well as their antiviral potential against COVID-19 infection. Three specific search fields used include "traditional medicine and COVID-19", natural products and COVID-19 "medicinal plants and COVID-19". The search involved appearance in the title, keywords as well as abstract. Only relevant papers (original research articles and reviews) were included while books, book chapters, and short communications were excluded. Various plant parts are being used in different formulations and mixtures as well as in diverse conditions in different communities. Therefore, these natural bioactives can be synergistically employed in combination with pharmacological treatments to effectively manage various disease

* Corresponding Author's Emails: idrisahmed@lincoln.edu.my; ahmedris1400@gmail.com.

In: Medicinal Plants and COVID-19
Editor: Asita Elengoe
ISBN: 979-8-88697-958-9

pathologies. Detailed phytochemical, pharmacological, and clinical studies are, however, recommended to authenticate the safety and efficacy of these plants. Government should also formulate relevant public health policies for the sustainable use of these resources in the adequate management and prevention of the COVID-19 pandemic and other health outcomes.

Keywords: COVID-19, medicinal plants, natural products, traditional medicine

Introduction

The appearance of the new severe acute respiratory syndrome (SARS-COV-2) was reported for the first time in the Hubei Province of Wuhan, China in December 2019 with a few pneumonia cases linked to the new strain. Several cases were subsequently reported following a rapid increase of cases throughout larger parts of China with the first 41 patients comprising 30 men and 11 women confirmed through laboratory testing in Wuhan in early January 2020. The World Health Organization (WHO) later declared it a pandemic on the 11th of March 2020 following the rapid spread and increase in the number of cases and death worldwide (Khan and Al-Balushi, 2021; Sen et al., 2021). Some patients usually report body aches, diarrhea, loss of taste and smell, sore throat, runny nose, and difficulty in breathing, in case of serious illness with those with underlying medical problems and aged people being more prone to develop serious illness. The rate of infection of SARS-COV-2 is usually accelerated by human-to-human transmission via contact or respiratory droplets as well as the consumption of infected animals (Oladele et al., 2020; Sen et al., 2021). Though no consensus yet concerning the first COVID-19 transmission to humans among the scientific community (Khan and Al-Balushi, 2021), nevertheless, genome sequencing together with other analyses suggested bats as the origin of the enveloped zoonotic virus before it was transmitted to humans through an unknown intermediate (Nguyen et al., 2022).

Older people and immune-compromised patients with certain underlying co-morbidities including cardiovascular diseases, diabetes and hypertension have been reported to have higher mortality rates (Khan and Al-Balushi, 2021; Parida et al., 2021). Immunodeficiency has also been linked to malnutrition and deficiency of certain single nutrients including zinc, manganese, selenium, folic acid and other vitamins. The supplementation of vitamin D, however, could lessen the infection and death from COVID-19 (Khabour and Hassanein, 2021).

Furthermore, some medications such as angiotensin receptor blockers, as well as statins that are commonly used for the treatment of atherosclerosis and hypertension co-morbidities, are thought to cause the upsurge in acute respiratory distress syndrome risk, severe pneumonia and death among people suffering from COVID-19 (Khan and Al-Balushi, 2021).

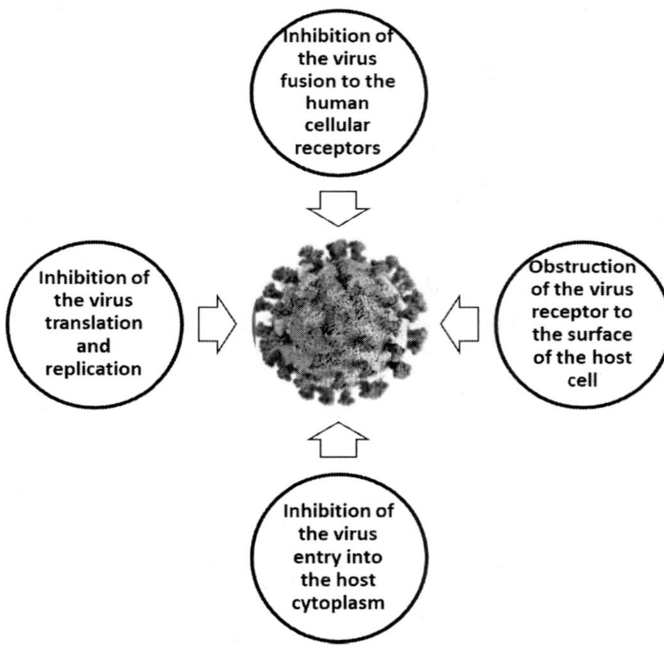

Figure 1. Common targets for antivirals.

Efforts have, thus, been intensified in the quest for efficient treatments and management of the new outbreak with the earliest therapies focused on drugs such as ivermectin, azithromycin, and hydroxychloroquine. The hydroxychloroquine strategies, however, failed to produce the desired impact, thus, it was unrecommended and challenged by WHO. Furthermore, favipravir and remdesivir were two commonly used antiviral drugs that reportedly reduced viral load in some patients (Sen et al., 2021). So far, vaccines have been the most commonly effective and preventive method to combat COVID-19 (Mukherjee et al., 2022). Thus, the identification of other antiviral drug candidates with the potent capability of interfering with at least one or more replication cycle stages of the coronavirus was suggested. The various known targets for antivirals (Figure 1) include (i) inhibition of the virus fusion to the receptors of human cells (ii) obstruction of the virus receptor to the surface of the host cell (ii) inhibition of the virus access into the host cytoplasm as well as (iv) inhibition of the replication and translation of the virus. Some repurposed drugs were, therefore, identified as potential anti-COVID-19 drugs in a few clinical trials (Khan and Al-Balushi, 2021).

Alternatively, the use of common natural products and medicinal plants has always been considered a global therapeutic strategy for the effective management of various health impacts including the recent COVID-19 (Cordoba-Tovar et al., 2022; Sen et al., 2021). According to the International Union for Conservation of Nature, about 80% of the world's population depends on medicinal plant use as the sole means of health management with at least 50,000 to 80,000 natural plant products worldwide having medicinal purposes (Córdoba-Tovar et al., 2022a; Nyagumbo et al., 2022; Sen et al., 2021). The invention of new therapeutic agents with low toxicity but immense pleiotropic biological activities such as antiviral potency is the key to tackling the ravaging pandemic. Several hundreds of natural products have been identified as possessing medicinal activities against various human viruses (Khan et al., 2021). Most medicinal plant bioactives have the capability to regulate immune and apoptotic pathways (Natesh et al., 2021). Furthermore, herbal formulations are very effective in relieving several COVID-19 symptoms and reducing patients' probability of severe conditions development (Mlozi, 2022; Mukherjee et al., 2022). One of the rich potential drug resources is plant polysaccharides with antiviral activity. Polysaccharides such as arabinoxylan, fucoidans, glucans, glucoarabinan, pectins, sulfated xylomannan, and xylan act through the inhibition of both the replication and binding of the virus to cells. The structural complexity and diversity of plant polysaccharides also contribute to their antiviral properties in diverse viral infection process stages (Sen et al., 2021).

In contrast to the European and North American Governments, health authorities in South Korea, India, and China swiftly issued guidelines to recognize the traditional medicine role in mitigating COVID-19 impacts (Córdoba-Tovar et al., 2022a; Sen et al., 2021). The Indian Government, under the Ministry of AYUSH, for instance, endorsed four medicinal herbs including Dalchini (*Cinnamomum zeylanicum*), Sunthi (*Zingiber officinale*), Krishna Marich (*Piper nigrum*), and Tulsi (*Ocimum sanctum*) as body immunity boosters as the first line of defense against COVID-19 (Sen et al., 2021).

Similarly, there was a creation of health programs to protect and save lives in 23 Chinese provinces through the timely provision of guidelines for medicinal plant use including *Atractylodis macrocephala* Koidz, *A. membranaceus*, *Saposhnikovia divaricata* (Turcz.) Schischk. and *G. uralensis* (Córdoba-Tovar et al., 2022a).

The divergence in history, philosophical and cultural views of medication and medical care in addition to discordant standards and policies in the legalization and regulation of traditional practices might have been the underlying issues for the non-recognition of traditional medicine in North America as well as Europe. Nevertheless, the objective of the current review is to evaluate the potential of some regional natural products and medicinal plants that have been proposed or are being used for COVID-19 therapy and counteracting SARS-COV-2 impacts.

Materials and Methods

Relevant information search was embarked upon from different databases including ScienceDirect, PubMed, and Scopus. Three explicit search fields used include "traditional medicine and COVID-19", natural products and COVID-19 "medicinal plants and COVID-19". The search involved appearance in the title, keywords, and abstract. Only relevant papers (original research articles and reviews) were included while books, book chapters, and short communications were excluded.

Pathogenesis of COVID-19

Coronaviruses are enveloped non-segmented positive-sense viruses with a characteristic single-stranded RNA genome (Oladele et al., 2020; Verma et al., 2021). They are grouped under the family Coronaviridae, (subfamily Orthocoronaviridae) and comprise mainly four genera (Alpha, Beta, Gamma, as well as Delta coronavirus). With genome size lengths ranging from 26 to 32 kilobases, coronaviruses are said to have the largest RNA virus genomes of diameter 80–120 nm (Oladele et al., 2020). Based on antigenic and genetic criteria, however, coronaviruses are grouped into three categories: α-CoVs, β-CoVs, as well as γ-CoVs. The novel SARS-COV-2 that is responsible for the viral COVID-19 disease belongs to β-CoVs genera (Tahir ul Qamar et al.,

2020; Verma et al., 2021). The viral genome consists of two main open reading frames (ORFs), namely ORF1a and ORF1ab forming 16 nonstructural proteins from proteolytical cleavage (Yang et al., 2022). The four key structural proteins identified in SARS-CoV-2 include nucleocapsid, envelope, membrane and spike proteins. The beta-coronaviruses usually generate a ~800 kilo Dalton polypeptide that is cleaved proteolytically to produce various proteins following genome transcription. The 3- chymotrypsin-like protease (3CLpro), as well as papain-like protease (PLpro), are the enzymes responsible for the proteolytic processing with the 3CLpro involved in the cleavage of the polyproteins at 11 unique sites to produce various non-structural proteins considered essential for the replication of the virus (Alamri et al., 2020; Tahir ul Qamar et al., 2020).

The average period of incubation for COVID-19 ranges from 3-5 days with its most common clinical features being very comparable to other pneumonia from viruses. The features include sore throat, sneezing, cough, fatigue, fever, hypoxemia shortness of breath as well as other complications while in severe and critical cases, it may include multiple organ failure as well as death (Divya et al., 2020; Oladele et al., 2020). Viral transmission among humans usually occurs through the permeation of infected aerosols from a sneeze, cough, or respiratory droplets via inhalation either through the mouth or nose into the lungs COVID-19, however, has a higher mortality rate (3.7%) in comparison with influenza [>1% mortality rate] (Oladele et al., 2020). Respiratory failure, heart failure, gastroenteritis and cytokine storm syndrome were reported in some severe COVID-19 cases with acute respiratory distress syndrome as the chief death cause (Oladele et al., 2020; Parida et al., 2021). Viral infections also major initiate secondary hemophagocytic lymphohistiocytosis otherwise known as macrophage activation syndrome which is a life-threatening health state comprising a heterogeneous hyper-inflammatory syndrome group. Its usual characteristics include a severe and sudden hyper-cytokinemia from untimely survival of cytotoxic T-lymphocytes and histiocytes which ultimately results in failure of multiple organs, hemophagocytosis and very high mortality as a result of persistent fever, cytopenias, hyper-ferritinemia and pulmonary in some patients (Oladele et al., 2020).

Effects of SARS-CoV-2 Virus on Different Human Body Parts

COVID-19 is a severe but acute respiratory viral infection caused by SARS-CoV-2 in humans. The viral structure is characterized by four major proteins namely; envelope, membrane, nucleocapsid and spike proteins (Sen et al., 2021; Verma et al., 2021). The spike proteins generate homotrimers which protrude on the viral surface to enable its binding to the host cells by enticing angiotensin-converting enzyme 2 (ACE2) in the cells of the lower respiratory tract. The nucleocapsid proteins are mostly involved in interacting with the viral genome while the membrane proteins are equally involved in the virion assembly. The membrane proteins also determine the viral envelope shape and stabilize the nucleocapsid protein to allow the viral assembly completion while the envelope proteins play an important role in both the production and maturation of the virus (Verma et al., 2021). Basically, the spike proteins bind to the ACE2 receptors in the epithelium of the nose, mouth, and lungs. Thus, they use this route to gain entrance into the body.

The viruses mainly affect the upper part of the respiratory tract (nose, sinuses as well as throat) and the lower part of the respiratory tract (lungs and windpipe), thus, resulting in the respiratory tract infection onset. SARS-CoV-2 affects major body organs such as the brain, gut, kidneys, heart and blood vessels (Sen et al., 2021). The commonest modes of action of the virus proposed earlier include hyper-inflammation and immunosuppression characterized by a swift and lethal hyper-cytokinemia with failure of multiple organs as well as ACE2 reduction to enhance the permeability of the pulmonary vascular and alveoli damage (Oladele et al., 2020). Several mental issues have also been reported from SARS-CoV-2 infection being both neuro-virulent and neuro-invasive. Neuropsychological problems have also been reported in one in three recovering patients while patients with pre-existing mental illnesses reportedly experience more deteriorating conditions (Mukherjee et al., 2022).

Susceptibility to COVID-19 Infection

COVID-19 symptoms are mostly non-specific and, sometimes, can be asymptomatic. Other characteristic symptoms range from severe pneumonia by coughing, anosmia, fever, headache, diarrhea, fatigue, sore throat, heightened sputum production, dyspnea, anorexia, rhinorrhea, skin sensitivity, pleurisy, myalgias, hemoptysis, shortness of breath and also death (Mukherjee et al., 2022). Anyone irrespective of age or sex from children to elder people can contract the SARS-CoV-2 virus and get COVID-19 infection. However, those with a deteriorated immune system from some common pre-existing infectious or chronic diseases such as chronic obstructive pulmonary disease, cancer, coronary artery disease, hypertension, HIV, kidney disease, liver disease, neurodegenerative disorders, and type 2 diabetes are more liable to COVID-19 infection and get serious illness (Khan and Al-Balushi, 2021; Khanal et al., 2022; Oladele et al., 2020; Sen et al., 2021). Several case reports demonstrated an increased mortality rate with advancing age while people above 80 years show the highest known mortality rate with males having a higher case fatality rate than females at any age (Mukherjee et al., 2022). Therefore, it is crucial to build adequate body resilience against viral infections through natural immunity supplements (Sen et al., 2021). The immunity system boosting involves appropriate modulation of the multiple proteins that are involved in homeostasis regulation (Khanal et al., 2022).

Medicinal Plants for Protection against SARS-CoV-2

Besides pharmacological treatment, humans have adopted various therapeutic approaches since time immemorial to prevent or manage different health ailments. Until today, about 80% of the global population still uses only herbal remedies for basic health care owing to the availability of different medicinal plant varieties across the globe for various purposes. The traditional approach to disease management is not only passed from one generation to the other worldwide, but their usage is also characterized by effectiveness, cultural preferences, affordability, safety, and ample accessibility when needed and at all times (Nyagumbo et al., 2022; Oladele et al., 2020). Medicinal plants and other natural products remain an enormous reservoir for the development of potential novel bioactives against both viral and non-viral diseases (Abubakar et al., 2022). Therefore, the antiviral, antioxidant, anti-inflammatory, antipyretic, cytoprotective properties and immunomodulatory properties of various medicinal plants have made them a suitable candidate for global COVID-19 management whether alone or in combination with other standard medicines (Khan and Al-Balushi, 2021). Plants are very rich in secondary compounds and metabolites including flavonoids, phenolics, saponins, tannins, triterpenes and alkaloids with competitive but dose-dependent inhibition of the SARS-CoV-2 main protease (3CLpro) activity which may help to improve patients' recovery (Ebadollahinatanzi et al., 2021; Murugesan et al., 2021; Natesh et al., 2021; Yang et al., 2022).

An *in-silico* study also reported the exhibition of good binding affinity by poly-methoxyflavones and poly-ethoxyflavones for the target spike proteins of SARS-CoV-2 making them potential candidates for anti-viral drugs against most of the SARS-CoV-2 variants (Solo and doss, 2021). Several researchers have also adopted computational screening methods for potential anti-COVID-19 agent discovery against SARS-CoV-2 infection (Alamri et al., 2020; Khanal et al., 2022; Murugesan et al., 2021; Tahir ul Qamar et al., 2020). The exploration of herbal preparations and bioassay-guided isolation of highly potent lead compounds from natural products and medicinal herbs remain an attractive sustainable approach to combatting COVID-19). The broad-spectrum antiviral activity of most natural products may be useful in inhibiting the various multiple steps in viral replication and infection (Mukherjee et al., 2022). Phytochemicals generally act by interfering with viral entry through attachment to the viral proteins, inhibiting viral replication, and hindering transcription and viral protein expression (Khan et al., 2021). The commonest families of medicinal plants identified for relevant biological and pharmacological activities include but are not limited to Asparagaceae, Alliaceae, Canellaceae, Chenopodiaceae, Cannabaceae, Chrysobalanaceae, Crassulaceae, Convolvulaceae, Cyperaceae, Cucurbitaeae, Flacourtiaceae, Hyacinthaceae, Heteropyxidaceae, Kirkiaceae, Myrsinaceae, Myrothamnaceae, Nymphaeaceae, Pittosporaceae, Phyllanthaceae, Poaceae, Plumbaginaceae, Polygonaceae, Polygalaceae, Proteaceae, Pteridaceae, Sapindaceae, Ranunculaceae, Ulmaceae, Tiliaceae, Zingiberaceae, Annonaceae, Acanthaceae, Bignoniaceae, Asclepiadaceae, Loganiaceae, Clusiaceae, Rhamnaceae, Thymelaeaceae, Rutaceae, Celastraceae, Verbenaceae, Loranthaceae, Ebenaceae, Amarilidaceae, Amaranthaceae, Commelinaceae, Phytovaceae, Meliaceae, Apiaceae, Moraceae, Asphodelaceae, Myrtaceae, Combretaceae, Solanaceae, Lamiaceae, Apocynaceae, Anacardiaceae, Euphorbiaceae, Asteraceae, Rubiaceae and Fabaceae. Among these, the families with the largest number of species identified are Rutaceae, Apocynaceae, Verbenaceae, Lamiaceae, Anacardiaceae, Euphorbiaceae, Asteraceae, Fabaceae, and Rubiaceae (Cordoba-Tovar et al., 2022; Nyagumbo et al., 2022).

Globally, about 390,000 plant species have been identified, with about 60,000 being exclusively employed as medicinal plants while about 26,000 reportedly have unswerving scientific evidence following confirmation from different cultural practices across various geographical regions. The net revenues from the use and international trade of medicinal plants have also tripled over time from 1.3 billion USD in 1998 to 3.3 billion USD in 2018 (Cordoba-Tovar et al., 2022). Some of the plants with very promising potential to fight major illnesses, including COVID-19, are listed in Table 1.

Table 1. Some common natural products and medicinal herbs being used for COVID-19 management

Natural products	Properties/activities	References
Nigella sativa L.,	Antiviral, antioxidant, anti-inflammatory, antibacterial, and immunomodulation	(Hossain et al., 2021; Mukherjee et al., 2022)
Zingiber officinale Roscoe	Antiviral, antioxidant, anti-inflammatory, antibacterial, and immunomodulation	(Cordoba-Tovar et al., 2022b; Mukherjee et al., 2022)
Glycyrrhiza glabra L.	Antiviral, antioxidant, anti-inflammatory, antibacterial, and immunomodulation	(Khan et al., 2021; Mukherjee et al., 2022)
Allium sativum L.	Antiviral, antioxidant, anti-inflammatory, antibacterial, and immunomodulation	(Khan et al., 2021; Mukherjee et al., 2022)
Rhus chinensis Mill.	Antiviral, antioxidant, anti-inflammatory, antibacterial, and immunomodulation	(Li et al., 2022; Mukherjee et al., 2022)
Isatis indigotica Fortune ex Lindl.	Antiviral, antioxidant, anti-inflammatory, antibacterial, and immunomodulation	(Chen et al., 2021; Mukherjee et al., 2022)
Cassia tora L.	Antiviral, antioxidant, anti-inflammatory, antibacterial, and immunomodulation	(Antonisamy et al., 2017; Mukherjee et al., 2022)
Calophyllum blancoi Planch.	Antiviral, antioxidant, anti-inflammatory, antibacterial, and immunomodulation	(Das et al., 2021; Mukherjee et al., 2022)
Tripterygium regelii Sprague	Antiviral, antioxidant, anti-inflammatory, antibacterial, and immunomodulation	(Liu et al., 2022; Mukherjee et al., 2022)
Cibotium barometz (L.) J. Sm.	Antiviral and anti-inflammatory	(Heng et al., 2020; Mukherjee et al., 2022)
Chrysanthemum indicum L.	Antiviral, anti-inflammatory and immunomodulation	(Mukherjee et al., 2022; Qiu et al., 2022)
Dryopteris crassirhizoma Nakai	Antiviral, antioxidant, anti-inflammatory, antibacterial, and immunomodulation	(Hai et al., 2022; Mukherjee et al., 2022)
Mucuna pruriens (L.) DC.	Antiviral, antioxidant, anti-inflammatory, antibacterial, and immunomodulation	(Elabd et al., 2022; Mukherjee et al., 2022)

Natural products	Properties/activities	References
Gentiana scabra Bunge	Antiviral, anti-inflammatory and immunomodulation	(Mukherjee et al., 2022; Zheng et al., 2021)
Scutellaria spp.	Antiviral, antioxidant, anti-inflammatory, antibacterial, and immunomodulation	(Maleki et al., 2022; Mukherjee et al., 2022)
Rosmarinus officinalis (L.) Scheid	Antiviral, antioxidant, anti-inflammatory, antibacterial, and immunomodulation	(Eid et al., 2022; Khan et al., 2021)
Salvia officinalis L.	Antiviral, antioxidant, anti-inflammatory, antibacterial, and immunomodulation	(Abdel Rahman et al., 2022; Mukherjee et al., 2022)
Cinnamomum zeylanicum L.	Antiviral, antioxidant, anti-inflammatory, antibacterial, and immunomodulation	(Mukherjee et al., 2022; Saki et al., 2020)
Psoralea corylifolia L.	Antiviral, antioxidant, anti-inflammatory, antibacterial, and immunomodulation	(Mahajan et al., 2022; Mukherjee et al., 2022)
Ecklonia cava	Antiviral, anti-inflammatory and immunomodulation	(Mukherjee et al., 2022; Zhang et al., 2021)
Taxillus chinensis (DC.)	Antiviral, antioxidant, anti-inflammatory, antibacterial, and immunomodulation	(Mukherjee et al., 2022; Qin et al., 2022)
Azadirachta indica A. Juss	Antiviral, anti-inflammatory and immunomodulation	(Mukherjee et al., 2022; Patil et al., 2022)
Toona sinensis (Juss.). Roem.	Antiviral, antioxidant, anti-inflammatory, antibacterial, and immunomodulation	(Mukherjee et al., 2022; Peng et al., 2019)
Broussonetia papyrifera (L.) L'Her. ex Vent	Antiviral, antioxidant, anti-inflammatory, antibacterial, and immunomodulation	(Chen et al., 2022; Mukherjee et al., 2022)
Andrographis paniculata (Burm.f.) Nees	Antiviral, anti-inflammatory and immunomodulation	(Hossain et al., 2021; Mukherjee et al., 2022
Thymus vulgaris L.	Antioxidant and antifibrotic	(Bahri et al., 2022; Ebadollahinatanzi et al., 2021; Tardugno et al., 2022)
Hyssopus officinalis L.	Antiviral and antifungal	(Ebadollahinatanzi et al., 2021; Sharifi-Rad et al., 2022)
Mentha spp.	Antimicrobial, antioxidant, and anti-inflammatory	(Cordoba-Tovar et al., 2022b; Ebadollahinatanzi et al., 2021; Stringaro et al., 2018)
Ziziphus jujube L.	Antiviral, anti-inflammatory and immunomodulation	(Ebadollahinatanzi et al., 2021; Lu et al., 2021)
Althea officinalis L.	Post-viral acute cough treatment	(Ciprandi and Tosca, 2021; Ebadollahinatanzi et al., 2021)
Malva sylvestris L.	Anti-inflammatory, antioxidant, and antiproliferative	(Batiha et al., 2022; Ebadollahinatanzi et al., 2021)

Table 1. (Continued)

Natural products	Properties/activities	References
Viola odorata L.	Antimicrobial, antioxidant, antiviral, and anti-inflammatory	(Ebadollahinatanzi et al., 2021; Feyzabadi et al., 2017)
Echinacea purpurea L.	Antiviral, anti-inflammatory and immunomodulation	(Ebadollahinatanzi et al., 2021; Mukherjee et al., 2022; Sharifi-Rad et al., 2018)
Matricaria chamomilla L.	Antiviral, antioxidant, anti-inflammatory, antibacterial, and immunomodulation	(Cordoba-Tovar et al., 2022b; Ebadollahinatanzi et al., 2021; Habibzadeh and Zohalinezhad, 2021)
Verbascum thapsus L.	Antiviral, antioxidant, anti-inflammatory, antibacterial, and immunomodulation	(Ebadollahinatanzi et al., 2021; A. Gupta et al., 2022)
Daucus carota L.	Antiviral, anti-inflammatory and immunomodulation	(Ebadollahinatanzi et al., 2021; Qi et al., 2021)
Eucalyptus spp.	Antiviral, anti-inflammatory and immunomodulation	(Cordoba-Tovar et al., 2022b; Faezeh Taghizadeh et al., 2022)
Citrus limon	Antiviral, antioxidant, anti-inflammatory, antibacterial, and immunomodulation	(Cordoba-Tovar et al., 2022b; Singh et al., 2021)
Cymbopogon citratus	Antiviral, antioxidant, anti-inflammatory, antibacterial, and immunomodulation	(Ben Ammar et al., 2022; Cordoba-Tovar et al., 2022b)
Garlic (*Allium sativum*)	Antiviral, anti-inflammatory and immunomodulation	(Ahmed and Wang, 2021; Cordoba-Tovar et al., 2022b)
Moringa oleifera	Antiviral, antioxidant, anti-inflammatory, antibacterial, and immunomodulation	(Cordoba-Tovar et al., 2022b; Mushtaq et al., 2021)
Orange	Antiviral, antioxidant, anti-inflammatory, antibacterial, and immunomodulation	(Cordoba-Tovar et al., 2022b; Honore et al., 2022)
Or´egano (*Origanum vulgare*)	Antiviral, antioxidant, anti-inflammatory, antibacterial, and immunomodulation	(Cordoba-Tovar et al., 2022b; Diniz do Nascimento et al., 2020)
Tinospora cordifolia	Antioxidant, anti-inflammatory, antiviral, and anticancer	(Hein et al., 2023; Kumar et al., 2022)
Oroxylum indicum (L.) Kurz	Antimicrobial, antiviral, anti-inflammatory, and antioxidant	(Cardoso Reis et al., 2022; Hein et al., 2023)
Aquilaria malaccensis Lam.	Anti-allergy, anti-inflammatory, antiviral, antioxidant, and cardioprotective	(Hein et al., 2023; Sarmah et al., 2022)

Natural products	Properties/activities	References
Chromolaena odorata (L.)	Antibacterial, anti-microbial and wound-healing properties	(Hein et al., 2023; Olawale et al., 2022)
Chloranthus elatior Link.	Antiviral, anti-inflammatory and immunomodulation	(Hein et al., 2023; Y.-y. Liu et al., 2022)
Polysaccharides from medicinal plants	Interference with virus life cycle and inhibition of virus's attachment to the cell of the host	(Sen et al., 2021)
Syzygium aromaticum (L.) Merr.	Antiviral, anti-inflammatory and immunomodulation	(Faqer et al., 2022; Mukherjee et al., 2022)
Piper nigrum L.	Antiviral, anti-inflammatory and immunomodulation	(Mukherjee et al., 2022; Zahin et al., 2021)
Camellia sinensis (L.) Kuntze	Antiviral, antioxidant, anti-inflammatory, antibacterial, and immunomodulation	(Mukherjee et al., 2022; Saeed et al., 2017)
Curcuma longa L.	Antiviral, anti-inflammatory and immunomodulation	(Mukherjee et al., 2022; Park et al., 2022)
Tribulus terrestris L.	Antiviral, antioxidant, anti-inflammatory, antibacterial, and immunomodulation	(Aldaddou et al., 2022; Mukherjee et al., 2022)
Piper betle L.	Antiviral, anti-inflammatory and immunomodulation	(Gupta et al., 2022; Zamakshshari et al., 2021)

Role of Ethnobotanical and Ethnopharmacological Studies

Ethnobotanical and ethnopharmacological studies have received serious attention widely in recent times, especially during the present COVID-19 pandemic in smaller and underdeveloped countries. The lack of effective treatment and non-availability of vaccines to combat the rapid expansion of SARS-CoV-2 throughout the early pandemic days forced people to explore and adopt commonly available medicinal plants as alternatives to treat, halt and prevent the COVID-19 virus spread. Furthermore, the ethnobotanical and ethnopharmacological knowledge associated with locally available medicinal plants in different regions of the world shows the potential benefits of these medicinal plants in addition to the growth of environmental awareness in facing future challenges (Cordoba-Tovar et al., 2022). Nevertheless, experimental and clinical studies are vital for the validation of the effectiveness as well as safety of these natural endowments. Furthermore, *in silico* (computational) studies, genetic engineering and drug repurposing could also be explored. In addition, government policies and guidelines should be reviewed time-to-time to advance and promote traditional and alternative medicine (Cordoba-Tovar et al., 2022).

Conclusion

This review shows the relevance and role of cultural and traditional-based approaches for the management as well as treatment of COVID-19. The study highlights some of the commonly used anti-viral and anti-COVID medicinal plants that have so far been reported in Malaysia and other regions of the world. Various plant parts are used in different formulations and mixtures as well as in diverse conditions in different communities. Significant attention is being given to traditionally employed medicinal plants in recent years as alternative therapeutic and healing solutions against novel communicable diseases. These natural bioactives can be synergistically employed in combination with pharmacological treatments for the effective management of various disease pathologies. Detailed phytochemical, pharmacological, and clinical studies are, however, recommended to authenticate the safety and efficacy of these plants. Government should also formulate relevant public health policies for the sustainable use of these resources in the adequate management and prevention of the COVID-19 pandemic and other health outcomes.

References

Abdel Rahman AN, Van Doan H, Elsheshtawy HM, Dawood A, Salem SM, Sheraiba NI, Masoud SR, Abdelnaeim NS, Khamis T, Alkafafy M, Mahboub HH. Dietary *Salvia officinalis* leaves enhances antioxidant-immune-capacity, resistance to *Aeromonas sobria* challenge, and growth of *Cyprinus carpio*. *Fish & Shellfish Immunology* (2022) 127:340-348.

Abubakar IB, Kankara SS, Malami I, Danjuma JB, Muhammad YZ, Yahaya H, Singh D, Usman UJ, Ukwuani-Kwaja AN, Muhammad A, Ahmed SJ. Traditional medicinal plants used for treating emerging and re-emerging viral diseases in northern Nigeria. *European Journal of Integrative Medicine* (2022) 49:102094.

Ahmed T, Wang CK. Black Garlic and its bioactive compounds on human health diseases: A Review. *Molecules* (2021) 26(16).

Alamri MA, Altharawi A, Alabbas AB, Alossaimi MA, Alqahtani SM. Structure-based virtual screening and molecular dynamics of phytochemicals derived from Saudi medicinal plants to identify potential COVID-19 therapeutics. *Arabian Journal of Chemistry* (2020) 13(9):7224-7234.

Aldaddou WA, Aljohani ASM, Ahmed IA, Al-Wabel NA, El-Ashmawy IM. Ameliorative effect of methanolic extract of *Tribulus terrestris* L. on nicotine and lead-induced degeneration of sperm quality in male rats. *Journal of Ethnopharmacology* (2022) 295:115337.

Antonisamy P, Dhanasekaran M, Kim HR, Jo SG, Agastian P, Kwon KB. Anti-inflammatory and analgesic activity of ononitol monohydrate isolated from *Cassia tora L.* in animal models. *Saudi Journal of Biological Sciences* (2017) 24(8):1933-1938.

Bahri S, Mlika M, Nahdi A, Ben Ali R, Jameleddine S. Thymus Vulgaris Inhibit Lung Fibrosis Progression and Oxidative Stress Induced by Bleomycin in Wistar Rats. *Nutrition Cancer* (2022) 74(4):1420-1430.

Batiha GE, Tene ST, Teibo JO, Shaheen HM, Oluwatoba OS, Teibo TK, Al-Kuraishy HM, Al-Garbee AI, Alexiou A, Papadakis M. The phytochemical profiling, pharmacological activities, and safety of *Malva sylvestris*: a review. *Naunyn-Schmiedeberg's Archives of Pharmacology* (2022):1-20.

Ben Ammar R, Mohamed ME, Alfwuaires M, Abdulaziz Alamer S, Bani Ismail M, Veeraraghavan VP, Sekar AK, Ksouri R, Rajendran P. Anti-inflammatory activity of geraniol isolated from lemon grass on Ox-LDL-stimulated endothelial cells by upregulation of heme oxygenase-1 via PI3K/Akt and Nrf-2 signaling pathways. *Nutrients* (2022) 14(22).

Cardoso Reis AC, Valente GM, Silva BdM, de Brito Magalhães CL, Kohlhoff M, Brandão GC. Anti-arboviral activity and chemical characterization of hispidulin and ethanolic extracts from *Millingtonia hortensis* L. and *Oroxylum indicum* (L.) Kurz (Bignoniaceae). *Natural Product Research* (2022):1-5.

Chen Q, Lan HY, Peng W, Rahman K, Liu QC, Luan X, Zhang H. *Isatis indigotica*: A review of phytochemistry, pharmacological activities and clinical applications. *Journal of Pharmacy and Pharmacology* (2021) 73(9):1137-1150.

Chen Y, Wang L, Liu X, Wang F, An Y, Zhao W, Tian J, Kong D, Zhang W, Xu Y, Ba Y. The genus Broussonetia: An updated review of phytochemistry, pharmacology and applications. *Molecules* (2022) 27(16).

Ciprandi G, Tosca MA. Non-pharmacological remedies for post-viral acute cough. *Monaldi Archives for Chest Disease* (2021) 92(1).

Córdoba-Tovar L, Barón PAR, Marrugo-Negrete J, Roa-Fuentes LL, Jonathan MP. Scientific production on medicinal plants and their efficacy against COVID-19: A review and scientometric analysis based on VOSviewer. *Acta Ecologica Sinica* (2022):1-9.

Cordoba-Tovar L, Ríos-Geovo V, Largacha-Viveros MF, Salas-Moreno M, Marrugo-Negrete JL, Ramos PA, Chaverra LM, Jonathan MP. Cultural belief and medicinal plants in treating COVID-19 patients of Western Colombia. *Acta Ecologica Sinica* (2022) 42(5):476-484.

Das A, Pandita D, Jain GK, Agarwal P, Grewal AS, Khar RK, Lather V. Role of phytoconstituents in the management of COVID-19. *Chemico-Biological Interactions* (2021) 341:109449.

Diniz do Nascimento L, Barbosa de Moraes AA, Santana da Costa K, Pereira Galúcio JM, Taube PS, Leal Costa CM, Neves Cruz J, de Aguiar Andrade EH, Guerreiro de Faria

LJ. Bioactive natural compounds and antioxidant activity of essential oils from spice plants: New findings and potential applications. *Biomolecules* (2020) 10(7).

Divya M, Vijayakumar S, Chen J, Vaseeharan B, Durán-Lara EF. South Indian medicinal plants can combat deadly viruses along with COVID-19? - A review. *Microbial Pathogenesis* (2020) 148:104277.

Ebadollahinatanzi A, Arabrahmatipour G, Abedi M. COVID-19, health recovery by medicinal plants (health solutions): An overview. *Toxicology Letters* (2021) 350: S136.

Eid AM, Jaradat N, Issa L, Abu-Hasan A, Salah N, Dalal M, Mousa A, Zarour A. Evaluation of anticancer, antimicrobial, and antioxidant activities of rosemary (*Rosmarinus Officinalis*) essential oil and its Nanoemulgel. *European Journal of Integrative Medicine* (2022) 55:102175.

Elabd H, Faggio C, Mahboub HH, Emam MA, Kamel S, El Kammar R, Abdelnaeim NS, Shaheen A, Tresnakova N, Matter A. *Mucuna pruriens* seeds extract boosts growth, immunity, testicular histology, and expression of immune-related genes of mono-sex Nile tilapia (*Oreochromis niloticus*). *Fish & Shellfish Immunology* (2022) 127:672-680.

Faezeh Taghizadeh S, Panahi A, Esmaeilzadeh Kashi M, Kretschmer N, Asili J, Ahmad Emami S, Azizi M, Shakeri A. Structural diversity of complex phloroglucinol derivatives from Eucalyptus species. *Chemistry Biodiversity* (2022) 19(6): e202200025.

Faqer OE, Bendiar S, Rais S, Elkoraichi I, Dakir M, Elouaddari A, El Amrani A, Oudghiri M, Mtairag EM. Phytochemical characterization and immunomodulatory effects of aqueous, ethanolic extracts and essential oil of *Syzygium aromaticum L.* on human neutrophils. *Scientific African* (2022) 18: e01395.

Feyzabadi Z, Ghorbani F, Vazani Y, Zarshenas MM. A Critical Review on Phytochemistry, pharmacology of *Viola odorata* L. and related multipotential products in traditional Persian medicine. *Phytotherapy Research* (2017) 31(11):1669-75.

Gupta A, Atkinson AN, Pandey AK, Bishayee A. Health-promoting and disease-mitigating potential of *Verbascum thapsus* L. (common mullein): A review. *Phytotherapy Research* (2022) 36(4):1507-1522.

Gupta RK, Guha P, Srivastav PP. Phytochemical and biological studies of betel leaf (*Piper betle* L.): Review on paradigm and its potential benefits in human health. *Acta Ecologica Sinica* (2022).

Habibzadeh S, Zohalinezhad ME. Antiviral activity of *Matricaria chamomilla* in the treatment of COVID-19: Molecular Docking study. *European Journal of Integrative Medicine* (2021) 48:101975.

Hai P, He Y, Wang R, Jian Y, Gao Y, Wu X, Chen N, Ye L, Li R. Antimicrobial acylphloroglucinol meroterpenoids and acylphloroglucinols from *Dryopteris crassirhizoma*. *Planta Medica* (2022).

Hein PP, Arunachalam K, Fu Y, Zaw M, Yang Y, Yang X. Diversity of medicinal plants and their therapeutic usages of Kachin people (Jinghpaw) in the central part of Kachin State, Myanmar. *Journal of Ethnopharmacology* (2023) 302:115921.

Heng YW, Ban JJ, Khoo KS, Sit NW. Biological activities and phytochemical content of the rhizome hairs of *Cibotium barometz* (Cibotiaceae*). Industrial Crops and Products* (2020) 153:112612.

Honore PM, Redant S, Djimafo P, Preseau T, Cismas BV, Kaefer K, Gutierrez LB, Anane S, Gallerani A, Attou R. High-dose vitamin C: How an orange can be a lemon. *Critical Care Medicine* (2022) 50(9): e719-e20.

Hossain MS, Sharfaraz A, Dutta A, Ahsan A, Masud MA, Ahmed IA, Goh BH, Urbi Z, Sarker MM, Ming LC. A review of ethnobotany, phytochemistry, antimicrobial pharmacology and toxicology of *Nigella sativa L. Biomedicine & Pharmacotherapy* (2021) 143:112182.

Hossain S, Urbi Z, Karuniawati H, Mohiuddin RB, Moh Qrimida A, Allzrag AM, Ming LC, Pagano E, Capasso R. *Andrographis paniculata* (Burm. f.) Wall. ex Nees: An updated review of phytochemistry, antimicrobial pharmacology, and clinical safety and efficacy. *Life (Basel)* (2021) 11(4).

Khabour OF, Hassanein SFM. Use of vitamin/zinc supplements, medicinal plants, and immune-boosting drinks during COVID-19 pandemic: A pilot study from Benha city, Egypt. *Heliyon* (2021) 7(3): e06538.

Khan SA, Al-Balushi K. Combating COVID-19: The role of drug repurposing and medicinal plants. *Journal of Infection and Public Health* (2021) 14(4):495-503.

Khan T, Khan MA, Mashwani Z-u-R, Ullah N, Nadhman A. Therapeutic potential of medicinal plants against COVID-19: The role of antiviral medicinal metabolites. *Biocatalysis and Agricultural Biotechnology* (2021) 31:101890.

Khanal P, Duyu T, Patil BM, Dey YN, Pasha I, Wanjari M, Gurav SS, Maity A. Network pharmacology of AYUSH recommended immune-boosting medicinal plants against COVID-19. *Journal of Ayurveda and Integrative Medicine* (2022) 13(1):100374.

Kumar M, Hasan M, Sharma A, Suhag R, Maheshwari C, Chandran D, Sharma K, Dhumal S, Senapathy M, Natarajan K, Punniyamoorthy S. *Tinospora cordifolia* (Willd.) Hook.f. & Thomson polysaccharides: A review on extraction, characterization, and bioactivities. *International Journal of Biological Macromole-cules* (2023) 229:463-475.

Li M, Wang A, Zhang Y, Han T, Guan L, Fan D, Liu J, Xu Y. A comprehensive review on ethnobotanical, phytochemical and pharmacological aspects of *Rhus chinensis* Mill. *Journal of Ethnopharmacology* (2022) 293:115288.

Liu Y, Wang AF, Naseem A, Ye HL, Jiang P, Li XM, Wang SY, Pan J, Guan W, Lan W, Yang BY. Phenylpropanoids and triterpenoids from *Tripterygium regelii* and their anti-inflammatory activities. *Phytochemistry Letters* (2022) 49:73-78.

Liu YY, Li YZ, Huang SQ, Zhang HW, Deng C, Song XM, Zhang DD, Wang W. Genus Chloranthus: A comprehensive review of its phytochemistry, pharmacology, and uses. *Arabian Journal of Chemistry* (2022) 15(11):104260.

Lu Y, Bao T, Mo J, Ni J, Chen W. Research advances in bioactive components and health benefits of jujube (*Ziziphus jujuba* Mill.) fruit. *Journal of Zhejiang University Science B* (2021) 22(6):431-449.

Mahajan N, Koul B, Gupta P, Shah BA, Singh J. *Psoralea corylifolia L.*: Panacea to several maladies. *South African Journal of Botany* (2022) 149:963-993.

Maleki S, Akaberi T, Emami SA, Akaberi M. Diterpenes of Scutellaria spp.: Phytochemistry and pharmacology. *Phytochemistry* (2022) 201:113285.

Mlozi SH. The role of natural products from medicinal plants against COVID-19: Traditional medicine practice in Tanzania. *Heliyon* (2022) 8(6): e09739.

Mukherjee PK, Efferth T, Das B, Kar A, Ghosh S, Singha S, Debnath P, Sharma N, Bhardwaj PK, Haldar PK. Role of medicinal plants in inhibiting SARS-CoV-2 and in the management of post-COVID-19 complications. *Phytomedicine* (2022) 98:153930.

Murugesan S, Kottekad S, Crasta I, Sreevathsan S, Usharani D, Perumal MK, Mudliar SN. Targeting COVID-19 (SARS-CoV-2) main protease through active phytocompounds of ayurvedic medicinal plants – *Emblica officinalis* (Amla), *Phyllanthus niruri* Linn. (Bhumi Amla) and *Tinospora cordifolia* (Giloy)-A molecular docking and simulation study. *Computers in Biology and Medicine* (2021) 136:104683.

Mushtaq BS, Hussain MB, Omer R, Toor HA, Waheed M, Shariati MA, Sergey P, Heydari M. *Moringa oleifera* in malnutrition: A comprehensive review. *Current Drug Discovery Technologies* (2021) 18(2):235-243.

Natesh J, Mondal P, Kaur B, Abdul Salam AA, Kasilingam S, Meeran SM. Promising phytochemicals of traditional Himalayan medicinal plants against putative replication and transmission targets of SARS-CoV-2 by computational investigation. *Computers in Biology and Medicine* (2021) 133:104383.

Nguyen TT, Abdelrazek M, Nguyen DT, Aryal S, Nguyen DT, Reddy S, Nguyen QV, Khatami A, Nguyen TT, Hsu EB, Yang S. Origin of novel coronavirus causing COVID-19: A computational biology study using artificial intelligence. *Machine Learning with Applications* (2022) 9:100328.

Nyagumbo E, Pote W, Shopo B, Nyirenda T, Chagonda I, Mapaya RJ, Maunganidze F, Mavengere WN, Mawere C, Mutasa I, Kademeteme E. Medicinal plants used for the management of respiratory diseases in Zimbabwe: Review and perspectives potential management of COVID-19. *Physics and Chemistry of the Earth, Parts A/B/C* (2022) 128:103232.

Oladele JO, Ajayi EI, Oyeleke OM, Oladele OT, Olowookere BD, Adeniyi BM, Oyewole OI, Oladiji AT. A systematic review on COVID-19 pandemic with special emphasis on curative potentials of Nigeria-based medicinal plants. *Heliyon* (2020) 6(9): e04897.

Olawale F, Olofinsan K, Iwaloye O. Biological activities of *Chromolaena odorata*: A mechanistic review. *South African Journal of Botany* (2022) 144:44-57.

Parida PK, Paul D, Chakravorty D. Nature's therapy for COVID-19: Targeting the vital non-structural proteins (NSP) from SARS-CoV-2 with phytochemicals from Indian medicinal plants. *Phytomedicine Plus* (2021) 1(1):100002.

Park YG, Cho JH, Choi J, Ju EM, Adam GO, Hwang DI, Lee JH, An SY, Choi HK, Park CB, Oh HG. Immunomodulatory effects of *Curcuma longa L.* and *Carthamus tinctorius L.* on RAW 264.7 macrophages and cyclophosphamide-induced immunosuppression C57BL/6 mouse models. *Journal of Functional Foods* (2022) 91:105000.

Patil SM, Shirahatti PS, Ramu R. *Azadirachta indica A.* Juss (neem) against diabetes mellitus: a critical review on its phytochemistry, pharmacology, and toxicology. *Journal of Pharmarcy Pharmacology* (2022) 74(5):681-710.

Peng W, Liu Y, Hu M, Zhang M, Yang J, Liang F, Huang Q, Wu C. *Toona sinensis*: A comprehensive review on its traditional usages, phytochemisty, pharmacology and toxicology. *Revista Brasileira de Farmacognosia* (2019) 29(1):111-124.

Qi K, Xia G, Huang G, Huang H. Extraction, chemical modification, and antioxidant activities of *Daucus carota* polysaccharide. *Chemical Biology & Drug Design* (2021) 98(6):1098-1103.

Qin M, Huang Q, Yang X, Yu L, Tang Y, Zhang C, Qin D, Zou W, Deng J, Liu J, Hu H. *Taxillus chinensis* (DC.) Danser: a comprehensive review on botany, traditional uses, phytochemistry, pharmacology, and toxicology. *Chinese Medicine* (2022) 17(1):136.

Qiu T, Shi Y, Wang R, Wang J, Wang W, Zhu J, Wang W, Wu Y, Li K, Liu J. Treatment effects of phosphorylated *Chrysanthemum indicum* polysaccharides on duck viral hepatitis by protecting mitochondrial function from oxidative damage. *Veterinary Microbiology* (2022) 275:109600.

Saeed M, Naveed M, Arif M, Kakar MU, Manzoor R, Abd El-Hack ME, Alagawany M, Tiwari R, Khandia R, Munjal A, Karthik K. Green tea (*Camellia sinensis*) and l-theanine: Medicinal values and beneficial applications in humans-A comprehensive review. *Biomedicine & Pharmacotherapy* (2017) 95:1260-1275.

Saki M, Seyed-Mohammadi S, Montazeri EA, Siahpoosh A, Moosavian M, Latifi SM. *In vitro* antibacterial properties of *Cinnamomum zeylanicum* essential oil against clinical extensively drug-resistant bacteria. *European Journal of Integrative Medicine* (2020) 37:101146.

Sarmah P, Das B, Saikia J, Konwar P, Mudoi KD, Saikia SP, Banik D. An insight into the immunomodulatory potential of wood oil of *Aquilaria malaccensis* Lam. with an emphasis into related phytomedicine, biomarkers, pharmacology, and toxicity. *South African Journal of Botany* (2022) 151:695-712.

Sen IK, Chakraborty I, Mandal AK, Bhanja SK, Patra S, Maity P. A review on antiviral and immunomodulatory polysaccharides from Indian medicinal plants, which may be beneficial to COVID-19-infected patients. *International Journal of Biological Macromolecules* (2021) 181:462-470.

Sharifi-Rad J, Quispe C, Kumar M, Akram M, Amin M, Iqbal M, Koirala N, Sytar O, Kregiel D, Nicola S, Ertani A. Hyssopus essential oil: An update of its phytochemistry, biological activities, and safety profile. *Oxidative Medicine and Cellular Longevity* (2022) 2022:8442734.

Sharifi-Rad M, Mnayer D, Morais-Braga MF, Carneiro JN, Bezerra CF, Coutinho HD, Salehi B, Martorell M, del Mar Contreras M, Soltani-Nejad A, Uribe YA. Echinacea plants as antioxidant and antibacterial agents: From traditional medicine to biotechnological applications. *Phytotherapy Research* (2018) 32(9):1653-1663.

Singh N, Yarla NS, Siddiqi NJ, de Lourdes Pereira M, Sharma B. Features, pharmacological chemistry, molecular mechanism and health benefits of lemon. *Medicinal Chemistry* (2021) 17(3):187-202.

Solo P, Doss MA. Potential inhibitors of SARS-CoV-2 (COVID 19) spike protein of the delta and delta plus variant: *In silico* studies of medicinal plants of North-East India. *Current Research in Pharmacology and Drug Discovery* (2021) 2:100065.

Stringaro A, Colone M, Angiolella L. Antioxidant, antifungal, antibiofilm, and cytotoxic activities of Mentha spp. essential oils. *Medicines (Basel)* (2018) 5(4).

Tahir ul Qamar M, Alqahtani SM, Alamri MA, Chen L-L. Structural basis of SARS-CoV-2 3CLpro and anti-COVID-19 drug discovery from medicinal plants. *Journal of Pharmaceutical Analysis* (2020) 10(4):313-319.

Tardugno R, Serio A, Purgatorio C, Savini V, Paparella A, Benvenuti S. *Thymus vulgaris L.* essential oils from *Emilia romagna* apennines (Italy): Phytochemical composition and antimicrobial activity on food-borne pathogens. *Natural Product Research* (2022) 36(3):837-842.

Verma D, Mitra D, Paul M, Chaudhary P, Kamboj A, Thatoi H, Janmeda P, Jain D, Panneerselvam P, Shrivastav R, Pant K. Potential inhibitors of SARS-CoV-2 (COVID-19) proteases PLpro and Mpro/ 3CLpro: molecular docking and simulation studies of three pertinent medicinal plant natural components. *Current Research in Pharmacology and Drug Discovery* (2021) 2:100038.

Yang F, Jiang XL, Tariq A, Sadia S, Ahmed Z, Sardans J, Aleem M, Ullah R, Bussmann RW. Potential medicinal plants involved in inhibiting 3CLpro activity: A practical alternate approach to combating COVID-19. *Journal of Integrative Medicine* (2022) 20(6):488-496.

Zahin M, Bokhari NA, Ahmad I, Husain FM, Althubiani AS, Alruways MW, Perveen K, Shalawi M. Antioxidant, antibacterial, and antimutagenic activity of *Piper nigrum* seeds extracts. *Saudi Journal of Biological Sciences* (2021) 28(9):5094-5105.

Zamakshshari N, Ahmed IA, Nasharuddin MNA, Mohd Hashim N, Mustafa MR, Othman R, Noordin MI. Effect of extraction procedure on the yield and biological activities of hydroxychavicol from *Piper betle L.* leaves. *Journal of Applied Research on Medicinal and Aromatic Plants* (2021) 24:100320.

Zhang W, An EK, Park HB, Hwang J, Dhananjay Y, Kim SJ, Eom HY, Oda T, Kwak M, Lee PC, Jin JO. *Ecklonia cava* fucoidan has potential to stimulate natural killer cells *in vivo*. *International Journal of Biology Macromolecules* (2021) 185:111-21.

Zheng Y, Fang D, Huang C, Zhao L, Gan L, Chen Y, Liu F. *Gentiana scabra* restrains hepatic pro-inflammatory macrophages to ameliorate non-alcoholic fatty liver disease. *Frontiers in Pharmacology* (2021) 12:816032.

Chapter 5

Traditional Plant-Based Therapies for SARS-CoV-2 Infections

Chun Hoe Tan*, **PhD**
Erlina Abdullah, PhD
Hema Priyaa Vijayan
and Rathimalar Ayakannu, PhD
Department of Biotechnology, Faculty of Applied Science, Lincoln University College, Selangor, Malaysia

Abstract

Coronavirus disease (COVID-19), which is caused by severe acute respiratory syndrome coronavirus-2 (SARS-CoV-2) infection has made a great negative impact on our healthcare system since 2019. To the best of the authors' knowledge, there is no specific and curable treatment for the disease and the current therapeutic strategies are only dealing with the symptoms of the disease (symptomatic treatment) with prevention aimed at reducing community transmission (vaccination). This situation has brought intensive research into traditional systems of medicine. Traditional medicine using herbal extracts has been developed to combat various illnesses since ancient times. The excellent and promising pharmacological properties of medicinal plants have paved the way for developing therapy against novel SARS-CoV-2 infection. The current book chapter aims to summarize the data of traditional plant-based therapies, i.e., traditional Chinese and Indian medicine to warrant their use in COVID-19 management. A total of four relevant traditional polyherbal Chinese formulations and two traditional Indian formulations

* Corresponding Author's Email: chunhoe@lincoln.edu.my.

In: Medicinal Plants and COVID-19
Editor: Asita Elengoe
ISBN: 979-8-88697-958-9

were described herein; all of which exhibit antiviral, antipyretic, and anticough properties against infection along with potent immunomodulatory activities.

Keywords: antipyretic, herbs, traditional Chinese medicine, traditional Indian medicine

Introduction

Traditional plant-based therapies have been playing an important role in the control of the epidemics caused by human coronaviruses, such as severe acute respiratory syndrome coronavirus (SARS) in 2002, swine flu (H1N1) in 2009, Middle Eastern Respiratory Syndrome (MERS) in 2012, and COVID-19 in 2019. Coronavirus disease (COVID-19), which is an infectious respiratory disease that is caused by SARS-CoV-2 has been reported to affect millions of people worldwide and has shown significantly high morbidity and mortality rates within a very short period. With the severe progression of the COVID-19 epidemic without a specific and curable drug therapy for this disease, there has been a broad range of efforts made by traditional medicine practitioners, clinicians, pharmaceutical companies, researchers, and governments to make traditional plant-based medicines as an alternative option in treating various stages of patients with COVID-19 (Girija et al., 2022; Sun, 2022). Two major traditional plant-based therapies have proven high efficacy in the treatment of COVID-19, VIZ., traditional Chinese medicine (TCM) and traditional Indian medicine (TIM) system, both of which are a vast reservoir of medicinal plants.

Traditional Chinese Medicine (TCM)

Originated from the Shang dynasty (14th–11th centuries BCE) in China, traditional Chinese medicine (TCM) has a long history of use in the East Asian cultural sphere, which encompasses many countries that were historically influenced by Chinese cultures, such as Japan, Korea, Malaysia (partly), etc. Herbal therapy is a major component of TCM which comprises the application of the Chinese herbal extract (e.g., decoction) and Chinese patent medicine during the treatment, the latter refers to herbal medicines that are innovated into an available form, such as tablets, oral solutions, and dry suspensions.

Several TCM prescriptions are derived from various ancient Chinese medical books, such as *Huang Di Nei Jing* (Inner Canon of the Yellow Emperor), *Shang Han Za Bing Lun* (Treatise on Febrile and Miscellaneous Diseases), and *Wen Yi Lun* (Theory of Epidemic Febrile Diseases) Collectively, these books described SARS-Cov-2 infection as an introduction of *evil/poisonous gas* to the lung which is highly infectious and epidemic. In addition, the disease is caused by a cold pathogen that leads to cold, body aches, fatigue, and severe internal injury (collateral disease) (Ren et al., 2021). According to the Pneumonia Treatment Protocol for Novel Coronavirus Infection in TCM, COVID-19 belongs to the category of "plague diseases" located in the lungs, causing "dampness, heat, toxicity, and blood stasis" in the body system (Ling et al., 2022). In this regard, the approach used by the TCM generally includes expelling dampness, heat, and toxins, repairing the injury to the viscera and tissues, as well as supplementing healthy *qi* (vital energy that circulates through the body at all times) (Jia and Wu, 2021). During the COVID-19 pandemic, TCM is often consumed by patients to treat and prevent the disease (Zhao et al., 2021), enhance immunity, and tune up their physiological condition during the recovery process by modifying the microenvironment of the tissues and cells (Lu, 2015; Xu et al., 2022).

A total of 40 representative clinical trials have been conducted on the efficacy of TCM in the treatment of COVID-19. In addition, plenty of clinical data provided strong evidence that combined TCM and Western medicine treatment have significantly improved the recovery of COVID-19 patients (Lyu et al., 2021). To the best of the authors' knowledge, there are more than 18 recommended TCM used in the prevention and treatment of COVID-19 at different stages of the disease (Luo et al., 2020). Among others, the highly recommended three TCM are Lianhua Qingwen capsule (granules), Jinhua Qinggan granules, and Xuebijing injection (Lyu et al., 2021).

***Lianhua Qingwen* (LHQW) Capsule/Granule**

Lianhua Qingwen (LHQW) is a Chinese patent medicine developed by Shijiazhuang Yiling Pharmaceutical for the treatment of colds and influenza. It is a capsule and granular form of a mixture composed of 13 herbs. During the SARS pandemic outbreak in 2003, it was approved expeditiously as a new drug in the same year by the China National Medical Products Administration. Since then, it has been a principal TCM used by the global community in the subsequent viral outbreak pandemic, e.g., the H1N1 influenza pandemic in

2009 and the COVID-19 epidemic in 2020 (Alam et al., 2021; Jia and Wu, 2021).

Table 1. The composition of LHQW

No.	Common name	Scientific name	Biological activities
1.	Forsythia fruit	*Forsythia suspensa* (Thunb.) Vahl.	To reduce heat, to relieve toxicity and inflammation (Sung et al., 2016)
2.	Honeysuckle flower	*Lonicera japonica* Thunb.	To reduce/treat inflammation (Du et al., 2020)
3.	Fried ephedra stem	*Ephedra sinica* Stapf.	To diffuse and descend the *Qi* in the lungs, moisten the lung, and stop wheezing (Yang and Al-Khafaji, 2010)
4.	Dryopteris root	*Dryopteris crassirhizoma* Nakai.	To expel worms, stop bleeding, and treat the common cold (Hwang et al., 2013)
5.	Isatis root	*Isatis indigotica* Fort.	To treat colds (Zhang et al., 2020)
6.	Gypsum	*Gypsum fibrosum*	To reduce fevers and alleviate thirst (Ikarashi et al., 2013)
7.	Patchouli	*Pogostemon cablin* (Blanco) Benth.	To eliminate heat and dampness, calm nerves, and alleviate fatigue (Kim et al., 2015)
8.	Integripetal rhodiola herb (root and rhizome)	*Rhodiola crenulata* (Hook. f. et Thoms.) H. Ohba	To clear heat in the lungs and eliminate toxins in the body (Lee et al., 2020)
9.	Houttuynia	*Houttuynia cordata* Thunb.	To clear heat, eliminate toxins, and treat pulmonary symptoms, e.g., lung abscess, dyspnea, phlegm, and cough (Lu et al., 2018)
10.	Rhubarb root and rhizome	*Rheum palmatum* L.	To clear heat (Kong et al., 2014)
11.	Fried apricot seed	*Prunus armeniaca* L. var. ansu Maxim.	To treat the common cold, coughs, bronchial asthma, rheumatism, and constipation (Koh and Tham, 2011)
12.	Licorice root	*Glycyrrhiza uralensis* Fisch.	To remove toxins (Gong et al., 2015)
13.	Menthol	-	To reduce pain, to treat cough and headache (Lin et al., 2002)

In general, LHQW is highly effective in reducing heat, ventilating the lungs, relieving cough, resolving phlegm, and removing the toxin, all of which are attributed to the herbal components (Table 1) that mainly exhibit anti-viral/anti-influenza, anti-inflammation, and antipyretic activities. Mechanical studies on LHQW show that it involves the inhibition of virus shedding, suppression of virus-induced NF-κB activation, down-regulation of virus-induced IL-6, IL-8, TNF-α, IP-10, and MCP-1, reduction of the nuclear export of the viral RNP, and decrease inflammatory cytokines level in the early stage of infection (reduce cytokine storm) (Zhang et al., 2020). Nonetheless, the most common side effects after consumption of LHQW are gastrointestinal system damage, and skin and appendage damage (Hu et al., 2022).

Jinhua Qinggan (JHQG) Granules

Jinhua Qinggan (JHQG) granule is composed of 12 Chinese herbs (Table 2) that were developed based on two classic formulae that are used to treat respiratory infections, i.e., Maxingshigan decoction and Yinqiao powder (Lyu et al., 2021). It was initially formulated for the treatment of swine flu (H1N1 influenza) in 2009 (Kageyama et al., 2022).

Table 2. The composition of JHQG

No.	Common name	Scientific name	Biological activities
1.	Honeysuckle flower	*Lonicera japonica* Thunb.	To reduce/treat inflammation (Du et al., 2020)
2.	Gypsum	*Gypsum fibrosum*	To reduce fevers and alleviate thirst (Ikarashi et al., 2013)
3.	Fried ephedra stem	*Ephedra sinica* Stapf	To diffuse and descend the *Qi* in the lungs, moisten the lung, and stop wheezing (Yang and Al-Khafaji, 2010)
4.	Siberian apricot	*Prunus sibirica* L.	To treat respiratory disorders, such as coughing, wheezing, asthma, emphysema, and bronchitis (Qin et al., 2019)
5.	Chinese skullcap	*Scutellaria baicalensis* Georgi.	To expel the heat of the upper energizer (Ye et al., 2021)
6.	Forsythia fruit	*Forsythia suspensa*	To reduce heat, to relieve toxicity and inflammation (Sung et al., 2016)
7.	Fritillaria bulb	*Fritillaria thunbergii* Miq	To remove toxins, eliminate phlegm, release stagnated lung-qi, and treat "wind-fire syndrome" (Nile et al., 2021)
8.	Anemarrhena rhizome	*Anemarrhena asphodeloides* Bunge	To clear heat and treat cough, with antiviral effects (Liu et al., 2023)
9.	Burdock	*Arctium lappa* L.	To treat fever, relieve throat pain, and detoxify blood (Yosri et al., 2023)
10.	Sweet wormwood	*Artemisia annua* L.	To treat fever (Lee et al., 2023)
11.	Peppermint	*Mentha haplocalyx* Briq.	To treat fever, cold, cough, and influenza (Zhao et al., 2018)
12.	Licorice	*Glycyrrhiza glabra* L.	To treat painful swellings, coughs, colds, and influenza (Bisht et al., 2022).

Following its potential therapeutic efficacy for H1N1 influenza, JHQG granules have been highly recommended for use in patients suspected of SARS-CoV-2 infection to prevent the onset and progression of the disease. In several randomized controlled trials on patients with SARS-CoV-2 infection, the co-administration of JHQG granules and conventional medicine has displayed significant improvement in viral clearance, respiratory symptoms alleviation, recovery time from pneumonia, moderation of psychological anxiety, reduction of fever, and enhancement of appetite compared with the administration of conventional medicine alone (An et al., 2021; Kageyama et al., 2022; Shah et al., 2022; Wang et al., 2011). In addition, computational studies unveiled that the mechanism of action might be attributed to its active components (e.g., kaempferol, baicalein, and oroxylin A) which regulate multiple signaling pathways and molecular targets by binding with SARS-CoV-2 3CL hydrolase, thereby downregulating angiotensin I converting enzyme 2, the viral entry receptor (Kageyama et al., 2022; Luo et al., 2020).

Table 3. The composition of XBJ

No.	Common name	Scientific name	Biological activities
1.	Safflower	*Carthamus tinctorius*	To promote blood circulation, remove blood stasis, and reduce pain (Zhou et al., 2014)
2.	Chinese peony	*Paeonia lactiflora*	To treat pain, inflammation, and immune disorders (Zhu et al., 2022b)
3.	Sichuan lovage rhizome	*Ligusticum chuanxiong*	To improve blood circulation by circulating "Qi," dispel wind, and relieve pain (Wang et al., 2022)
4.	Female ginseng	*Angelica sinensis*	To detoxify the blood and activate blood circulation (Ping et al., 2022)
5.	Red sage	*Salvia miltiorrhiza*	To promote blood circulation and remove blood stasis (Jiao et al., 2022)

Xuebijing Injection

The critical stage of SARS-CoV-2 infection leads to cytokine storm and immune disorder, causing inflammatory injury, lung fibrosis, and sepsis. At this critical stage, Xuebijing (XBJ) injection combined with antiviral drugs will be prescribed to reduce lung lesions and provide protection to vital organs by regulating immune inflammatory response (Wang et al., 2022; Xing et al., 2020; Zhu et al., 2022b). The Xuebijing (XBJ) injection is prepared by water extraction from five medicinal plants (Table 3) based on the XueFuZhuYu (XFZY) decoction (Tianyu and Liying, 2021; Zhu et al., 2022b). Particularly, mechanistic studies conducted by Xing et al. (2020) indicated that quercetin, luteolin, apigenin, and other components in XBJ could affect TNF, MAPK1, and IL6, thereby reducing the inflammation of severe COVID-19 patients.

Table 4. The composition of TJQW

No.	Common name	Scientific name	Biological activities
1.	Forsythia fruit	*Forsythiae fructus*	To reduce heat from the lung (Ye et al., 2021)
2.	Edible tulip	*Pseudobulbus cremastrae seu Pleiones*	To clear heat, remove the plague toxin, remove phlegm, and open the orifices (Ye et al., 2021; Zhu et al., 2022a)
3.	Japanese honeysuckle flower	*Lonicera japonica*	To remove the heat of the lungs (Ye et al., 2021)
4.	Chinese skullcap	*Scutellariae baicalensis* Georgi.	To remove the heat of the lungs (Ye et al., 2021)
5.	Isatis Leaf	*Folium isatidis*	To remove the heat of the lungs (Ye et al., 2021), to improve the symptom of sore throat, cold, and fever (Chen et al., 2022)
6.	Bupleurum root	*Bupleurum chinense* DC. and *Bupleurum scorzonerifolium* Willd	To treat influenza, fever, and inflammation (Yang et al., 2017)
7.	Sweet wormwood	*Artemisia annua*	To treat fever (Lee et al., 2023)
8.	Cicada slough	*Periostracum cicadae*	To treat soreness of the throat and reduce fever (Yang et al., 2015)
9.	Peucedanum root	*Peucedani radix*	To treat a cough with thick sputum and dyspnea, non-productive cough, and upper respiratory infections (Song et al., 2014)
10.	Yellow Himalayan fritillary	*Fritillaria cirrhosa*	To treat respiratory diseases, such as cough (Wang et al., 2016)
11.	Fritillaria bulb	*Fritillaria thunbergii*	To remove toxins, eliminate phlegm, release stagnated lung-qi, and treat "wind-fire syndrome" (Nile et al., 2021)
12.	Poria mushroom	*Poria cocos*	To treat inflammation (Yu et al., 2022)
13.	Japanese Apricot	*Fructus mume*	To relieve cough, treat ulceration, and improve digestive function (Chen et al., 2011)
14.	Figwort root	*Radix scrophulariae*	To treat phthisis with cough and sore throat (Huang et al., 2012)
15.	Mongolian milkvetch	*Astragalus propinquus*	To prevent respiratory tract infection (Sivakumar et al., 2021)
16.	Prince ginseng	*Radix pseudostellariae*	To improve immunity and appetite, nourish vitality, and moisten lungs (Cai et al., 2020)

Toujie Quwen (TJQW) Granules

Another TCM that is deemed worthy of description is Toujie Quwen (TJQW) granule, which was the first TCM used by China Guangzhou eighth people's hospital to treat the COVID-19 patient (Ye et al., 2021). TJQW includes 16 TCM components (Table 4) (Huang et al., 2020), and some of them are also found in LHQW capsules.

Similar to LHQW, TJQW exhibited toxin removal and heat-clearing effects attributed to the detoxification and antipyretic activities of individual components in the granules. The pharmacological targets of TJQW in modulating COVID-19 symptoms include caspase 3, STAT3, ESR1, FPR2, BCL2L1, BDKRB2, myeloperoxidase, and angiotensin-converting enzyme. In addition, TJQW was found to be able to manipulate the estimated glomerular filtration rate. A case series study involving 121 mild-stage patients revealed that TJQW treatment for six days significantly improved the proportion of patients ranging from 71-84% (Luo et al., 2021).

Traditional Indian Medicine (TIM)

The TIM, which comprises Ayurveda, Yoga, Unani, Siddha, and Homoeopathy (AYUSH) is one of the oldest medical practices globally and has been providing holistic healthcare service to human civilization since its inception (Adhikari and Paul, 2018). In particular, Ayurveda and Siddha were developed through daily skills of utilizing natural resources, such as herbal extracts for maintaining good health and treating various diseases. During the COVID-19 pandemic, TIM practitioners have been using herbal extracts to treat respiratory clinical indications, as they are effective, safe (minimal adverse reactions), well-tolerated, and widely accepted across all ages (Mahaboob Ali et al., 2022; Srivastava et al., 2021). According to Siddha Medical Literature, the symptoms and signs of COVID-19 are similar to Kaba Suram, including cold, cough, and fever (Srivastava et al., 2021).

In India, Kabasura Kudineer (KSK) and Nilavembu Kudineer (NVK) are the two most infamous formulations that are currently used in the treatment of COVID-19. Many of the studies focus on computational analyses and the potentiality of their phytoconstituents to interact with SARS-CoV-2 proteins (Mahaboob Ali et al., 2022). The currently available TIM (both alone and combinatory treatment with Western medicine) is effective in shortening the recovery period of asymptomatic and mildly symptomatic COVID-19 (Mahaboob Ali et al., 2022).

Kabasura Kudineer (KSK)

Kabasura Kudineer (KSK) is a Siddha formulation described in the Siddha manuscript 'Citta Vaittiyattirattu'. It is currently used for the treatment of asymptomatic, mild to moderate COVID-19 in India (Tamil Nadu) by improving immunity and treating several respiratory symptoms, e.g., flu-like fever (Jeyanthi and Kumar, 2020; Kiran et al., 2022; Mahaboob Ali et al., 2022; Natarajan et al., 2021). The formulation contains 15 plants (Table 5) and 32 phytoconstituents, in which the majority of these plants possess a bitter taste.

The main objective of the consumption of this formulation is to increase and normalize *pitham* (heat) and eliminate *kapham* (phlegm) from the lungs, attributed to the antipyretic and anticough effects of the plant components in the formulation. An *in silico* study conducted by Kiran et al. (2022) unveiled that nine phytoconstituents (from six plants extracts) of KSK, such as magnoflorine, 5-hydroxy-7,8-dimethoxyflavanone, tinosponone, cirsimaritin, chrysoeriol, 6-methoxygenkwanin, vasicinone, quercetin, and luteolin have a potent binding affinity with SARS-CoV-2 spike protein. In addition, the calculation conducted by the same research group has also shown that vasicine, andrographolide, ursolic acid, quercetin, and medicine were able to inhibit the ACE2 receptor (Kiran et al., 2022).

Table 5. The composition of KSK

No.	Common name	Scientific name	Biological activities
1.	Ginger	*Zingiber officinale* Rosc.	To treat the common cold, fever, sore throat, and pain (Mbaveng and Kuete, 2017)
2.	Long pepper	*Piper longum* L.	To remove endotoxin (Yadav et al., 2020)
3.	Clove	*Syzygium aromaticum* (L.) Merr. and Perry	Anti-inflammatory effect (Habtemariam, 2019)
4.	Atlas daisy	*Anacyclus pyrethrum* (L.) Lag.	To treat respiratory infections (Alliouche Kerboua et al., 2021)
5.	Indian stinging nettle	*Tragia involucrata* L.	To treat fever and cough (Pallie et al., 2020)
6.	Neermulli in Tamil	*Hygrophila auriculata* (Schum.) Heine Lam.	To treat cough (Sethiya et al., 2018)

Table 5. (Continued)

No.	Common name	Scientific name	Biological activities
7.	Myrobalan	*Terminalia chebula* Retz.	To relieve fever with antiviral effects (Hassan Bulbul et al., 2022)
8.	Malabar nut	*Justicia adhatoda* L.	To treat coughs and colds (Gulati et al., 2016)
9.	Mexican mint	*Plectranthus amboinicus* (Lour.) Spreng.	To treat headaches, cough, fever, and cold (Gonçalves et al., 2012)
10.	Crepe ginger	*Costus speciosus* (J. Koenig) Sm.	To treat fever (Sohrab et al., 2021)
11.	Malabar gulbel	*Tinospora sinensis* (Lour.) Merr.	To treat fever (Chi et al., 2016)
12.	Blue fountain bush	*Clerodendrum serratum* (L.) Moon	To treat cough, fever, inflammation, pain, and some respiratory disorders (Patel et al., 2014)
13.	King of bitters	*Andrographis paniculata* (Burm.f.) ex Nees	To treat fever, upper respiratory infection, and flu (Kumar et al., 2021)
14.	Nut grass	*Cyperus rotundus* L.	To treat fever (Peerzada et al., 2015)
15.	Velvetleaf	*Cissampelos pareira* L.	To treat fever, cough, sore throat, and headache (Kumari et al., 2021)

Table 6. The composition of NVK

No.	Common name	Scientific name	Biological activities
1.	Ginger	*Zingiber officinale* Rosc.	To treat the common cold, fever, sore throat, and pain (Mbaveng and Kuete, 2017)
2.	Indian sandalwood	*Santalum album* Linn	To treat fever (Kumar et al., 2015)
3.	Black pepper	*Piper nigrum* Linn	To treat colds, coughs, and fevers (Ashokkumar et al., 2021)
4.	King of bitters	*Andrographis paniculata* (Burm) Wall. ex Nees	To treat fever, upper respiratory infection, and flu (Kumar et al., 2021)
5.	Nut grass	*Cyperus rotundus* Linn	To treat fever (Peerzada et al., 2015)
6.	Vetiver grass	*Vetiveria zizanioide* (Linn) Nash	To treat fever and headache (Chou et al., 2012)
7.	Flat-top mille graines	*Hedyotis corymbosa* (Linn.) Lam.	To treat colds and coughs (Das et al., 2019)
8.	Vetiver grass	*Plectranthus vettiveroides* (Linn.) Nash.	To treat fever (Velvizhi et al., 2020)
9.	Snake tomato	*Trichosanthes cucumerina* Linn.	To treat fever (Arawwawala et al., 2020)

Nilavembu Kudineer (NVK)

Nilavembu Kudineer (NVK) is Siddha medicine and was used in the treatment of chikungunya and dengue (Srivastava et al., 2021). It consists of nine plants to treat respiratory disorders due to their anti-inflammatory, anti-analgesic, and antipyretic effects (Table 6) (Anbarasu et al., 2011). For the management of COVID-19, NVK prevents the entry of the SARS-Cov-2 virus into the host body through interaction with the ACE2 receptor (Anbarasu et al., 2011).

Conclusion

The current book chapter summarized the traditional plant-based therapies from the Chinese and Indian medicine systems. It is well-noted that most of the traditional formulations consist of medicinal plant extracts that exhibit antipyretic, anti-cough, anti-inflammation, and immunomodulatory effects. Various clinical results showed that these formulations can effectively modulate COVID-19 symptoms, enhance patients' recovery, prevent patients from progressing into critical conditions, and decrease the mortality rate. In addition, these clinical results also provided verification that the precious knowledge and wisdom bequeathed by the ancestors are relevant and practically useful even in the modern era.

References

Adhikari PP, Paul SB. History of Indian traditional medicine: A medical inheritance. *Asian Journal of Pharmaceutical and Clinical Research* (2018) 11(1):421-426.

Alam S, Sarker MMR, Afrin S, Richi FT, Zhao C, Zhou J-R, Mohamed IN. Traditional herbal medicines, bioactive metabolites, and plant products against COVID-19: Update on clinical trials and mechanism of actions. *Frontiers in Pharmacology* (2021) 12.

Alliouche Kerboua K, Benosmane L, Namoune S, Ouled-Diaf K, Ghaliaoui N, Bendjeddou D. Anti-inflammatory and antioxidant activity of the hot water-soluble polysaccharides from *Anacyclus pyrethrum* (L.) Lag. roots. *Journal of Ethnopharmacology* (2021) 281: 114491.

An X, Xu X, Xiao M, Min X, Lyu Y, Tian J, Ke J, Lang S, Zhang Q, Fan A, Liu B. Efficacy of Jinhua Qinggan granules combined with Western medicine in the treatment of confirmed and suspected COVID-19: A randomized controlled trial. *Frontiers in Medicine* (2021) 8:728055.

Anbarasu K, Manisenthil KKT, Ramachandran S. Antipyretic, anti-inflammatory and analgesic properties of Nilavembu Kudineer choornam: a classical preparation used in the treatment of chikungunya fever. *Asian Pacific Journal of Tropical Medicine* (2011) 4(10):819-823.

Arawwawala L, Thabrew M, Arambewela L. A review of the pharmacological properties of *Trichosanthes cucumerina* Linn of Sri Lankan origin. *Unique Journal of Pharmaceutical and Biological Sciences* (2020) 1(1):3-6.

Ashokkumar K, Murugan M, Dhanya MK, Pandian A, Warkentin TD. Phytochemistry and therapeutic potential of black pepper [*Piper nigrum* (L.)] essential oil and piperine: a review. *Clinical Phytoscience* (2021) 7(1): 52.

Bisht D, Rashid M, Arya RKK, Kumar D, Chaudhary SK, Rana VS, Sethiya NK. Revisiting liquorice (*Glycyrrhiza glabra* L.) as anti-inflammatory, antivirals and immunomodulators: Potential pharmacological applications with mechanistic insight. *Phytomedicine Plus* (2022) 2(1):100206.

Cai X, Yang Q, Weng Q, Wang S. pH Sensitive doxorubicin-loaded nanoparticle based on Radix pseudostellariae protein-polysaccharide conjugate and its improvement on HepG2 cellular uptake of doxorubicin. *Food and Chemical Toxicology* (2020) 136:111099.

Chen J, Zhu Z, Gao T, Chen Y, Yang Q, Fu C, Zhu Y, Wang F, Liao W. *Isatidis radix* and *Isatidis folium*: A systematic review on ethnopharmacology, phytochemistry and pharmacology. *Journal of Ethnopharmacology* (2022) 283:114648.

Chen Y, Wong RWK, Seneviratne CJ, Hägg U, McGrath C, Samaranayake LP, Kao R. The antimicrobial efficacy of *Fructus mume* extract on orthodontic bracket: A monospecies-biofilm model study *in vitro*. *Archives of Oral Biology* (2011) 56(1):16-21.

Chi S, She G, Han D, Wang W, Liu Z, Liu B. Genus *Tinospora*: Ethnopharmacology, phytochemistry, and pharmacology. *Evidence-based Complementary and Alternative Medicine* (2016) 2016: 9232593.

Chou S-T, Lai C-P, Lin C-C, Shih Y. Study of the chemical composition, antioxidant activity and anti-inflammatory activity of essential oil from *Vetiveria zizanioides*. *Food Chemistry* (2012) 134(1): 262-268.

Das S, Mondal N, Mondal S, Ghosh P, Ghosh C, Das C, Chatterjee S. Botanical features, phytochemical and pharmacological overviews of *Oldenlandia corymbosa* Linn.: A brief review. *The Pharma Innovation Journal* (2019) 8(2):464-468.

Du L, Li D, Zhang J, Du J, Luo Q, Xiong J. Elicitation of *Lonicera japonica* thunb suspension cell for enhancement of secondary metabolites and antioxidant activity. *Industrial Crops and Products* (2020) 156:112877.

Girija PLT, Sivan N, Naik P, Murugavel YA, Ravindranath TM, Krishnaswami CV. Standalone ayurvedic treatment of high-risk COVID-19 patients with multiple co-morbidities: A case series. *Journal of Ayurveda and Integrative Medicine* (2022) 13(1):100466.

Gonçalves TB, Braga MA, de Oliveira FF, Santiago GM, Carvalho CB, e Cabral PB, de Melo Santiago T, Sousa JS, Barros EB, do Nascimento RF, Nagao-Dias AT. Effect of subinihibitory and inhibitory concentrations of *Plectranthus amboinicus* (Lour.)

Spreng essential oil on *Klebsiella pneumoniae*. *Phytomedicine* (2012) 19(11):962-968.

Gong H, Zhang BK, Yan M, Fang PF, Hu CP, Yang Y, Cao P, Jiang P, Fan XR. A protective mechanism of licorice (*Glycyrrhiza uralensis*): Isoliquiritigenin stimulates detoxification system via Nrf2 activation. *Journal of Ethnopharmacology* (2015) 162:134-139.

Gulati K, Rai N, Chaudhary S, Ray A. (2016). Chapter 6 - Nutraceuticals in Respiratory Disorders. In Gupta, R. C. (Ed.), *Nutraceuticals* (pp. 75-86). Boston: Academic Press.

Habtemariam S. (2019). Chapter 16 - The chemical and pharmacological basis of cloves (*Syzygium aromaticum* (L.) Merr. and L.M.Perry) as potential therapy for type 2 diabetes and associated diseases. In Habtemariam, S. (Ed.), *Medicinal Foods as Potential Therapies for Type-2 Diabetes and Associated Diseases* (pp. 551-578): Academic Press.

Hassan Bulbul MR, Uddin Chowdhury MN, Naima TA, Sami SA, Imtiaj MS, Huda N, Uddin M G. A comprehensive review on the diverse pharmacological perspectives of *Terminalia chebula* Retz. *Heliyon* (2022) 8(8):e10220.

Hu C, He B, Gong F, Liang M, Zhao D, Zhang G. The adverse reactions of Lianhua Qingwen capsule/granule compared with conventional drug in clinical application: A meta-analysis. *Frontiers in Pharmacology* (2022) 13: 764774.

Huang XY, Chen CX, Zhang XM, Liu Y, Wu XM, Li YM. Effects of ethanolic extract from *Radix scrophulariae* on ventricular remodeling in rats. *Phytomedicine* (2012) 19(3):193-205.

Huang Y, Zheng WJ, Ni YS, Li MS, Chen JK, Liu XH, Tan XH, Li JQ. Therapeutic mechanism of Toujie Quwen granules in COVID-19 based on network pharmacology. *BioData Mining* (2020) 13(1):1-21.

Hwang Y-H, Ha H, Ma JY. Acute oral toxicity and genotoxicity of *Dryopteris crassirhizoma*. *Journal of Ethnopharmacology* (2013) 149(1):133-139.

Ikarashi N, Ogiue N, Toyoda E, Nakamura M, Kon R, Kusunoki Y, Aburada T, Ishii M, Tanaka Y, Machida Y, Ochiai W. Elucidating the mechanism by which *Gypsum fibrosum*, a traditional Chinese medicine, maintains cutaneous water content. *Biological and Pharmaceutical Bulletin* (2013) 36(10):1615-1621.

Jeyanthi V, Kumar G. COVID-19 outbreak: An overview and India's perspectives on the management of infection. *Indian Journal of Science and Technology* (2020) 13(36): 3716-3724.

Jia Z, Wu Y. Clinical applications and pharmacological research progress of Lianhua Qingwen capsules/granules. *Journal of Traditional Chinese Medical Sciences* (2021) 8(2):101-109.

Jiao T, Liu Y, Han M, Wang S, Zhang X, Zhang X, Zhao Z. Protective effect and possible mechanisms of *Salvia miltiorrhiza* Bge. for the treatment of diabetic nephropathy: A systematic review and meta-analysis of animal studies. *Journal of Traditional Chinese Medical Sciences* (2022) 9(4):351-364.

Kageyama Y, Aida K, Kawauchi K, Morimoto M, Ebisui T, Akiyama T, Nakamura T. Jinhua Qinggan granule, a Chinese herbal medicine against COVID-19, induces rapid changes in the neutrophil/lymphocyte ratio and plasma levels of IL-6 and IFN-γ: An open-label, single-arm pilot study. *World Academy of Sciences Journal* (2022) 4(1): 2.

Kim KH, Beemelmanns C, Clardy J, Cao S. A new antibacterial octaketide and cytotoxic phenylethanoid glycosides from *Pogostemon cablin* (Blanco) Benth. *Bioorganic and Medicinal Chemistry Letters* (2015) 25(14):2834-2836.

Kiran G, Karthik L, Shree Devi MS, Sathiyarajeswaran P, Kanakavalli K, Kumar KM, Ramesh Kumar D. *In silico* computational screening of Kabasura Kudineer - Official siddha formulation and JACOM against SARS-CoV-2 spike protein. *Journal of Ayurveda and Integrative Medicine* (2022) 13(1):100324.

Koh KH, Tham FY. Screening of traditional Chinese medicinal plants for quorum-sensing inhibitors activity. *Journal of Microbiology, Immunology and Infection* (2011) 44(2):144-148.

Kong X, Wan H, Su X, Zhang C, Yang Y, Li X, Yao L, and Lin N. *Rheum palmatum* L. and *Coptis chinensis* Franch., exert antipyretic effect on yeast-induced pyrexia rats involving regulation of TRPV1 and TRPM8 expression. *Journal of Ethnopharmacology* (2014) 153(1):160-168.

Kumar R, Anjum N, and Tripathi Y. Phytochemistry and pharmacology of *Santalum album* L.: A review. *World Journal of Pharmaceutical Research* (2015) 4(10):1842-1876.

Kumar S, Singh B, and Bajpai V. *Andrographis paniculata* (Burm.f.) Nees: Traditional uses, phytochemistry, pharmacological properties and quality control/quality assurance. *Journal of Ethnopharmacology* (2021) 275:114054.

Kumari S, Anmol Bhatt V, Suresh PS, and Sharma U. *Cissampelos pareira* L.: A review of its traditional uses, phytochemistry, and pharmacology. *Journal of Ethnopharmacology* (2021) 274: 113850.

Lee BJ, Weyers M, Haynes RK, van der Kooy F. Discovery of artemisinin in *Artemisia annua*, its current production, and relevance to sub-Saharan Africa. *South African Journal of Botany* (2023) 153:21-27.

Lee JW, Kim YS, Dong X, Park JS, Shin WB, Kim SJ, Go EA, Park PJ. Anti-inflammatory effect of *Rhodiola crenulata* extracts through the down-regulation of MyD88 dependent pathway and induction of autophagy. *Journal of Functional Foods* (2020) 64:103703.

Lin R, Tian J, Huang G, Li T, Li F. Analysis of menthol in three traditional Chinese medicinal herbs and their compound formulation by GC-MS. *Biomedical Chromatography* (2002) 16(3): 229-233.

Ling Y, Dong Y, Li Q, Huang D, Li Z, Chen M. Treatment of COVID-19 by monoclonal antibodies and the traditional Chinese medicine. *Medicine in Novel Technology and Devices* (2022) 15:100159.

Liu C, Cong Z, Wang S, Zhang X, Song H, Xu T, Kong H, Gao P, and Liu X. A review of the botany, ethnopharmacology, phytochemistry, pharmacology, toxicology and quality of *Anemarrhena asphodeloides* Bunge. *Journal of Ethnopharmacology* (2023) 302:115857.

Lu DY. (2015). Chapter 7 - Assistant chemotherapy. In Lu DY. (Ed.), *Personalized Cancer Chemotherapy* (pp. 43-48). Oxford: Woodhead Publishing.

Lu Y, Jiang Y, Ling L, Zhang Y, Li H, Chen D. Beneficial effects of *Houttuynia cordata* polysaccharides on "two-hit" acute lung injury and endotoxic fever in rats associated with anti-complementary activities. *Acta Pharmaceutica Sinica B* (2018) 8(2):218-227.

Luo H, Gao Y, Zou J, Zhang S, Chen H, Liu Q, Tan D, Han Y, Zhao Y, Wang S. Reflections on treatment of COVID-19 with traditional Chinese medicine. *Chinese Medicine* (2020) 15(1): 94.

Luo H, Yang M, Tang QL, Hu XY, Willcox ML, Liu JP. Characteristics of registered clinical trials on traditional Chinese medicine for coronavirus disease 2019 (COVID-19): A scoping review. *European Journal of Integrative Medicine* (2021) 41:101251.

Lyu M, Fan G, Xiao G, Wang T, Xu D, Gao J, Ge S, Li Q, Ma Y, Zhang H, Wang J. Traditional Chinese medicine in COVID-19. *Acta Pharmaceutica Sinica B* (2021) 11(11): 3337-3363.

Mahaboob Ali AA, Bugarcic A, Naumovski N, Ghildyal R. Ayurvedic formulations: Potential COVID-19 therapeutics? *Phytomedicine Plus* (2022) 2(3):100286.

Mbaveng AT, and Kuete V. (2017). Chapter 30 - *Zingiber officinale*. In Kuete V. (Ed.), *Medicinal Spices and Vegetables from Africa* (pp. 627-639): Academic Press.

Natarajan S, Anbarasi C, Sathiyarajeswaran P, Manickam P, Geetha S, Kathiravan R, Prathiba P, Pitchiahkumar M, Parthiban P, Kanakavalli K, Balaji P. Kabasura Kudineer (KSK), a poly-herbal Siddha medicine, reduced SARS-CoV-2 viral load in asymptomatic COVID-19 individuals as compared to vitamin C and zinc supplementation: findings from a prospective, exploratory, open-labeled, comparative, randomized controlled trial, Tamil Nadu, India. *Trials* (2021) 22(1):623.

Nile SH, Su J, Wu D, Wang L, Hu J, Sieniawska E, Kai G. *Fritillaria thunbergii* Miq. (Zhe Beimu): A review on its traditional uses, phytochemical profile and pharmacological properties. *Food and Chemical Toxicology* (2021) 153:112289.

Pallie MS, Perera PK, Kumarasinghe N, Arawwawala M, Goonasekara CL. Ethnopharmacological use and biological activities of *Tragia involucrata* L. *Evidence-Based Complementary and Alternative Medicine* (2020) 2020: 8848676.

Patel JJ, Acharya SR, Acharya NS. *Clerodendrum serratum* (L.) Moon. – A review on traditional uses, phytochemistry and pharmacological activities. *Journal of Ethnopharmacology* (2014) 154(2):268-285.

Peerzada AM, Ali HH, Naeem M, Latif M, Bukhari AH, Tanveer A. *Cyperus rotundus* L.: Traditional uses, phytochemistry, and pharmacological activities. *Journal of Ethnopharmacology* (2015) 174:540-560.

Ping Z, Chen X, Fang L, Wu K, Liu C, Chen H, Jiang X, Ma J, Yu W. Effect of *Angelica sinensis* extract on the angiogenesis of preovulatory follicles (F1-F3) in late-phase laying hens. *Poultry Science* (2022):102415.

Qin F, Yao L, Lu C, Li C, Zhou Y, Su C, Chen B, Shen Y. Phenolic composition, antioxidant and antibacterial properties, and in vitro anti-HepG2 cell activities of wild apricot (*Armeniaca sibirica* L. Lam) kernel skins. *Food and Chemical Toxicology* (2019) 129:354-364.

Ren W, Liang P, Ma Y, Sun Q, Pu Q, Dong L, Luo G, Mazhar M, Liu J, Wang R, Yang S. Research progress of traditional Chinese medicine against COVID-19. *Biomedicine and Pharmacotherapy* (2021) 137:111310.

Sethiya NK, Ahmed NM, Shekh RM, Kumar V, Kumar Singh P, Kumar V. Ethnomedicinal, phytochemical and pharmacological updates on *Hygrophila auriculata* (Schum.) Hiene: An overview. *Journal of Integrative Medicine* (2018) 16(5):299-311.

Shah MR, Fatima S, Khan SN, Ullah S, Himani G, Wan K, Lin T, Lau JY, Liu Q, Lam DS. Jinhua Qinggan granules for non-hospitalized COVID-19 patients: A double-blind, placebo-controlled, and randomized controlled trial. *Frontiers in Medicine* (2022) 9:928468.

Sivakumar TR, Surendhiran D, Chen K, LV P, Vinothkanna A, Prathiviraj R, Sethupathy S, Sirajunnisa AR. Network pharmacology based analysis of *Astragalus propinquus* components for the treatment of rheumatoid arthritis and diabetes. *South African Journal of Botany* (2021) 139:92-105.

Sohrab S, Mishra P, Mishra S K. Phytochemical competence and pharmacological perspectives of an endangered boon-*Costus speciosus* (Koen.) Sm.: a comprehensive review. *Bulletin of the National Research Centre* (2021) 45(1):209.

Song YL, Jing WH, Chen YG, Yuan YF, Yan R, Wang YT. 1^H Nuclear magnetic resonance based-metabolomic characterization of *Peucedani radix* and simultaneous determination of praeruptorin A and praeruptorin B. *Journal of Pharmaceutical and Biomedical Analysis* (2014) 93:86-94.

Srivastava A, Rengaraju M, Srivastava S, Narayanan V, Gupta V, Upadhayay R, Kumar J, Parameswaran S. Efficacy of two siddha polyherbal decoctions, Nilavembu Kudineer and Kaba Sura Kudineer, along with standard allopathy treatment in the management of mild to moderate symptomatic COVID-19 patients-a double-blind, placebo-controlled, clinical trial. *Trials* (2021) 22(1): 570.

Sun J. Promoting self-healing power and balancing immune response: a holistic, effective strategy of traditional Chinese medicine in treating COVID-19. *Pharmacological Research - Modern Chinese Medicine* (2022) 5:100199.

Sung Y-Y, Lee AY, Kim HK. *Forsythia suspensa* fruit extracts and the constituent matairesinol confer anti-allergic effects in an allergic dermatitis mouse model. *Journal of Ethnopharmacology* (2016) 187:49-56.

Tianyu Z, Liying G. Identifying the molecular targets and mechanisms of xuebijing injection for the treatment of COVID-19 via network parmacology and molecular docking. *Bioengineered* (2021) 12(1):2274-2287.

Velvizhi D, Nartunai G, Susikumar S, Ilavarasan R. Morpho-anatomical, phytochemical and TLC/HPTLC studies on whole plant of *Plectranthus vettiveroides* (Jacob) N.P. Singh and B.D. Sharma. *Pharmacognosy Journal* (2020) 12(1): 58-65.

Wang C, Cao B, Liu QQ, Zou ZQ, Liang ZA, Gu L, Dong JP, Liang LR, Li XW, Hu K, He XS. Oseltamivir compared with the Chinese traditional therapy Maxingshigan–Yinqiaosan in the treatment of H1N1 Influenza. *Annals of Internal Medicine* (2011) 155(4):217-225.

Wang D, Yang J, Du Q, Li H, Wang S. The total alkaloid fraction of bulbs of *Fritillaria cirrhosa* displays anti-inflammatory activity and attenuates acute lung injury. *Journal of Ethnopharmacology* (2016) 193:150-158.

Wang J, Wang L, Zhou H, Liang XD, Zhang MT, Tang YX, Wang JH, Mao JL. The isolation, structural features and biological activities of polysaccharide from *Ligusticum chuanxiong*: A review. *Carbohydrate Polymers* (2022) 285:118971.

Xing Y, Hua YR, Shang J, Ge WH, Liao J. Traditional Chinese medicine network pharmacology study on exploring the mechanism of Xuebijing Injection in the treatment of coronavirus disease 2019. *Chinese Journal of Natural Medicines* (2020) 18(12):941-951.

Xu J, Liu H, Fan Y, and Ji B. Traditional Chinese Medicine is effective for COVID-19: A systematic review and meta-analysis. *Medicine in Novel Technology and Devices* (2022) 16:100139.

Yadav V, Krishnan A, Vohora D. A systematic review on *Piper longum* L.: Bridging traditional knowledge and pharmacological evidence for future translational research. *Journal of Ethnopharmacology* (2020) 247:112255.

Yang F, Dong X, Yin X, Wang W, You L, Ni J. *Radix bupleuri*: A review of traditional uses, botany, phytochemistry, pharmacology, and toxicology. *BioMed Research International* (2017) 2017:7597596.

Yang L, Li G Y, Wang H Y, Zhang K, Zhu Y, Zhao W, Wang H, Wang J H. N-Acetyldopamine derivatives from *Periostracum cicadae*. *Phytochemistry Letters* (2015) 11:275-279.

Yang Y, Al-Khafaji M. (2010). Chapter Two - Herbs that release the Exterior. In Yang Y. and Al-Khafaji M. (Eds.), *Chinese Herbal Medicines (Second Edition)* (pp. 25-39). Edinburgh: Churchill Livingstone.

Ye M, Luo G, Ye D, She M, Sun N, Lu Y-J, Zheng J. Network pharmacology, molecular docking integrated surface plasmon resonance technology reveals the mechanism of Toujie Quwen Granules against coronavirus disease 2019 pneumonia. *Phytomedicine* (2021) 85:153401.

Yosri N, Alsharif SM, Xiao J, Musharraf SG, Zhao C, Saeed A, Gao R, Said NS, Di Minno A, Daglia M, Guo Z. *Arctium lappa* (Burdock): Insights from ethnopharmacology potential, chemical constituents, clinical studies, pharmacological utility and nanomedicine. *Biomedicine and Pharmacotherapy* (2023) 158:114104.

Yu J, Hu Q, Liu J, Luo J, Liu L, Peng X. Metabolites of gut microbiota fermenting *Poria cocos* polysaccharide alleviates chronic nonbacterial prostatitis in rats. *International Journal of Biological Macromolecules* (2022) 209:1593-1604.

Zhang ZJ, Morris-Natschke SL, Cheng YY, Lee KH, Li RT. Development of anti-influenza agents from natural products. *Medicinal Research Reviews* (2020) 40(6):2290-2338.

Zhao B T, Kim T I, Kim Y H, Kang J S, Min B S, Son J K, Woo M H. A comparative study of *Mentha arvensis* L. and *Mentha haplocalyx* Briq. by HPLC. *Natural Product Research* (2018) 32(2):239-242.

Zhao Z, Li Y, Zhou L, Zhou X, Xie B, Zhang W, Sun J. Prevention and treatment of COVID-19 using traditional Chinese medicine: A review. *Phytomedicine* (2021) 85:153308.

Zhou X, Tang L, Xu Y, Zhou G, Wang Z. Towards a better understanding of medicinal uses of *Carthamus tinctorius* L. in traditional Chinese medicine: A phytochemical and pharmacological review. *Journal of Ethnopharmacology* (2014) 151(1):27-43.

Zhu J, Liu Z, Pu Y, Xu J, Zhang S, Bao Y. Green synthesized gold nanoparticles from *Pseudobulbus Cremastrae seu Pleiones* show efficacy against hepatic carcinoma potentially through immunoregulation. *Drug Delivery* (2022a) 29(1):1983-1993.

Zhu LX, Tang L, Feng PF. Xuebijing injection combined with antibiotics for severe pneumonia: A meta analysis of randomized clinical trials. *Phytomedicine Plus* (2022b) 2(4):100344.

Chapter 6

Medicinal Plants for SARS-CoV-2 Infections

Erlina Abdullah*, PhD
Chun Hoe Tan, PhD
Hema Priyaa Vijayan, MSc
and Rathimalar Ayakannu, PhD
Department of Biotechnology, Faculty of Applied Science, Lincoln University College, Petaling Jaya, Selangor, Malaysia

Abstract

Medicinal plants are claimed to be less toxic, inexpensive and effective compared to commercial pharmaceutical drugs. The coronavirus disease 2019 (COVID-19) is a respiratory viral disease caused by severe acute respiratory syndrome coronavirus 2 (SARS-CoV-2). It has had a huge impact on global human health and economic development. The implementation of vaccination programs around the world has shown a tremendous effect of protection as well as reducing the risk of SARS-CoV-2 transmission among the communities, thus lowering mortality cases and hospitalization. However, vaccination also faces challenges in fighting against the new emerging variants which can evade the immune system, increase transmissibility and cause severe diseases. Therefore, in this chapter, five medicinal plants potential for the treatment of SARS-CoV-2 infection were discussed. Selected medicinal plants include *Azadirachta indica* A. Juss (neem), *Allium sativum* L. (Garlic), *Zingiber officinalis* (Ginger), *Syzigium aromaticum* (Clove), and *Camellia sinensis* (L.) Green Tea; have been proven to contain various biological properties to treat diseases and COVID-19-related symptoms with potent

---

* Corresponding Author's Email: erlina@lincoln.edu.my.

In: Medicinal Plants and COVID-19
Editor: Asita Elengoe
ISBN: 979-8-88697-958-9

antiviral and immunomodulatory effects. Literature studies reveal that most of those active plant compounds bind to the main proteins of the SARS-CoV-2 and possibly inhibit the activities of the protein in the host cells, therefore suppressing the spread of SARS-COV-2 infections. The proposed phytoconstituents could be further considered to be examined in *in-vitro* and clinically to confirm their effectiveness in human bodies. A future experimental study should also be conducted to determine the suitable dosage of those compounds to ensure their safety for administration.

Keywords: cytotoxicity, infection, medicinal plants, SARS-CoV-2

Introduction

Malaysia is known as a country that is rich in medicinal plant species. It is also endowed with multi-ethnic cultures offering a unique combination of folk and traditional medication such as Ayurveda, Traditional Chinese Medicine (TCM), etc. for the development of the herbal industry. Medicinal plants are claimed to be less toxic, inexpensive and effective compared to commercial pharmaceutical drugs (Sooi and Keng, 2013). Severe acute respiratory syndrome coronavirus 2 (SARS-CoV-2) is a causal factor of the coronavirus disease 2019 (COVID-19) which generates a major health issue, economic and social well-being for the human population globally. The implementation of vaccination programs around the world has shown a tremendous effect of protection as well as reducing the risk of SARS-CoV-2 transmission among the communities, thus lowering mortality cases and hospitalization. However, recent studies showed that the vaccine efficacy is reduced within six months, thus requiring a need of booster injection. Vaccination also faces the challenge of fighting against the new emerging variants which can evade the immune system, increase transmissibility and cause severe diseases (WHO, 2021; Shastri and Parikh, 2021).

Apart from that, drug discovery for targeting the host cells is crucial for the treatment of SARS-CoV-2-infected patients. Drugs such as remdesivir, favipiravir, hydroxychloroquine and azithromycin have been utilized against COVID-19 which causes several side effects, yet still, there are no significant therapies in combating the SARS-CoV-2 infection specifically (Awadasseid et al., 2021). Researchers focus on the study of medicinal plants to fight against SARS-CoV-2 infection. They are targeting the functional protein of SARS-CoV-2 and the host immune modulation. SARS-CoV-2 gains entry into the host cells by binding to the cell receptor of angiotensin-converting enzyme 2 (ACE-2). The antihypertensive activities through ACE inhibition have been demonstrated in various medicinal plant types of research. Hence, the blocking of this receptor could be of potential in the management of SARS-CoV-2 infection (Yan et al., 2020; Patten et al., 2013).

Previous research in molecular docking analysis has identified three molecules (Crocin, Digitoxigenin, and β-Eudesmol) from medicinal plants as potential inhibitors for SARS-CoV-2 main protease based on their binding energy and interaction with the active site of protein studied (Aanouz et al., 2021). Therefore, this paper is aimed to provide a literature review on five selected medicinal plants' potential for the treatment of SARS-CoV-2 infection. Medicinal plants were selected according to their therapeutic uses and bioactive properties such as antiviral and immunomodulatory activities.

List of Medicinal Plants Potential for SARS-CoV-2 Infection

Azadirachta indica A. Juss (Neem)

Azadirachta indica A. Juss also known as neem has been traditionally used as a treatment for several diseases including viral infections such as animals suffering from bovine and avian poxvirus infections by applying a paste of neem leaves directly on the infected skin (Biswas et al., 2002). Scientific reports have proven a broad spectrum of antiviral effects of different parts (leaves, fruits, seeds, and barks) of neem extracts including inhibition against smallpox virus, chicken pox virus, poliovirus, dengue virus and HIV (Udeinya et al., 2008; Kumar, 2020; Kumar and Navaratnam, 2013). Additionally, neem extracts displayed antiviral and immunomodulatory effects against Herpes Simplex Virus type 1 and SARS-CoV-2 (Ogidigo et al., 2022; Baildya et al., 2021; Sudhakaran et al., 2022; Kumar, 2020; Kumar and Navaratnam, 2013; Lim et al., 2021).

Positive antiviral effects of neem-derived compounds were also demonstrated in a molecular docking study where nimbolin A, nimocin, and cycloartenol showed strong binding capabilities towards the SARS-CoV-2 membrane (M) and envelope (E) protein structures indicating their potential to inhibit these compounds functionality (Borkotoky and Banerjee, 2021).

The mutation of SARS-CoV-2 in the receptor binding domain (RBD) of the spike protein has emerged new variants of SARS-CoV-2 which are known as delta and omicron. The mutations caused changes in the structural properties of the proteins wherein the omicron variant possesses a higher fraction of alpha-helix structure as compared to the delta variant. As a consequence, the same ligands may bind at different locations within the binding pockets of the receptor protein of omicron and delta variants (Lupala et al., 2022). Further study was conducted using the computational method to investigate the potency of neem extract compounds against those SARS-CoV-2 variants. It was found that nimbolinin B12-methyl-ether and nimbidinin have demonstrated the highest negative docking score against both the S-protein of delta and omicron variants. Both compounds are considered as significant inhibitory agents with high binding efficacy within the pockets of S-glycoproteins of those variants (Ansari et al., 2022).

Allium sativum L. (Garlic)

Garlic or *Allium sativum* from the Liliaceae family is one of the most cultivated plants in the world. It has been used as food spice and flavoring agent in cooking as well as providing vitamins, minerals and innumerable nutrients to human health being (Alam et al., 2016). Literature surveys showed garlic is an efficient natural antibiotic against the wide spectrum of bacteria and viruses, thus its various biological compounds are the potential source of antiviral agents against SARS-CoV-2 infections (Khubber et al., 2020).

Numerous studies have demonstrated the immune system boosting the capability of garlic by stimulating certain cell types, such as macrophages, lymphocytes, natural killer (NK) cells, dendritic cells, and eosinophils, by mechanisms including modulation of cytokine secretion, immunoglobulin production, phagocytosis, and macrophage activation (Arreola et al., 2015). Essential oils from garlic, specifically diallyl sulfide (DAS), are able to block the virus entry into the host cells while activating a molecular pathway that lower the secretions of culprit pro-inflammatory-inflammatory cytokines (Asif et al., 2020). For example, the garlic essential oils were observed to inhibit the ACE-2 protein activity in SARS-CoV-2 leading to a loss of this viral receptor in the host cell and attacking the receptor in the human body and SARS-CoV-2 main protease (PDB ID 6LU7) (Jamal, 2022).

SARS-CoV-2 main proteases consist of papain-like protease (PL-Pro) and 3-chymotrypsin (C)-like cysteine protease (3CL-Pro) which act as regulators for the virus replication activity. The blocking of this activity by targeting these two proteins can inhibit the replication of the virus in humans, however, the 3CL-Pro (Mpro) is more desirable to be targeted for SARS-Cov-2 infection as the formation of 3CL-Pro inhibitors are not toxic to the human being (Yan and Wu, 2021; Zhang et al., 2021; Roe et al., 2021; Zhang et al, 2020). Molecular docking analysis was performed between 20 bioactive compounds from garlic to investigate the interaction stability level between those compounds and the target protein, Mpro. The lower binding affinity indicates the more stable ligand-protein complex interaction, thus resulting in higher inhibition activity performed by the ligand compounds. Based on the study, 5 compounds were identified as the best candidates for therapeutic agents against SARS-CoV-2, those are squalene, 1,4-dihydro-2,3-benzoxanthiin 3-oxide, 1,2,3-propanetriyl ester, trans-13-octadecenoic acid and methyl-11-hexadecenoate with the binding affinity of -7, -6.5, -5.9, -5.7 and -5.6 kcal/mol, respectively (Listiyani et al., 2022). Previously, Pandey et al, 2021, had reported that alliin compound in garlic is the most promising antiviral agents compared to other organosulfur compounds (S-(allyl/methyl/ethyl/propyl)-cysteine, S-propyl L-cysteine, and S-ally-mercapto cysteine) based on the inhibition of the Mpro of SARS-CoV-2 through H-bonds with protease.

Zingiber officinalis (Ginger)

Zingiber officinalis has been used in the treatment of respiratory diseases. The traditional uses of *Z. officinalis* rhizome are proven as much related to the treatment of COVID-19-associated symptoms including enhancement of appetite, clearing the throat, alleviating cough, acting as a painkiller, antipyretic, asthmatic conditions, vomiting, stomach ache, heart disorders, exhaustion and physical weakness, and headaches (Riazur et al., 2011). The consumption of ginger for COVID-19 patients in Saudi Arabia has indicated a reduced percentage of hospitalization to 28% for users compared to 38% for non-users (Aldwihi et al., 2021).

Experimental studies showed that ginger extracts can inhibit the replication of the influenza A virus (H1N1, H5N1 and H9N2) by stimulating the production of the first line of defense against virus infection, TNF-α (Wang et al., 2020; Seo and Webster, 2002). The antiviral effects of ginger administration are also supported by the inhibition of attachment and internalization of the human respiratory syncytial virus (RSV) to both human lung cells and liver cells. In addition, the extract stimulates the mucosa cells to increase sections of interferon-beta (IFN-β) secretions (Chang et al., 2012).

An in-silico study (molecular docking) found that 2 compounds present in ginger, which is chlorogenic acid and hesperidin, had high binding affinities for Mpro with binding energies of -7.5 and -8.3 kcal/mol, hence possibly inhibiting the Mpro activities of the SARS-CoV-2 (Jahan et al., 2021). Prasanth et al., 2020, in their study had discovered 9 phytoconstituents from ginger as potential Mpro inhibitors for the SARS-CoV-2, which include [6]-Paradol,[6]-Shogaol, Cedr-8-ene, Copaene, Gingerenone A, Isogingerenone B, Shogasulfonic acid A and Zonarenebe. Moreover, other ginger-derived compounds such as gingerol, geraniol, shogaol, zingiberene, zingiberenol, zingerone and sesquiphellandrene are able to bind to SARS-CoV-2 S-protein thus promoting interference of S protein-ACE-2 interaction. Therefore, the attachment of the SARS-CoV-2 virus to its target host cells is inhibited (Ahkam et al., 2020; Joshi et al., 2020a).

Syzigium aromaticum (Clove)

Syzigium aromaticum or cloves is widely used as a spice that contains many bioactive compounds for medicinal benefits since a long time ago. Literature studies have demonstrated that the *S. aromaticum* flower bud is an effective remedy for various ailments, especially in toothache and tooth infections as well as other treatments such as stomach disorder, bowel disorder, wound infections, vomiting, nausea, cholera, malaria, tuberculosis, etc. (Batiha et al., 2020). Based on their traditional uses, cloves have been implemented in the prevention and treatment of SARS-CoV-2 infections. The preparation of a decoction is by boiling the cloves with other plant materials for 15 minutes and the released volatile are then inhaled by patients for five minutes. The same protocol also applied to a drinkable decoction (Kanyinda, 2020).

Eugenol (4-allyl-2-methoxyphenol), a major constituent purified from cloves, was utilized for its anti-viral activity in an anti-herpes simplex virus (HSV) property. Pure eugenol was found to exert a higher antiviral effect than the ethanol extracts of the whole clove buds (Tragoolpua and Jatisatienr, 2007). In a different study, eugenol was reported to inhibit the HSV-1 and HSV-2 replication with 50% inhibitory concentration (IC_{50}) at 25.6 μg/mL and 16.2 μg/mL, respectively (Benencia and Courreges, 2000). Eugenol also exhibited antiviral activities against the influenza A virus (IAV) and Ebola virus (Dai et al, 2003; Lane et al., 2019). A very recent study discovered that the clove water and ethanol extracts had potentiated the suppression of SARS-

CoV-2 spike protein binding to ACE-2 and therefore inhibits the ACE-2 activity that initiates the spread of SARS-CoV-2 infection (Li et al., 2022).

Several computational studies have shown the anti-COVID-19 effects of clove extracts (Pandey et al., 2020; Joshi et al., 2020b). For example, an in-silico study observed that kaempferol extracted from cloves was able to bind the substrate binding pocket of the main protease of SARS-CoV-2 with high affinity interacting with the active site residue such as Cys145 and Hs41 through hydrophobic interactions and hydrogen bonding (Rehman et al., 2021). Other clove compounds such as bicornin and biflorin also show their potential inhibitory activities through molecular docking studies with high binding affinities for Mpro, (-9.2 kcal/mol) and (-8.5 kcal/mol), respectively (Rehman et al., 2021).

***Camellia sinensis* (L.) Green Tea**

Tea leaves have been consumed all over the world as one of the most popular beverages to date. The intake of tea is also recommended for a broad spectrum of health benefits such as headaches, body aches, pains, indigestion, depression, detoxification and as an energizer (Ferrara et al., 2001). Tea is classified into three categories which is green tea, black tea, and oolong tea. Their difference is in terms of their processing where the green tea is made from unfermented leaves, black tea is fully fermented and oolong tea is partially fermented (Alic, 1999).

Catechin is a class of phenolic compounds abundantly present in *C. sinensis*. Epigallocatechin-3-gallate (EGCG) is a major component of catechin which displayed inhibition against the influenza virus and direct antiviral effects against Retroviridae, Orthomyxoviridae, and Flaviviridae. EGCG also exerts interference with the human immunodeficiency virus (HIV), blocking the enzymatic activity of the HIV-1 reverse transcriptase, hepatitis C virus (HCV), and herpes simplex virus 1 and 2 (Mahmood et al., 2016; Zakaryan et al., 2017; Lee et al., 2018; Song et al, 2005). The *in-vitro* study revealed that the catechin-separated compound (fraction 5) resulted in the highest inhibitory activity against the SARS-CoV-2 PLpro with IC_{50} of 0.125 μg/ml (Montone et al, 2021). Also, the EGCG and other catechins, theasinensin A (TSA) and theaflavin 3,3′-di-*O*-gallate (TFDG) significantly inactivate the SARS-CoV-2 infection by interacting with the S protein for the prevention of ACE-2 attachment on the cell surface (Ohgitani et al., 2021).

Furthermore, the polyphenols from green tea are reported as potential matrix metalloproteinase (MMP) inhibitors for SARS-CoV-2 main protease. MMPs play an essential role in immunity response, inflammation, cell growth, organ morphogenesis, wound healing, angiogenesis, apoptosis and embryonic development. Overexpression of MMPs can initiate various diseases, therefore the inhibition of MMPs overexpression could arrest the spread of several diseases. For example, epicatechin, epicatechingallate, epigallocatechin, and epigallocatechingallate in the tea inhibit MMP-2, and MMP-9 (Kanbarkar and Mishra, 2021). MMP-2 overexpression was reported to be responsible for chronic lung diseases (Kong et al., 2009), while MMP-9 overexpression activates viral lung or pulmonary infections, angiogenesis, and metastasis (Dabo et al., 2015).

Conclusion

Five medicinal plants reviewed here are among the most potential for the treatment against COVID-19 disease based on their biological properties to treat various ailments and COVID-19-related symptoms as well as their scientifically proven antiviral activities and immunomodulatory effects. A lot of computational studies through molecular docking analysis also demonstrated the binding abilities of different compounds from the medicinal plants to the SARS-CoV-2 proteins which promote the inhibition and replication of the virus and interference of its enzyme activities that block their spread into the host cells. The phytoconstituents proposed as antiviral agents against SARS-CoV-2 infection could be further considered to be examined *in-vitro* and clinically to confirm their effectiveness in human bodies. Moreover, future experimental studies should be conducted to determine the suitable dosage of those compounds to ensure their safety for administration.

References

Aanouz I, Belhassan A, El-Khatabi K, Lakhlifi T, El-Ldrissi M, Bouachrine M. Moroccan Medicinal plants as inhibitors against SARS-CoV-2 main protease: Computational investigations. *J Biomol Struct Dyn* (2021) 39(8):2971-2979.

Ahkam AH, Hermanto FE, Alamsyah A, Aliyyah H. Virtual prediction of antiviral potential of ginger (*Zingiber officinale*) bioactive compounds against spike and M^{pro} of SARS-CoV-2 protein. *J Biol Res* (2020) 25(2):52-57.

Alam K, Hoq O, Uddin S. Medicinal plant *Allium sativum*. A review. *J Med Plant Stud* (2016) *4*(6):72-79.

Aldwihi LA, Khan SI, Alamri FF, AlRuthia Y, Alqahtani F, Fantoukh OI, Assiri A, Almohammed OA. Patients' behavior regarding dietary or herbal supplements before and during COVID-19 in Saudi Arabia. *Int J Environ Res Publ Health* (2021)18(10):5086.

Alic M. Green tea for remission maintenance in Crohn's disease. *Am J Gastroenterol* (1999) 94 (6):1710.

Ansari WA, Rizvi F, Khan MA, Khan ZA, Khan MF. Computational Study Reveals the Inhibitory Effects of Chemical Constituents from *Azadirachta indica* (Indian Neem) Against Delta and Omicron Variants of SARS-CoV-2. *Coronaviruses* (2022) 3(5): 62-72.

Arreola R, Quintero-Fabián S, López-Roa RI, Flores-Gutiérrez EO, Reyes-Grajeda JP, Carrera-Quintanar L, Ortuño-Sahagún D. Immunomodulation and anti-inflammatory effects of garlic compounds. *J Immunol Res* (2015).

Asif M, Saleem M, Saadullah M, Yaseen HS, al Zarzour R. COVID-19 and therapy with essential oils having antiviral, anti-inflammatory, and immunomodulatory properties. *Inflammopharmacology* (2020) 28(5):1153-1161.

Awadasseid, A, Wu Y, Tanaka Y, Zhang W. Effective drugs used to combat SARS-CoV-2 infection and the current status of vaccines. *Biomed Pharmacother* (2021) 137:111330.

Baildya N, Khan AA, Ghosh NN, Dutta T, Chattopadhyay AP. Screening of potential drug from *Azadirachta indica* (Neem) extracts for SARS- CoV-2: An insight from molecular docking and MD-simulation studies. *J Mol Struct* (2021) 1227:129390.

Batiha GES, Alkazmi LM, Wasef LG, Beshbishy AM, Nadwa EH, Rashwan EK. *Syzygium aromaticum* L. (Myrtaceae): Traditional uses, bioactive chemical constituents, pharmacological and toxicological activities. *Biomolecules* (2020) 10(2).

Benenci F, Courreges M. *In vitro* and *in vivo* activity of eugenol on human herpesvirus. *Phytother Res* (2000) 14:495-500.

Biswas K, Chattopadhyay I, Banerjee RK, Bandyopadhyay U. Biological activities and medicinal properties of neem (*Azadirachta indica*). *Curr Sci* (2002) 82:1336-1345.

Borkotoky S, Banerjee M. A computational prediction of SARS-CoV-2 structural protein inhibitors from *Azadirachta indica* (neem). *J Biomol Struct Dyn* (2021) 39(11):4111-4121.

Chang J, Wang K, Yeh C, Shieh D, Chiang LC. Fresh ginger (*Zingiber officinale*) has anti-viral activity against human respiratory syncytial virus in human respiratory tract cell lines. *J Ethnopharmacol* (2012) 145: 146-151.

Dabo AJ, Cummins N, Eden E, Geraghty P. Matrix metalloproteinase 9 exerts antiviral activity against respiratory syncytial virus. *PLoS One* (2015) 10(8):1.

Dai JP, Zhao XF, Zeng J, Wan, QY, Yang JC, Li WZ, Chen XX, Wang GF, Li KS. Drug screening for autophagy inhibitors based on the dissociation of Beclin1-Bcl2 complex using BiFC technique and mechanism of eugenol on anti-influenza A virus activity. *PLoS One* (2013) 8:e61026.

Ferrara L, Montesano D, Senatore A. The distribution of minerals and flavonoids in the tea plant (*Camellia sinensis*). *Il Farmaco* (2001) 56: 397-401.

Jahan R, Paul AK, Bondhon TA, Hasan A, Jannat K, Mahboob T, Nissapatron V, Pereira ML, Wiart C, Wilairatana P, Rahmatullah M. *Zingiber officinale*: Ayurvedic uses of the plant and *in silico* binding studies of selected phytochemicals with Mpro of SARS-CoV-2. *Natl Prod Commun* (2021) 16(10): 1934578X211031766.

Jamal QMS. Antiviral Potential of Plants against COVID-19 during Outbreaks-An Update. *Int J Mol Sci* (2022) 23(21): 13564.

Joshi A, Sunil Krishnan G, Kaushik V. Molecular docking and simulation investigation: effect of beta-sesquiphellandrene with ionic integration on SARS-CoV2 and SFTS viruses. *J Genet Eng Biotechnol* (2020a) 18(1):78.

Joshi T, Joshi T, Sharma P, Mathpal S, Pundir H, Bhatt V, Chandra S. *In silico* screening of natural compounds against COVID-19 by targeting Mpro and ACE2 using molecular docking. *Eur Rev Med Pharmacol Sci* (2020b) 24:4529-4536.

Kanbarkar N, Mishra S. Matrix metalloproteinase inhibitors identified from Camellia sinensis for COVID-19 prophylaxis: an *in silico* approach. *Adv Tradit Med* (2021) 21(1):173-188.

Kanyinda JNM. Coronavirus (COVID-19): A protocol for prevention and treatment (Covalyse®). *Eur J Med Health Sci* (2020) 2:1-4.

Khubber S, Hashemifesharaki R, Mohammadi M, Gharibzahedi SMT. Garlic (*Allium sativum* L.): a potential unique therapeutic food rich in organosulfur and flavonoid compounds to fight with COVID-19. *Nutr J* (2020) 19(1):1-3.

Kong MY, Gaggar A, Li Y, Winkler M, Blalock JE, Clancy JP. Matrix metalloproteinase activity in paediatric acute lung injury. *Int J Res Med Sci* (2009) 6(1):9-17.

Kumar VS, Navaratnam V. Neem (*Azadirachta indica*): Prehistory to contemporary medicinal uses to humankind. *Asian Pac J Trop Biomed* (2013) 3(7):505-514.

Kumar AHS. Molecular docking of natural compounds from Tulsi (*Ocimum sanctum*) and neem (*Azadirachta indica*) against SARS-CoV-2 protein targets. *Biol Eng Med Sci Rep* (2020) 6(1):11-13.

Lane T, Anantpadma M, Freundlich JS, Davey RA, Madrid PB, Ekins S. The natural product eugenol is an inhibitor of the ebola virus *in vitro*. *Pharm Res* (2019) 36:1-6.

Lee YH, Jang YH, Kim YS, Kim J, Seong BL. Evaluation of green tea extract as a safe personal hygiene against viral infections. *J Biol Eng* (2018) 12:1.

Li Y, Liu Z, Zeng M, El Kadiri A, Huang J, Kim A, He X, Sun J, Chen P, Wang TTY, Zhang Y, Gao B, Xie Z, Yu LL. Chemical compositions of clove (*Syzygium aromaticum* (L.) Merr. & L.) extracts and their potentials in suppressing SARS-CoV-2 spike protein-ACE2 binding, inhibiting ACE2, and scavenging free radicals. *J Agric Food Chem* (2022) 70(45):14403-14413.

Lim XY, Teh BP, Tan TYC. Medicinal plants in COVID-19: potential and limitations. *Front Pharmacol* (2021) 12:611408.

Listiyani P, Kharisma VD, Ansori ANM, Widyananda MH, Probojati RT, Murtadlo AAA, Turista DDR, Ullah ME, Jakhmola V, Zainul R. *In silico* phytochemical compounds screening of *Allium sativum* targeting the Mpro of SARS-CoV-2. *Pharmacognosy J* (2022) 14(3).

Lupala CS, Ye Y, Chen H, Su XD, Liu H. Mutations on RBD of SARS-CoV-2 Omicron variant result in stronger binding to human ACE2 receptor. *Biochem Biophys Res Commun* (2022) 590:34-41.

Mahmood MS, Mártinez JL, Aslam A, Rafique A, Vinet R, Laurido C, Hussain I, Abbas RZ, Khan A, Ali S. Antiviral effects of green tea (*Camellia sinensis*) against pathogenic viruses in human and animals (a mini-review). *Afr J Tradit Complement Altern Med* (2016) 13:176.

Montone CM, Aita SE, Arnoldi A, Capriotti AL, Cavaliere C, Cerrato A, Lammi C, Piovesana S, Ranaldi G, Laganà A. Characterization of the trans-epithelial transport of green tea (*C. sinensis*) Catechin extracts with *in vitro* inhibitory effect against the SARS-CoV-2 papain-like protease activity. *Molecules* (2021) 26(21):6744.

Ogidigo JO, Iwuchukwu IA, Ibeji CU, Okpalefe O, Soliman MES. Natural phyto, compounds as possible noncovalent inhibitors against SARS-CoV-2 protease: computational approach. *J Biomol Struct Dyn* (2022) 40(5):2284-2301.

Ohgitani E, Shin-Ya M, Ichitani M, Kobayashi M, Takihara T, Kawamoto M, Kinugasa H, Mazda O. Significant inactivation of SARS-CoV-2 in vitro by a green tea catechin, a catechin-derivative, and black tea galloylated theaflavins. *Molecules* (2021) 26(12):3572.

Pandey P, Singhal D, Khan F, Arif M. An in silico screening on *Piper nigrum*, *Syzygium aromaticum* and *Zingiber officinale* roscoe derived compounds against SARS-CoV-2: A drug repurposing approach. *Biointerface Res Appl Chem* (2020) 11:11122-11134.

Pandey P, Khan F, Kumar A, Srivastava A, Jha NK. Screening of potent inhibitors against 2019 novel coronavirus (Covid-19) from *Allium sativum* and *Allium cepa*: An *in silico* approach. *Biointerface Res Appl Chem* (2021) 11(1):7981-7993.

Patten GS, Abeywardena MY, Bennett LE. Inhibition of angiotensin converting enzyme, angiotensin II receptor blocking, and blood pressure lowering bioactivity across plant families. *Crit Rev Food Sci Nutr* (2013) 56:181-214.

Prasanth DSNBK, Panda SP, Rao AL, Chakravarti G, Teja N, Vani VBN, Sandhya T. *In-silico* strategies of some selected phytoconstituents from *Zingiber officinale* as SARS-CoV-2 main protease (COVID-19) inhibitors. *Indian J Pharm Educ Res* (2020) 54(3):s552-559.

Riazur R, Naveed A, Qaiser J, Tariq S, Khalil A, Ghazala S, Not A. *Zingiber officinale* roscoe (pharmacological activity). *J Med Plant Res* (2011) 5(3):344-348.

Rehman M, AlAjmi MF, Hussain A. Natural compounds as inhibitors of SARS-CoV-2 main protease (3CLpro): A molecular docking and simulation approach to combat COVID-19. *Curr Pharm Des* (2021) 27(33):3577-3589.

Roe MK, Junod NA, Young AR, Beachboard DC, Stobart CC. Targeting novel structural and functional features of coronavirus protease nsp5 (3CLpro, Mpro) in the age of COVID-19. *J Gen Virol* (2021) 102(3):001558.

Seo SH, Webster R. Tumor necrosis factor alpha exerts powerful anti-influenza virus effects in lung epithelial cells. *J Virol* (2002) 76:1071-1076.

Shastri J, Parikh S, Aggarwal V, Agrawal S, Chatterjee N, Shah R, Devi P, Mehta P, Pandey R. Severe SARS-CoV-2 breakthrough reinfection with Delta variant after recovery from breakthrough infection by alpha variant in a fully vaccinated health worker. *Front Med* (2021) 8:1-13.

Song JM, Lee KH, Seong BL. Antiviral effect of catechins in green tea on influenza virus. *Antivir Res* (2005) 68:66-74.

Sooi LK, Keng SL. Herbal medicines: Malaysian women's knowledge and practice. *J Evid Based Complementary Altern Med* (2013):438139.

Sudhakaran G, Prathap P, Guru A, Rajesh R, Sathish S, Madhavan T, Arasu MV, Al-Dhabi NA, Choi KC, Gopinath P, Arockiaraj J. Anti-inflammatory role demonstrated both *in vitro* and *in vivo* models using nonsteroidal tetranortriterpenoid, Nimbin (N1) and its analogs (N2 and N3) that alleviate the domestication of alternative medicine. *Cell Biol Int* (2022) 46 (5):771-791.

Tragoolpua Y, Jatisatienr A. Anti-herpes simplex virus activities of *Eugenia caryophyllus* (Spreng.) Bullock & SG Harrison and essential oil, eugenol. *Phytother Res* (2007) 21:1153-1158.

Udeinya JI, Shu EN, Quakyi I, Ajayi FO. An antimalarial neem leaf extract has both schizonticidal and gametocytocidal activities. *Am J Ther* (2008) 15(2):108-110.

Wang J, Prinz RA, Liu XF, Xu XL. *In vitro* and *in vivo* antiviral activity of gingerenone A on influenza a virus is mediated by targeting Janus kinase 2. *Viruses* (2020) 12(10):1141.

World Health Organization (WHO). (2021). *Data for action: achieving high uptake of COVID-19 vaccines: gathering and using data on the behavioural and social driver of vaccination: a guidebook for immunization programmes and implementing partners: interim guidance.* Accessed 20th December 2022. http://www.who.int/publication/ite/WHO-2019-nCoV-vaccinationdemand-planning-2021.1.

Yan, R, Zhang Y, Li Y, Xia L, Guo Y, Zhou Q. Structural basis for the recognition of SARS-CoV-2 by full-length human ACE2. *Science* (2020) 367:1444-1448.

Yan S, Wu G. Potential 3-chymotrypsin-like cysteine protease cleavage sites in the coronavirus polyproteins pp1a and pp1ab and their possible relevance to COVID-19 vaccine and drug development. *FASEB J* (2021) 35(5):e21573.

Zakaryan H, Arabyan E, Oo A, Zandi K. Flavonoids: Promising natural compounds against viral infections. *Arch Virol* (2017) 162:2539-2551.

Zhang L, Lin D, Sun X, Curth U, Drosten C, Sauerhering L, Becker S, Rox K, Hilgenfeld R. Crystal structure of SARS-CoV-2 main protease provides a basis for design of improved α-ketoamide inhibitors. *Science* (2020) 368(6489):409-412.

Zhang Q, Xiang R, Huo S, Zhou Y, Jiang S, Wang Q, Yu F. Molecular mechanism of interaction between SARS-CoV-2 and host cells and interventional therapy. *Signal Transduct Target Ther* (2021) 6(1):233.

Chapter 7

SARS-CoV-2 and Medicinal Plants

Abdul Haleem Mohamad Wahhaj[1], BSc and Asita Elengoe[2,*], PhD

[1]School of Biotechnology, Manipal International University (MIU), Putra Nilai, Negeri Sembilan, Malaysia
[2]Department of Biotechnology, Faculty of Applied Science, Lincoln University College, Petaling Jaya, Selangor, Malaysia

Abstract

Severe acute respiratory syndrome coronavirus 2 (SARS-CoV-2) was discovered in December 2019. The World Health Organization (WHO) has declared a pandemic as a result of the rapid spread of the virus worldwide. It is a highly pathogenic virus. However, it is still incurable. The hunt for efficient and non-conventional therapeutic options, like antiviral medication compounds produced from plants, has also been strengthened by the toxicities and inadequate responses to resistant strains of synthetic antiviral treatments. Currently, traditional plants and their phytoconstituents have become one of the main public interests to hunt for efficient and reasonably priced medicines to meet the demands of the present. They have been proven to treat a variety of chronic and infectious diseases. Thus, in the present review, an attempt has been taken to summarize the SARS-CoV-2's morphology, pathogenic mechanism; signs and symptoms; clinical manifestations; different types of variants; medicinal plants and their mode of action. This information will help in developing an efficient and affordable drug against SARS-CoV-2.

Keywords: COVID-19, phytoconstituents, SARS-CoV-2

* Corresponding Author's Email: asitaelengoe@yahoo.com.

In: Medicinal Plants and COVID-19
Editor: Asita Elengoe
ISBN: 979-8-88697-958-9

Introduction

In 2003, severe acute respiratory syndrome (SARS) is a major health threat to the public. This zoonotic respiratory illness is brought on by the SARS coronavirus (SARS-CoV). It was the main reason for the outbreak of the epidemic in China's Guangdong province. According to research studies, the virus spread from bats to humans (Elengoe and Selvam, 2022). A novel coronavirus strain was initially identified in Wuhan, China, in December 2019. The virus was SARS-CoV-2. It is far more virulent than the SARS-CoV strain (Elengoe et al., 2022). The World Health Organization (WHO) recognized the outbreak of COVID-19 and classified it as a Public Health Emergency of International Concern in January 2020. In March 2020, the epidemic was classified as a pandemic. Based on the next gene sequencing results, the Chinese researchers reported that SARS-CoV-2 had shared 96.3% genetic similarity with a Yunnan bat coronavirus RaTG13. As of January 12th, 2023, COVID-19 had been linked to almost 669 million illnesses and 6.72 million deaths, making it one of the deadliest pandemics in history (Worldometer, 2022).

Currently, there are several variants identified from the mutation of the COVID-19 strain. These variants are namely Alpha (B.1.1.7 and Q lineages), Beta (B.1.351 and descendent lineages), Gamma (P.1 and descendent lineages), Delta (B.1.617.2 and AY lineages), Epsilon (B.1.427 and B.1.429), Eta (B.1.525), Iota (B.1.526), Kappa (B.1.617.1), 1.617.3, Mu (B.1.621, B.1.621.1) and Zeta (P.2). These variants are continued to be monitored in the United States. Moreover, no SARS-CoV-2 variants have been classified as a variant of high consequence (VOHC) as of yet (Centers for Disease Control and Prevention, 2023; Duong, 2021).

It is a disease that is spread by the air. It spreads quickly from one person to another through the air when people are close to one another. As people breathe, cough, sneeze, or speak, the virus passes from one infected person to another through the lips, nose, or eyes. It consequently also transmits through water droplets. It could also spread through contact with infected surfaces. A healthy person can contract the virus from asymptomatic carriers. They have up to two weeks to disseminate it. Asymptomatic carriers have the COVID-19 virus inside them for two weeks but do not exhibit any symptoms.

As its major organ, the lung is targeted by SARS-CoV-2. However, clinical research has shown that the virus can also infect other organs like the heart, blood vessels, kidneys, stomach, and brain. The symptoms of COVID-19 are extremely varied, ranging from none to serious illness (Guo et al.,

2020). Elderly persons, especially those over the age of 70, people with chronic conditions such as those that affect the respiratory system, diabetes, hypertension, and kidneys, and people with impaired immune systems (such as cancer and HIV patients) are all severely impacted (Elengoe, 2020).

Morphology

The SARS-CoV-2 is a large, spherical, enveloped, non-segmented, positive-sensed virus with a genome of nearly 30 kb (29,903 nucleotides) in length (Elengoe, 2020). The virus has unique club-like projections on the surface and proteomic studies have shown 2/3 of the genome encodes the replicase genes encode for two large polyproteins, pp1a and pp1ab that are processed into 15-16 non-structural proteins (nsp) via proteolytic cleavage and the remaining 1/3 contains the ORFs for the structural proteins. Structural proteins of the spike (S), envelope (E), membrane (M), and nucleocapsid (N) proteins are seen in the genome sequence of SARS-CoV-2 (Ou et al., 2020).

Phylogenetic analysis of SARS-CoV-2 genome sequences has shown about 50% sequence homology to MERS-CoV and about 79% sequence homology to SARS-CoV viruses (Singhal, 2020). SARS-CoV-2 and SARS-CoV belong to the β-coronavirus genera but the SARS-CoV-2 virus has shown more than 85% sequence homology to bat-derived SARS viruses proving that SARS-CoV-2 is a zoonotic virus (Xu et al., 2020).

Pathogenic Mechanism

The pathogenic mechanism of SARS-CoV-2 is very helpful in determining the viral infectivity and pathogenesis which also helps in immune surveillance and human intervention strategies towards the virus. Studies on the pathogenesis of SARS-CoV have shown that the spike (S) structural protein and N protein which is also a structural protein involved in virion assembly, maintain the stability of the virus particle increasing virus transcription and assembly efficiency. The S protein in the SARS-CoV-2 binds to receptors on cells to mediate infection in host cells. The S protein and angiotensin-converting enzyme 2 (ACE2) which is the host receptor play a major role in the development of infection.

SARS-CoV-2 affects the lungs mostly and alveolar epithelial cells are the first to be infected by respiratory droplets. The immune organs and the blood circulation are the next targets that are infected and the spread of infection to a serious content would cause extensive pulmonary consolidation, diffuse alveolar damage, and the formation of a transparent membrane, finally deteriorating into respiratory distress.

The entry to the host cells is mainly by respiratory droplets where the virus binds to a cell surface receptor and subsequently enters endosomes to fuse viral and lysosomal membranes. A trimer embedded with three receptor-binding S1 heads on top of a S2 protein stalk can be seen clearly on a mature virus the spike protein. The S1 contains a receptor-binding domain (RBD) that recognizes the angiotensin-converting enzyme 2 (ACE2) receptor specifically and constant structural change of S2 and S2 contributes to immune evasion and rapid infection mediation. The downregulation of the ACE2 enzyme creates excessive angiotensin II which leads to increased pulmonary vascular permeability facilitating effective viral transfection and replication. Other than receptor binding, protease activators such as TMPRSS2 and lysosomal proteases also have been observed to give better entry mechanisms for SARS-CoV-2 (Harrison et al., 2020).

SARS-CoV-2 is a respiratory virus mainly transmitted through contact with respiratory droplets or contaminated surfaces. More than 20 modes of transmission have been found to be infective creating replication, transcription, assembly and release of the virus within the host body causing cellular damage and leading to acute inflammatory and immune responses. Infection results in epithelial and endothelial cell apoptosis with increased vascular permeability that causes the release of pro-inflammatory cytokines and chemokines. This leads to severe immune damage starting from the lungs and spreads to other organs eventually (Xie et al., 2020).

Antibody-secreting cells, follicular helper T cells, activated $CD4^+$ and $CD8^+$ cells and IgM and IgG were the main antibodies that were detected in the blood of recovering patients and were identified that pro-inflammatory cytokines and chemokines were significantly decreased due to the immune response (Harrison et al., 2020).

It is observed that the binding ability of S protein with ACE2 and other receptors and specific recombination and insertion of RBD are the major factors that aid the pathogenic mechanism of SARS-CoV-2. The pathogenic mechanism of SARS-CoV-2 needs to be studied in order to understand how the virus creates immune evasion and potential cell infectivity through different strategies to create intervention strategies targeting cell entry of the virus. The fact that SARS-CoV-2 uses a cytokine overflow with renin-angiotensin system (RAS) and antibody-dependent enhancement (ADE) to create an unbreakable treatment and how SARS-CoV-2 invades human immune cells and influences the immune system to bring favorable balance among antiviral, anti-inflammatory and immune responses is still under research.

Signs and Symptoms

The clinical symptoms of SARS-CoV-2 have been finally accessed from data taken from infected persons across the world. The symptoms are mostly non-specific because the appearance of symptoms takes at least five days after infection and varies from one person to another. Patients infected with SARS-CoV-2 can show a variety of symptoms ranging from being asymptomatic to severe pneumonia. The severity condition is scaled initially by clinical symptoms and laboratory tests. Initial symptoms include fever or chills, dry cough, shortness of breath, muscle pain, headache, body aches, congestion or runny nose, sore throat, nausea or vomiting, diarrhea, fatigue and loss of taste and/or smell. Laboratory tests that prove elevated C-reactive protein, decreased lymphocyte count, increased lactate dehydrogenase, elevated cardiac enzymes, abnormal liver function tests, leukopenia, elevated D-dimer, elevated erythrocyte sedimentation rate, leukocytosis, elevated procalcitonin, and abnormal renal function are also considered as infected conditions because some people do not show clear symptoms but can be asymptomatic carriers. Most patients show clear symptoms of chest pain, diarrhea and upper respiratory tract symptoms because the SARS-CoV-2 affects the lungs as they are mainly transferred through respiratory droplets. Critical stages of SARS-CoV-2 can lead to death due to severe pneumonia, organ failure, acute respiratory tract infection and septic shock and people with chronic disease conditions like hypertension, diabetes, cardiovascular problems, kidney and liver diseases, cancer, HIV and auto-immune disorders are more susceptible to be infected by the SARS-CoV-2 virus as their immunity condition differs due to many medical reasons. Other than this elderly people, young children, pregnant ladies and smokers are also at an equal risk due to decreased immunity during these stages (Elengoe, 2020).

The SARS-CoV-2 virus incubation period is from 1-5 days whereas, most clinical symptoms appear after 5-11 days after infection. Patients are mostly asymptomatic within the first few days and become paucisymptomatic or critically ill within 2 weeks of infection. Diagnosis of patients infected with SARS-CoV-2 is a bit tricky as most patients are silent carriers of the virus. It depends on various factors such as the patient's age, the severity of illness, existing or recovered medical conditions, response to treatment and symptom presentation that vary from person to person. Moreover, SARS-CoV-2 is an evolving virus that evades host responses and adapts to the host environment by producing new biochemical mechanisms and different strains of SARS-CoV-2 that have been identified have rather astonishing mechanisms that bring about contrasting symptoms.

Clinical Manifestations

Pulmonary Manifestations

The respiratory system is the common organ affected by SARS-CoV-2. Pulmonary manifestations are considered the major clinical symptoms that identify SARS-CoV-2 infection. Pulmonary symptoms range from cough, shortness of breath, and sputum production to severe pneumonia, respiratory failure, and ARDS. Multiple studies have proven that cough, one of the common symptoms is the predominant pulmonary symptom observed in SARS-CoV-2 patients (Guo et al., 2020; Cascella et al., 2022). Currently, pulmonary symptoms of SARS-CoV-2 are classified according to the below:

Asymptomatic

Individuals without any clinical symptoms tested positive for the SARS-CoV-2 test.

Mild Illness

Positive individuals present symptoms like fever, cough, sore throat, malaise, headache, muscle pain, nausea, vomiting and diarrhea without shortness of breath or abnormal chest imaging.

Moderate Illness

Positive individuals show clear shortness of breath and radiographic results of respiratory tract infection.

Severe Illness

Positive individuals show lower pulse oximetry reading (SpO_2) $\leq$ 94%, (PaO_2/FiO_2) <300-200 ratio of partial pressure of arterial oxygen to fraction of inspired oxygen with a respiratory frequency of less than 30 breaths per minute and pulmonary infiltrate level less than 50% and oxygen will be supplied to the patient.

Critical Illness

Individuals show acute respiratory failure, septic shock, and/or multiple organ dysfunction.

Acute respiratory distress syndrome or ARDS is a severe pulmonary complication that causes decreased oxygen intake. Patients who experience severe hypoxia and will be given oxygen therapy. Some critically ill patients can develop severe pneumonia requiring intensive medical care followed by oxygen therapy due to life-threatening acute respiratory failure.

Extra-Pulmonary Manifestations

Although SARS-CoV-2 is a viral infection that affects the respiratory system, severe conditions have the possibility to develop extra-pulmonary manifestations such as organ dysfunction and septic shock. Many extra-pulmonary organs like the blood circulatory system, gastrointestinal system, cardiovascular system, nervous system and other organs are vulnerable to manifestations due to SARS-CoV-2 infection.

Renal Manifestations

In severe cases of SARS-CoV-2 people are susceptible to developing kidney injury. Patients develop acute kidney injury (AKI) due to an increase of fluid in blood or hypervolemia, drug and drug-related injury, vascular injury, and direct cytotoxicity of the virus. Other renal problems like proteinuria, hematuria, electrolyte abnormalities such as hyperkalemia, hyponatremia, and acid-base balance disturbance such as metabolic acidosis are also possible.

Cardiac Manifestations

- Myocardial ischemia/infarction (MI)
- ACS
- Arrhythmias
- Cardiomyopathy
- Cardiogenic shock

Hematologic Manifestations

- Lymphopenia
- Thrombocytopenia
- Leukopenia
- Elevated ESR levels
- C-reactive protein (CRP) level
- Lactate dehydrogenase (LDH)
- Leukocytosis
- Thromboembolic events such as PE, DVT, MI, ischemic strokes, and arterial thromboses
- Fibrinogen levels imbalance
- Prolonged prothrombin time (PT)
- Partial thromboplastin time(aPTT)

Gastrointestinal Manifestations

(Other than diarrhea, nausea and/or vomiting and abdominal pain)

- Anorexia
- Acute mesenteric ischemia
- Portal vein thrombosis

Hepatobiliary Manifestations

- Increase in aspartate transaminase(AST)
- Increase in alanine transaminase(ALT)
- Hepatic dysfunction

Endocrinologic Manifestations (Diabetes Mellitus Patients)

- Abnormal blood glucose levels
- Euglycemic ketosis
- Diabetic ketoacidosis

Neurologic Manifestations

- Anosmia
- Ageusia
- Impairment of consciousness
- Seizure disorder
- Toxic metabolic encephalopathy

Cutaneous Manifestations

- Acral lesions
- Erythematous maculopapular rash
- Urticarial rashes

Types of SARS-CoV-2 Variants

Table 1 shows the different types of SARS-CoV-2 variants such as alpha, beta, gamma and delta (Duong, 2021).

Table 1. Different types of SARS-CoV-2 variants

WHO Label	Alpha(α)	Beta(β)	Gamma(α)	Delta(α)
Variant Name	B.1.1.7	B.1.351 B.1.351 B.1.351.2 B.1.351.3	P.1 P.1.1 P.1.2	B.1.617.2 AY.1 AY.2 AY.3
First recorded	September 2020	October 2020	December 2020	December 2020
Country of the first Detection	United Kingdom	South Africa	Brazil	India
Key mutation (In the spike protein)	N501Y, D614G, P681H	K417N, E484K, N501Y, D614G, A701V	K417T, E484K, N501Y, D614G, H655Y	L452R, T478K, D614G, P681R
Concerns	Binds more readily to human angiotensin-converting enzyme 2 (hACE2) receptors 50% more transmissible Causes more severe disease Vaccines are effective.	Binds more readily to human angiotensin-converting enzyme 2 (hACE2) receptors Increased infectivity 50% more transmissible Causes more severe disease Vaccines are less effective.	Binds more readily to human angiotensin-converting enzyme 2 (hACE2) receptors Increased infectivity Believed to be more transmissible Unknown if it causes more severe disease Unknown if vaccines are less effective.	Higher possibility of reinfections 20% more transmissible May cause more severe disease Vaccines are still effective.
Major concern	Transmissibility Potential to infect cells is high.	Severity of illness Increased risk of death	Vaccine efficiency Prevent antibody efficiency	Transmissibility Potential to infect cells is high. Reduced neutralization
Countries reported	82	40	18	44
Possibility to evade the vaccine	Minimal	Moderate	Moderate	Minimal
Severity of Disease	High	High	High	High

Table 2. Medicinal plants and their therapeutic properties (Benarba and Pandiella, 2020; Hafez Ghoran et al., 2020; Anand et al., 2021; Sailah et al., 2021; Roy et al., 2021; Ahammad et al., 2021; Singh et al., 2021; Parmar et al., 2021; Wajs-Bonikowska et al., 2019; Batiha et al., 2020)

No.	Plant(s)	Scientific Name	Family	Part of the Plant	Active Constituents	Therapeutic properties
1.	Garlic	*Allium Sativum* L.	*Alliaceae*	Bulbs	Myrtenyl Acetat β-Thujone Camphre α-Thujone Diallyl trisulfide	Reduce the risk for cardiovascular diseases, have anti-tumor and anti-microbial effects, to aid respiration and digestion and to treat leprosy and parasitic infestation, antioxidant effect, antimicrobial effect, enhancement of detoxification foreign compound and hepatoprotection
2.	Cang Zhu	*Atractylodes lancea*	*Asteraceae*	Rhizome	Sesquiterpenes, sesquiterpenoids, polyethylene alkynes, phytosterols	Treat rheumatic diseases, digestive disorders, night blindness, and influenza anti-cancer, anti-obesity and anti-inflammatory
3.	Bai Zhu	*Atractylodis macrocephalae Koidz.*	*Asteraceae*	Rhizome	Sesquiterpenoids, polysaccharides, amino acids, vitamins, resins, and lactones	Anti-inflammatory, antitumor, antioxidant, antiviral and problems related to the gastrointestinal tract
4.	Chaihu	*Bupleurum chinense*	*Apiaceae*	Aerial parts	Flavonoids, dihydrochalcone, phenylpropanoid glycosides, lignan, sterols, stilbenoid	Antipyretic, anti-cancer immunomodulatory, gastrointestinal tract, and hepatoprotective effects
5.	Himalayan Thorowax	*Bupleurum* spp.	*Apiaceae*	Aerial parts	β-caryophyllene, β-caryophyllene oxide and spathulenol hexanal, heptanal (E)-2-nonenal (E,E)-2,4-decadienal, hexanoic acid, heptanoic acid, octanoic acid and hexadecanoic acid	Anti-inflammatory, antitumor, antioxidant, antimicrobial, antifungal, and analgesic activities

Table 2. (Continued)

No.	Plant(s)	Scientific Name	Family	Part of the Plant	Active Constituents	Therapeutic properties
6.	Tea	*Camellia sinensis* (L.) *Kuntze*	*Theaceae*	Leaves	Epigallocatechin gallate (EGCG), epigallocatechin (EGC), epicatechin gallate (ECG), and epicatechin (EC)	Antidiarrheal, antiemetic, cardiotonic, diuretic, expectorant, stimulant and astringent
7.	Yellow cypress	*Chamaecyparis obtusa var. formosana*	*Cupressaceae*	Heartwood	Flavonoids and proanthocyanidins	Anti-oxidant and anti-hyperglycemic activities
8.	Black Cohosh, Black Snakeroot, Bugbane	*Cimicifuga racemosa* (L.) *Nutt.*	*Ranunculaceae*	Rhizome		Anti-inflammatory, and increase immunity to relieve symptoms of menopause
9.	Cinnamon	*Cinnamomum Cassia*	*Lauraceae*	Bark	Tricyclene β-carophyllene	Remedy for respiratory, digestive and gynecological ailments used for chest pain, kidney disorders, high blood pressure, cramps, and cancer
10.	Sweet orange	*Citrus sinensis*	*Rutaceae*	Peels	Limonene myrcene, α-pinene; linalool; octanal; decanal,	Antithyroid, hypoglycemic, and insulin stimulatory properties,
11.	Chinese goldthread	*Coptis chinensis Franch.*	*Ranunculaceae*	Rhizomes	Berberine, coptisine, palmatine, and jatrorrhizine	Anti-inflammatory neuroprotective, neuroregenerative, anti-diabetic, anti-oxidative properties
12.	Autumn Crocus	*Crocus Sativus* L.	*Iridaceae*	dried red stigma	D-Camphor Camphene Borneol α-Pinene	Analgesic, sedative, fever reducer, treatment for pertussis and asthma, expectorant, antispasmodic, aphrodisiac, digestive, and carminative
13.	Sugi	*Cryptomeria japonica*	*Cupressaceae*	Heartwood	β-sitosterol, ferruginol, xanthoperol	Antifungal antioxidant, anti-lipase, antibacterial and melanin-biosynthesis-inhibition

No.	Plant(s)	Scientific Name	Family	Part of the Plant	Active Constituents	Therapeutic properties
14.	Holly Fern	*Cyrtomium fortunei*	*Dryopteridaceae*	Leaves	6’-methylglucuronide-5-hydroxy-chromone, ethyl α-d-glactopyranosid, neoechinulin A, 9,12,13-trihydroxyoctadeca-10(E),15(Z)-dienoic acid, phellopterin	Analgesic, anthelmintic, antibacterial, anticoagulant, antiviral, depurative, febrifuge and hemostatic properties
15.	Purple coneflower	*Echinacea purpurea*	*Asteraceae*	Aerial parts	Alkamides, caffeic acid derivatives, and polysaccharides	Immunostimulatory, anti-inflammatory, antioxidant, antibacterial, antiviral, and larvicidal properties
16.	Clove	*Eugenia Caryophyllus*	*Myrtaceae*	flower buds	Eugenol Eugenol acetate Tyranton Caryophyllene 1,4,7-Cycloundecatriene, 1,5,9,9-tetramethyl-,Z,Z, 1-8 Cineol, Cubenol	Used to treat roundworms and tapeworms, asthma, toothache, soar throat, dental, respiratory disorders, digestive system ailments, dyspepsia, gastritis, diarrhea, as antipyretic, aphrodisiac, carminative, appetizer, expectorant, antiemetic, anxiolytic, myorelaxant, analgesic, decongestant, anti-inflammatory, hypnotic, vermifuge, antibacterial agent
17.	Fleshy spurge, Oleander spurge Indian spurge tree	*Euphorbia neriifolia* L.	*Euphorbiaceae*	Leaves	Neriifolin-S, neriifolin, neriifoliene, euphol, neriifolione, cycloartenol, nerifoliol, lectin, euphonerins A-G, 3-O-acetyl-8-O-tigloylingol, taraxerol, antiquorin	Used as laxative, purgative, rubefacient, carminative and expectorant
18.	Fennel	*Foeniculum Vulgare*	*Apiaceae*	seeds and oil	Acetyleugenol Terpine-4-ol Sabinene	To treat a wide range of ailments related to the digestive, endocrine, reproductive, and respiratory systems

Table 2. (Continued)

No.	Plant(s)	Scientific Name	Family	Part of the Plant	Active Constituents	Therapeutic properties
19.	Lavender	*Lavandula Stoechas*	*Lamiaceae*	Flower and oil	Pulegone, menthol, menthone, Sabinene Eucalyptol Trans-Cinnamylacetat Trans-Cinnamaldehyde Undecan Undecane-2-6-dimethyl 2-carboxylic acid,3- methyl thiophene Picrocrocin	Antibacterial, antifungal, and antioxidant properties
20.	Weeping forsythia, Golden-bell	*Forsythia suspensa*	*Oleaceae*	Leaves	Steroids, triterpenoids, lignans, and phenylethanoid glycosides	Anti-inflammatory, diuretic, drainage, and antimicrobial
21.	Jasmine	*Gardenia jasminoides J. Ellis.*	*Rubiaceae*	Leaves	Glycosides, Alkaloids, Flavanoids, Tannins, Saponin, Steroids	Antioxidant, anti-inflammatory properties, and improve the immune system
22.	Liquorice	*Glycyrrhiza glabra* L.	*Fabaceae*	Roots	Saponin, flavonoids, isoflavonoids, stilbenoids, and coumarins	To treat respiratory disorders, hyperdipsia, epilepsy, fever, sexual debility, paralysis, stomach ulcers, rheumatism, skin diseases, hemorrhagic diseases, and jaundice with antioxidant, anticancer, anti-inflammatory, and antidiabetic properties

No.	Plant(s)	Scientific Name	Family	Part of the Plant	Active Constituents	Therapeutic properties
23.	Parsley tree	*Heteromorpha* spp.	*Apiaceae*	Aerial parts	α-pinene and 2,5-dimethoxy-p-cymene (thymohydroquinone dimethyl ether	Anti-inflammatory, anti-tumor properties
24.	Chameleon plant	*Houttuynia cordata*	*Saururaceae*	Aerial parts	Flavanoids and lipopolysaccharides,	Anti-inflammatory properties
25.	Perforate St John's-wort	*Hypericum perforatum* L.	*Hypericaceae*	Aerial parts	Naphthodianthrones, phloroglucinols, flavonoids, biflavones, phenylpropanes, proanthocyanidins	Antidepressant, antioxidant, anticancer, antiviral and neuroprotective properties
26.	Oleander	*Nerium Oleander*	*Apocynaceae*	Seeds and leaves	Dodecan Digitoxigenine Crocin Crocetin Calarene 1,2,4-Trithiolane,3,5-dimethyl Trans-Anethole	Used to treat hemorrhoids, ulcers, leprosy, and as an abortifacient
27.	Myrtle	*Myrtus Communis* L.	*Myrtaceae*	Seeds and leaves	Neriine γ-Terpinene pi-Cymene Picrocrocin	Used to treat inflammatory and infectious processes such as prostatitis, bronchitis, sinusitis, and colds, with antimicrobial, anti-inflammatory, and antioxidant activities
28.	Onion	*Allium Cepa* L.	*Amaryllidaceae*	Whole	Estragol β-Phellandren β-Eudesmol Amorphan α-Terpinyl acetate α-Terpeneol	Antioxidant, antimicrobial, antidiabetic, hyperlipidaemia, used to treat hypertension and diabetes, obesity

Table 2. (Continued)

No.	Plant(s)	Scientific Name	Family	Part of the Plant	Active Constituents	Therapeutic properties
29.	Lindera	*Lindera aggregata*	*Lauraceae*	Roots	(1'S)-12'-hydroxyl-linderegatine (1), (1S)-5'-O-p-hydroxybenzoyl norreticuline (2), (1R, 1'R)-11,11'-biscoclaurine (3)	Used to treat urinary system diseases such as enuresis and urinary stones, chronic gastritis and rheumatoid arthritis
30.	Chamomile Dog Fennel	*Anthemis hyalina* DC.	*Asteraceae*	Flowers	Carvacrol α-pinene	Used to treat hay fever, inflammation, muscle spasms, menstrual disorders, insomnia, ulcers, wounds, gastrointestinal disorders, rheumatic pain, and hemorrhoids
31.	Bay laurel	*Lauris Nobilis* L.	*lauráceas*	Leaves	Eugenol, methyl eugenol, pinene,cineole, Diallyl trisulfide, Trans-Anethole, Pulegone Linalool Fenchone pi-Cymene	Anti-inflammatory, antitumor, antioxidant, antispasmodic, antimicrobial, insecticidal, antimalarial, antifungal, and antioxidant activities
32.	Felt Fern Tongue Fern	*Pyrrosia lingua*	*Ploypodiaceae*	Leaves	3,'5,'5,7-Tetrahydroxy-6-C-β-D-glucopyranosyl-flavanone	Beneficial effect in pain, urinary stone formation and vaginal bleeding, stranguria, bloody stranguria, urolithiasis, urinary obstruction, difficulty and pain in micturition, hematemesis, rhinorrhagia, hematuria, metrorrhagia and coughing antibacterial, hyperglycaemic and antitumor activities

No.	Plant(s)	Scientific Name	Family	Part of the Plant	Active Constituents	Therapeutic properties
33.	Horse chestnut	*Aesculus hippocastanum* L.	*Sapindaceae*	Seeds	Saponins Flavonoids Aescin	Antioxidant, antiinflammatory, Anti-cancer chronic venous diseases and male infertility
34.	Korean mint Wrinkled giant hyssop Purple giant hyssop Indian mint	*Agastache rugosa*	*Lamiaceae*	Aerial parts	Methyl chavicol Linalool Pinene Anethole Anisaldehyde Limonene	Analgesic and antipyretic effects anti-inflammatory, and for its antibacterial to relieve nausea, vomiting and poor appetite
35.	Mongolian milkvetch	*Astragalus mongholicus*	*Fabaceae*	Leaves roots	Isoquercitrin, isorhamnetin-3-b-glucoside Ononin, calycosin, formononetin	Antibacterial and anti-inflammatory properties to protect and support the immune system, prevent colds and upper respiratory infections, lowering blood pressure, treat diabetes, and protect the liver.
36.	Mugwort	*Artemisia vulgaris*	*Asteraceae*	Whole plant	Allyl-methyl disulfide Limonene Diallyl sulfide Diallyl disulfide Acetat Bornyl Bicyclogermacrene D-Fenchol	Anti-inflammatory, antitumor, antioxidant, antispasmodic, antimicrobial, insecticidal, antimalarial, antifungal, and antioxidant activities
37.	Sweet wormwood Annual wormwood	*Artemisia annua* L.	*Asteraceae*	Whole plant	Artemisinin	Antimalarial for the treatment of multi-drug resistant strains of falciparum malaria

Table 2. (Continued)

No.	Plant(s)	Scientific Name	Family	Part of the Plant	Active Constituents	Therapeutic properties
38.	Toothpick-plant	*Ammi visnaga* L.	*Apiaceae*	Dried umbels	Isoamyl-2-methyl butyrate Cis- linalool oxide α-Terpinene Amyl valerate Amyl isobutyrate	Irritable stomachs and the womb; it stimulates urine and women's cycles (menstruation). The seeds of this plant were at one time used to treat sterility, convulsions in the urethra and kidney stones
39.	Chinese rhubarb, or Indian rhubarb	*Rheum officinale*	*Polygonaceae*	Roots	Hydroxyanthraquinones, tannins and gallic acid	Cathartic, anti-inflammatory, and antioxidative, and it is well known to promote excretion and decrease reabsorption of bilirubin
40.	Siler	*Saposhnikovia divaricata*	*Apiaceae*	Flowers and root	Chromones, coumarins, acid esters, and polyacetylenes	Anti-inflammatory, analgesic, antioxidant, antiproliferative, antitumor, and immunoregulatory properties
41.	Common sage	*Sauge officinal*	*Lamiaceae*	Leaves	2methyl 6-Isopropenyl-4,8a-Dimethyl-1,2,3,5,6,7,8, 8a-Octahydronaphthalene-2,3-diol α-Terpinene	Anti-inflammatory, antioxidant, antiseptic and antibiotic properties
42.	Ginger	*Zingiber officinalis*	*Zingiberaceae*	Roots	6-gingerol, 6-shogaol, and 6-paradol	Used to treat arthritis, cramps, sprains, sore throats, rheumatism, muscular aches, pains, vomiting, constipation, indigestion, hypertension, dementia, fever and infectious diseases

No.	Plant(s)	Scientific Name	Family	Part of the Plant	Active Constituents	Therapeutic properties
43.	Fennel flower	*Nigella sativa* L.	*Ranunculaceae*	Seeds	Thymoquinone, thymohydroquinone, dithymoquinone (nigellone), p-cymene, carvacrol, 4-terpineol, t-anethole, sesquiterpene longifolene, α-pinene, and thymol	Used as antihypertensive, liver tonics, diuretics, digestive, anti-diarrheal, appetite stimulant, analgesics, anti-bacterial and in skin disorders
44.	Formosan juniper	*Juniperus formosana Hayata.*	*Cupressaceae*	Heartwood	Leaves and trunk	Used to treat abdominal spasm, asthma, diarrhea, fever, gonorrhea, headache, respiratory problems, gastrointestinal infections, cardiovascular and/or blood disorders, and as astringent, antihypertensive, diuretic, carminative, appetizer, anticonvulsant, and flavoring agent
45.	Cade, cade juniper, prickly juniper, prickly cedar, or sharp cedar	*Juniperus oxycedrus* ssp. oxycedrus	*Cupressaceae*	Berry	Alpha-pinene, sylvestrene, p-cymene, and 13epi- manoyl oxide	Antibacterial, antimicrobial, insecticidal
46.	Japanese honeysuckle	*Lonicerae japonica*	*Capripoliaceae*	Fruits	Flavanoids	Antibacterial, antiviral, anti-inflammatory, antipyretic, liver and gall bladder-protective, anti-oxidative, anti-fertility, anti-platelet aggregation, hypolipidemic, anti-allergic and immunoregulatory effects with anti-upper respiratory tract infection

Table 2. (Continued)

No.	Plant(s)	Scientific Name	Family	Part of the Plant	Active Constituents	Therapeutic properties
47.	Red spider lily	*Lycoris radiata*	*Amaryllidaceae*	Stem cortex	Lycorine, Crinine, Galanthamine, Tazettine, Narciclasine, Lycorenine, Homolycorine and Montanine	Anti-acetylcholinesterase effect and anti-tumor activity, strong inhibition effect of flu virus, measles virus, poliovirus and SARS virus
48.	Chinaberry	*Melia azedarach* L.	*Meliaceae*	Barks and leaves	Benzyl 3-O-β-D-glucopyranosyl-7-hydroxybenzoat, spathulenol, 1,7,8-trihydroxy-2-naphtaldehyde, quercetin, astragalin and 2-methoxy-4-(2-propenyl)phenyl β-D-glucoside	Antioxidative, analgesic, anti-inflammatory, insecticidal, rodenticidal, antidiarrhoeal, deobstruent, diuretic, antidiabetic, cathartic, emetic, antirheumatic and antihypertensive
49.	Mint	*Mentha Rotundifolia*	*Lamiaceae*	Leaves	Carvone limonene 1,8-cineole β-pinene cis-dihydrocarvone, dihydrocarveol, Safranal	Used to treat antimicrobial, gastrointestinal, and nervous system disorders
50.	Moutan peony	*Paeonia suffruticosa*	*Paeoniaceae*	Roots, Bark and Whole plant	Catechin, paeonidanin, paeoniflorigenone, 2, 5-dihydroxy-4-methoxyacetophenone, paeonol, gallic acid	To treat blood flow, pain, hyperpigmentation, and mood disorders, autoimmune conditions and increase estrogen levels, used to treat fevers, boils, menstrual disorders, nosebleeds, ulcers, irritability and gastrointestinal infection
51.	Empress tree	*Paulownia tomentosa*	*Paulowniaceae*	Fruits	Prenylated flavonoids, lignans, phenolic glycosides, quinones, terpenoids, glycerides, phenolic acids	To relieve bronchitis, coughing, asthma, and phlegm, used to treat conjunctivitis, dysentery, enteritis, erysipelas, gonorrhea, hemorrhoids, parotitis, traumatic bleeding, and tonsillitis

No.	Plant(s)	Scientific Name	Family	Part of the Plant	Active Constituents	Therapeutic properties
52.	Cork tree	*Phellodendron amurense*	*Rutaceae*	Barks	Flavonoids (diosmin), alkaloids (berberine, jatrorrhizine, palmatine), saponins, and coumarins	Used to treat osteoarthritis, weight loss and obesity, diarrhea, abdominal pain, gastroenteritis, urinary tract infections, diabetes, meningitis, pneumonia, eye infections, tuberculosis, and cirrhosis of the liver
53.	Balloon flower	*Platycodon grandifiorus*	*Campanulaceae*	Roots and flowers	Flavonoids, phenolic acids, triterpenoid saponins, polyacetylene, and sterols	Anti-inflammatory, antitussive, antitumor, antioxidation, anti-inflammatory, hypoglycemic, anti-obesity, and immune enhancement, analgesic, sedative, and fever agent. effective in reducing cholesterol, blood-related diseases, high blood pressure, treatment of coughs with profuse phlegm, colds, bronchitis, pleurisy, pulmonary abscesses and throat infections
54.	Tuber fleeceflower	*Polygonum multiflorum*	*Polygonaceae*	Roots	stilbenes and anthraquinones 2,3,5,4′-tetrahydroxystilbene-2-O-β-glucoside (TSG), emodin-8-O-β-D-glucoside (EMG), and physcion-8-O-β-D-glucoside (PG)	Anti-tumor, anti-oxidative, anti-bacterial, anti-hyperlipidemia, anti-atherosclerosis, immunomodulating and hepatoprotective properties
55.	Indian snakeroot	*Rauwolfia* spp.	*Apocynaceae*	Roots, leaves, and stem	Alkaloids reserpine, rescinnamine, and deserpidine phytosterol, fatty acids, unsaturated alcohols and sugars	Used to treat convulsions, fever, weakness, inability to sleep, mental disorders, pain, arthritis, cancer, high blood pressure, diabetes, stomach, intestines, and liver

Table 2. (Continued)

No.	Plant(s)	Scientific Name	Family	Part of the Plant	Active Constituents	Therapeutic properties
56.	Balm-leaved Figwort	*Scrophularia scorodonia* L.	*Scrophulariaceae*	Aerial parts	Iridoids, phenylpropanoids, phenolic acids, flavonoids and saponins	Anti-inflammatory, antibacterial, hepatoprotective, immuno-modulator, cardiovascular, diuretic, protozoocidal, fungicidal, molluscicidal, cytotoxic, cytostatic, antitumor properties
57.	Chinese skullcap	*Scutellaria baicalensis*	*Lamiaceae*	Aerial parts	Baicalin	Anti-inflammatory, antioxidant, and antiapoptotic properties, and supports hepatoprotective properties
58.	Agrimony	*Stephania tetrandra*	*Menispermaceae*	Leaves	Tetrandrine, fangchinoline, cyclanoline and dimethyltetrandrine iodide	Anti-inflammatory, anti-tumor, antibacterial, and diuretic properties. used to manage water retention and edema, eliminate stagnant bronchial mucous, detoxify the body, promote urination, and expel heat in painful joint

Medicinal Plants and Their Mode of Actions

Indigenous traditional herbal therapy has a long history of treating a wide range of chronic and infectious diseases. As a result, the search for new antiviral medicines concentrates on both synthetic combinations and metabolites originating from plants. Numerous plant metabolites have the ability to prevent the replication of viruses with little or no negative effects on the host's physiology. Based on Kurokawa et al., 2010 study, it has been reported that the bioactive metabolites may have the ability to alter the host's immunological responses to viral infections. The mangrove plant (*Acanthus ilicifolius L.*), which belongs to the Acanthaceae family, has many therapeutic qualities, including anti-inflammatory, antioxidant, and hepatoprotective actions. Strong antiviral action against the hepatitis B virus is displayed by this medicinal herb. A study using a duck model showed that an alcoholic extract of the entire plant can lower the viral load by preventing DNA replication. However, the precise mechanism was not well understood (Wei et al., 2015). Table 2 explains the medicinal plants and their therapeutic properties.

Conclusion

Plant metabolites have a variety of therapeutic uses. They may have a synergistic impact, leading to better treatment results. In this review, pathogenic mechanisms; signs and symptoms; clinical manifestations; different types of variants; medicinal plants and their mode of action were discussed. Bioinformatics approach and laboratory experiments should be carried out to elucidate the mechanism of antiviral activities of the bioactive metabolites as well as the relationship between SARS-CoV-2 target protein three-dimensional (3-D) structure and activity for the synthesis of desired antiviral medication molecules from these medicinal plants.

References

Ahammad F, Alam R, Mahmud R, Akhter S, Talukder EK, Tonmoy AM, Fahim S, Al-Ghamdi K, Samad A, Qadri I. Pharmacoinformatics and molecular dynamics simulation-based phytochemical screening of neem plant (*Azadiractha indica*) against human cancer by targeting MCM7 protein. *Briefings in Bioinformatics* (2021) 22(5):bbab098.

Anand AV, Balamuralikrishnan B, Kaviya M, Bharathi K, Parithathvi A, Arun M, Senthilkumar N, Velayuthaprabhu S, Saradhadevi M, Al-Dhabi NA, Arasu MV. Medicinal plants, phytochemicals, and herbs to combat viral pathogens including SARS-CoV-2. *Molecules* (2021) 26(6):1775.

Batiha GE, Alkazmi LM, Wasef LG, Beshbishy AM, Nadwa EH, Rashwan EK. *Syzygium aromaticum L.* (Myrtaceae): Traditional uses, bioactive chemical constituents, pharmacological and toxicological activities. *Biomolecules* 2020 10(2):202.

Benarba B, Pandiella A. Medicinal plants as sources of active molecules against COVID-19. *Frontiers in Pharmacology* (2020) 11:1189.

Cascella M, Rajnik M, Aleem A, Dulebohn SC, Di Napoli R. *Features, evaluation, and treatment of coronavirus (COVID-19).* Statpearls (2022). Accessed 20 December 2022. https://www.ncbi.nlm.nih.gov/books/NBK554776/.

Centers for Disease Control and Prevention. *SARS-CoV-2 variant classifications and definitions (2022).* Accessed 12 January 2023. https://www.cdc.gov/coronavirus/2019-ncov/variants/variant-classifications.html.

COVID-19 Coronavirus pandemic. *Worldometer.* (2022). Accessed 15 March 2022. https://www.worldometers.info/coronavirus.

Duong D. Alpha, Beta, Delta, Gamma: What's important to know about SARS-CoV-2 variants of concern? *The Canadian Medical Association Journal* (2021).

Elengoe A. Indian spices boost the immune system against COVID-19. *The Annals of the University Dunarea de Jos of Galati. Fascicle VI-Food Technology* (2020) 44(2):189-206.

Elengoe A, Selvam S. Overview on origin, transmission, treatment and clinical researches of SARS-COV-2. *Journal of Health Sciences & Surveillance System* (2022) 10(1):2-12.

Elengoe A, Selvam S, Selvarajah M, Manikam H, Vijaya R. Sequence analysis and structure prediction of Malaysia SARS-CoV-2 strain's structural and accessory proteins. *Biointerface Research in Applied Chemistry* (2022) 12(1):3259-3304.

Guo YR, Cao QD, Hong ZS, Tan YY, Chen SD, Jin HJ, Tan KS, Wang DY, Yan Y. The origin, transmission and clinical therapies on coronavirus disease 2019 (COVID-19) outbreak-an update on the status. *Military Medical Research* (2020) 7(1):1-0.

Hafez Ghoran S, El-Shazly M, Sekeroglu N, Kijjoa A. Natural products from medicinal plants with anti-human coronavirus activities. *Molecules* (2021) 26(6):1754.

Harrison AG, Lin T, Wang P. Mechanisms of SARS-CoV-2 transmission and pathogenesis. *Trends in Immunology* (2020) 41(12):1100-1115.

Kurokawa M, Shimizu T, Watanabe W, Shiraki K. Development of new antiviral agents from natural products. *The Open Antimicrobial Agents Journal* (2010) 2(1):49-57.

Ou X, Liu Y, Lei X, Li P, Mi D, Ren L, Guo L, Guo R, Chen T, Hu J, Xiang Z. Characterization of spike glycoprotein of SARS-CoV-2 on virus entry and its immune cross-reactivity with SARS-CoV. *Nature Communications* (2020) 11(1):1-2.

Parmar DR, Soni JY, Guduru R, Rayani RH, Kusurkar RV, Vala AG, Talukdar SN, Eissa IH, Metwaly AM, Khalil A, Zunjar V. Discovery of new anticancer thiourea-azetidine hybrids: Design, synthesis, *in vitro* antiproliferative, SAR, *in silico* molecular docking against VEGFR-2, ADMET, toxicity, and DFT studies. *Bioorganic Chemistry* (2021) 115:105206.

Roy N, Davis S, Narayanankutty A, Nazeem PA, Babu TD, Abida PS, Valsala PA, Raghavamenon A. Garlic phytocompounds possess anticancer activity by specifically targeting breast cancer biomarkers-an *in silico* study. *Asian Pacific Journal of Cancer Prevention* (2016) 17(6):2883-2888.

Sailah I, Tumilaar SG, Lombogia LT, Celik I, Tallei TE. Molecular docking and dynamics simulations study of selected phytoconstituents of "Pangi" (*Pangium edule Reinw*) leaf as anti-SARS-CoV-2. *Philippine Journal of Science* (2021) 150:925-937.

Singh R, Bhardwaj VK, Sharma J, Kumar D, Purohit R. Identification of potential plant bioactive as SARS-CoV-2 Spike protein and human ACE2 fusion inhibitors. *Computers in Biology and Medicine* (2021) 136:104631.

Singhal T. A review of coronavirus disease-2019 (COVID-19). *The Indian Journal of Pediatrics* (2020) 87(4):1-6.

Xie Y, Wang Z, Liao H, Marley G, Wu D, Tang W. Epidemiologic, clinical, and laboratory findings of the COVID-19 in the current pandemic: systematic review and meta-analysis. *BMC infectious diseases* (2020) 20(1):1-2.

Xu J, Zhao S, Teng T, Abdalla AE, Zhu W, Xie L, Wang Y, Guo X. Systematic comparison of two animal-to-human transmitted human coronaviruses: SARS-CoV-2 and SARS-CoV. *Viruses* (2020) 2(2):244.

Wajs-Bonikowska A, Malarz J, Stojakowska A. Composition of essential oils from roots and aerial parts of *Carpesium divaricatum*, a traditional herbal medicine and wild edible plant from south-east asia, grown in Poland. *Molecules* (2019) 24(23):4418.

Wei PH, Wu SZ, Mu XM, Xu B, Su QJ, Wei JL, Yang Y, Qin B, Xie ZC. Effect of alcohol extract of *Acanthus ilicifolius L.* on anti-duck hepatitis B virus and protection of liver. *Journal of Ethnopharmacology* (2015) 60:1-5.

Chapter 8

Phytochemicals against SARS-CoV-2 Target Proteins

Hitheshi Saluwadana[1], BSc
Asita Elengoe[2,*], PhD
and Shanaya Amath[1], BSc
[1]School of Biotechnology, Manipal International University (MIU), Putra Nilai, Negeri Sembilan, Malaysia
[2]Department of Biotechnology, Faculty of Applied Science, Lincoln University College, Petaling Jaya, Selangor, Malaysia

Abstract

The COVID-19 pandemic infection was discovered in December 2019 after an epidemic outbreak in Wuhan City, Hubei Province, China. In 2020, the World Health Organization (WHO) named this disease as COVID-19 because it was caused by the severe acute respiratory syndrome coronavirus 2 (SARS-CoV-2). Vaccines have been developed to control the spread of the virus but no efficient antiviral treatment is available until now. Traditional herbs such as garlic, ginger, turmeric, and others have been used for thousands of years to treat a variety of chronic diseases. Therefore, the public show interest in medicinal plants to cure COVID-19. This review provides information about the SARS-CoV-2 virus's structural components; diagnostics tests, clinical manifestations; SARS-CoV-2 target proteins; medicinal plants and their biological activities. Understanding the target sites of SARS-CoV-2 will help the researchers to design active site-specific drugs. Phytochemicals can be used as safe drugs against SARS-CoV-2 target proteins.

* Corresponding Author's Email: asitaelengoe@yahoo.com.

In: Medicinal Plants and COVID-19
Editor: Asita Elengoe
ISBN: 979-8-88697-958-9

Keywords: COVID-19, phytochemicals, SARS-CoV-2

Introduction

SARS-CoV-2, which was initially discovered in Wuhan City, Hubei Province, China in December 2019, has the potential to become a fatal zoonotic illness that travels across several continents by the year 2020 (Elengoe, 2020). The cause of the coronavirus illness 2019 or COVID-19 outbreak is the beta coronavirus SARS-CoV-2, which is a member of the Coronaviridae family. On March 11, 2020, the World Health Organization (WHO) designated COVID-19 to be a worldwide pandemic, a disease to which people are not immune and which spreads across broader regions (such as many continents) (Elengoe, 2020).

Coronaviruses seldom spread from person to person. This new strain is believed to have originated in bats, but one study suggests that pangolins may have been the source, according to Trusted Source; because during the pandemic, the Wuhan market did not sell bats. For this reason, it was previously believed that pangolins, also known as scaly anteaters, were being sold illegally in some Chinese markets. SARS-CoV-2 is comparable to a number of pangolin-affecting coronaviruses. How the virus first infected people is still a mystery. Some accounts claim that the first occurrences were found in a seafood and animal market in Wuhan. It's conceivable that this is where SARS-CoV-2 first started to spread to people (Wu et al., 2020).

The disease is most contagious when an individual's symptoms are at their worst. However, it is possible for someone who is symptomless to spread the infection. Additionally, the infection can still be disseminated by someone who shows no symptoms. Droplets that carry viruses may fall onto nearby surfaces or objects (Wu et al., 2020). Other persons may contract the virus by coming into contact with these surfaces or items. The person is prone to get an infection if they later touch their mouth, nose, or eyes. As SARS-CoV-2 spread both inside and outside of China, individuals who had no direct contact with animals became afflicted. The sickness spreads from one person to the next, in other words (Kumar et al., 2020).

The coronavirus is currently expanding around the world, which implies that people are unintentionally acquiring and spreading it. This expanding international diffusion has resulted in the emergence of a pandemic. With more than 194 million confirmed cases and more than four million fatalities in approximately 200 countries, COVID-19 is currently spreading throughout the world (Worldometer, 2020).

Structural Components of SARS-CoV-2

Table 1 shows the structural components of SARS-CoV-2 (Khailany et al., 2020; Wang et al., 2020).

Table 1. The structural components of the SARS-CoV-2 and their function

Structural Component	Characterization	Function(s)
Open reading fame(orf)	Orf1a	Orf1a protein and nsps 4, 5, and 10 form a compartment together at the location of the coding sequence of Orf1ab
	Orf3	Proteins 3a and 3b (p3a and p3b, formerly known as X1 and X2) contain 274 and 154 amino acids, respectively. ORF3a and ORF3b encode them, and together they make up the second biggest sub-genomic RNA in the SARS-CoV genome
	Orf6	Indicating a large number of transport pathways for small ligands influencing their reactivity (with combining M protein)
	Orf8b	Trigger intracellular stress pathways and enact NOD-like receptor family pyrin region containing-3 (NLRP3) (SRAS-CoV)
	Orf10	Can be used to distinguish the infection more rapidly than PCR-based strategies
	Orf9	Help to localize Orf9b inside the particles but closer to the envelope
	Orf14	Catalytic activity, metal ion binding, ion binding
Structural proteins	Spike glycoprotein (S)	Involved in binding the host cell receptor and the virion. It allows the viral particles to enter the host cell
	Envelope protein (E)	Involved in the assembly and release of virions
	Membrane glycoprotein (M)	Maintain the shape of the virion. It enhances the curvature of the membrane and helps in binding to the nucleocapsid
	Nucleocapsid protein (N)	Essential for RNA synthesis and packaging of the genome into virions
Accessory proteins	Orf3a	Interacts with caveolin-1 during virus uptake and release, activates PERK pathway in the UPR and triggers apoptosis, activates p38 kinase to induce cytochrome-c mediated apoptosis, activates NF-κB and JNK, induces RANTES and IL-8 production

Table 1. (Continued)

Structural Component	Characterization	Function(s)
Accessory proteins (continued)	Orf3b	Induces necrosis and apoptosis, activates AP-1 through JNK/ERK pathway
	Orf6	Stimulates DNA synthesis, suppresses the expression of co-transfected plasmids
	Orf7a	Activates NF-κB and JNK for IL-8 and RANTES production, activates p38 and inhibits the translation of cellular proteins
	Orf7b	Required for Golgi localization (Maybe an attenuating factor *in vivo*)
	Orf8a	Induces caspase-dependent apoptosis (p8ab-activates the ATF6 branch of unfolded protein response)
	Orf8b	Stimulates cellular DNA synthesis
	Orf9b	Induces caspase-dependent apoptosis

Clinical Manifestations

The COVID-19 clinical manifestations are varied in the general population (Table 2). Importantly there are symptoms such as fever, cough, dyspnoea, malaise, fatigue and sputum/secretion. Sneezing, sore throat, goosebumps, headache, chest pain and diarrhea were other common symptoms (Singhal, 2020). Acute respiratory distress syndrome or ARDS is a severe pulmonary complication that causes decreased oxygen intake. Patients who experience severe hypoxia will be given oxygen therapy. Some critically ill patients can develop severe pneumonia requiring intensive medical care followed by oxygen therapy due to life-threatening acute respiratory failure (Viner et al., 2021).

Diagnostics Test

Table 3 shows the list of the diagnostics tests such as molecular, antigen, antibody and clinical tests.

Table 2. Clinical manifestation of SARS-CoV-2 (Wu et al., 2020; Mason, 2020)

Asymptomatic	Mild illness	Moderate illness	Severe illness	Critical illness
Individuals who test positive for SARS-CoV-2 while having no clinical symptoms	Positive people have symptoms such as fever, cough, sore throat, malaise, headache, muscle soreness, nausea, vomiting, and diarrhea, but no shortness of breath or abnormal chest imaging	Positive individuals show clear shortness of breath and radiographic results of respiratory tract infection	Positive individuals show lower pulse oximetry reading (SpO_2) $\leq$ 94%, (PaO_2/FiO_2) <300-200 ratio of partial pressure of arterial oxygen to fraction of inspired oxygen with a respiratory frequency of less than 30 breaths per minute and pulmonary infiltrate level less than 50% and oxygen will be supplied to the patient	Individuals show acute respiratory failure, septic shock, and/or multiple organ dysfunction

Table 3. List of the diagnostics tests (Wu et al., 2020; Hossain et al., 2020)

Testing type	Diagnostic test(s)	Specimen	Detection	Time takes	Advantages	Disadvantages
Molecular tests	rRT-PCR (Same with other NAAT tests. LAMP TMA NEAR SDA)	Nasal swab, sputum, respiratory trach aspirates/washes	SARS-CoV-2 nucleic acid	0-7 days (depends on the location.	• Early detection • Quick processing (hours) • Highly selective and specific	• Does not tell you if the virus is still active • Cannot detect those who've been infected and later recovered • Skilled personnel and equipment necessary

Table 3. (Continued)

Testing type	Diagnostic test(s)	Specimen	Detection	Time takes	Advantages	Disadvantages
Antigen test	Rapid diagnostic tests	Nasal or throat swab	The presence of viral antigen	1 hour or less	• Low cost • Fast results, usually within a few hours of taking the test • Highly accurate for positive results	• Reduced accuracy may lead to false results
Antibody tests	ELISA CLIA CGIA	Serum	The presence of antibodies created by an individual's immune system	0-3 weeks (depends on the location)	• Rapid test available at point-of-care • Low cost • Can be used to test large populations • Reliably detects exposure	• Cannot detect if the patient is contagious or if the infection is still present • Prone to false negatives if the patient has deferred immune response
Clinical tests	CT scan	-	CT- detection of radiological features	Usually within 24 hours	• Fast, painless, non-invasive, accurate	• Relatively expensive, cannot perform for pregnant women

SARS-CoV-2 Target Proteins

Table 4 shows the list of the target proteins with Protein Data Bank ID (PDB ID) and their functional roles (Berman et al., 2003).

Table 4. Target proteins (PDB ID) and their functional roles

Protein Name	Function	PDB id(s)	Ligand
Nsp 1	Inhibits translation machinery and production of immune defense factors	6ZLW	Ribosomal Complex
		6ZM7	Ribosomal Complex
		6ZME	Ribosomal Complex
		6ZMI	Ribosomal Complex
		6ZMO	Ribosomal Complex
		6ZN5	Ribosomal Complex
		6ZON	Ribosomal Complex
		6ZP4	Ribosomal Complex
		6ZOJ	Ribosomal Complex
Nsp 2	Putative role in apoptosis induction	-	-
Nsp 3	Involved in viral replication Nsp3-4-6 complex for ER modification and formation of DMV's Papin-like protease domain	6YWK	HEPES
		6YWL	ADP-ribose
		6YWM	MES
		6W6Y	AMP
		6WUU	Peptide inhibitor VIR250 ubiquitin proparglyamide
		6XAA	ISG15 C-terminal domain proparglyamide
		6XA9	ISG15 ADP-ribose MES
		6W9C	-
		6VXS	-
		6WO2	ADP-ribose
		6WCF	-
		6WEN	Peptide inhibitor VIR251
		6WEY	-
		6WOJ	-
		6WRH	-
		6WX4	-
		6WZV	-
		6XG3	-
Nsp 4	Nsp3-4-6 complex for ER modification and formation of DMV's	-	-
Nsp 5	Main protease	6W63	Inhibitor X77
		6WTT	Inhibitor GC-376
		6XCN	Leupeptin

Table 4. (Continued)

Protein Name	Function	PDB id(s)	Ligand
Nsp 5 (continued)	Main protease	6WQF	-
		7C8R	TG-0203770
		7C8T	TG-0205221
		6WNP	Boceprevir
		6WTK	Feline coronavirus drug
		6WTM	Feline coronavirus drug
		6WTJ	Feline coronavirus drug
		6XB1	-
		6XBO	-
		6XB2	-
		7BRR	GC376
		7BRO	-
		7BRP	Boceprevir
		6M03	-
		6XKF	-
		6XHM	7V-[(2S)-I-({(2S,3S)-3,4-dihydroxy-1-[(3S)-2-oxopyrrolidin-3-yl] butan-2-yllamino)-4-methyl-1-oxopentan-2-yI]-4-methoxy-1H-indole-2-carboxamide
		6XHU	-
		6XFN	UAW243
		6XBG	UAW246
		6XBH	UAW247
		6XBI	UAW248
		6XOA	-
		6XMK	Inhibitor 7J
		6YZ6	Leupeptin
		6XA4	UAW241
		7BQY	Inhibitor N3
		6LU7	Inhibitor N3
		6Z2E	Biotin-PEG(4)-Abu-Tle-Leu-Gln- vinylsulfone
		6XKH	-
		7BUY	Carmofur
		6M2N	-
		6M2Q	-
		6Y2G	Alpha-ketoamide 13b
		6Y2E	-
		6YB7	-
		6Y2F	Alpha-ketoamide 13b
		6LZE	Inhibitor llb
		6MOK	Inhibitor llb
		6Y84	

Protein Name	Function	PDB id(s)	Ligand
Nsp 5 (continued)	Main protease	7C8U	GC376
		6YVA	mISG15
		6YVF	AZD6482
		6YT8	Pyrithione zinc
		6YNQ	AZD6482
Nsp 6	Nsp3-4-6 complex for ER modification and formation of DMVs	-	-
Nsp 7	Part of the replication-transcription complex, involved in virus replication	7BTF	Nsp8, Nsp12
		6M71	Nsp8, Nsp12
		7BV1	Nsp8, Nsp12
		7BV2	Nsp8, Nsp12, Remdesivir+Nsp8, Nsp12
		7BW4	Nsp8, Nsp12
		6YYT	Nsp8
		6M51	Nsp8 C-term
		6W10	Nsp8 C-term
		6WQD	Nsp8 C-term
		6WTC	Nsp8 C-term
		6XIP	Nsp8, Nsp12
		6X2G	Nsp8, Nsp12
		7C2K	Nsp8, Nsp12
		7BZF	Nsp8, Nsp12
		6YHU	Nsp8
		6XEZ	Nsp8, Nsp12, Nsp13
		6XQB	Nsp8, Nsp12
Nsp 8		7BTF	Nsp7, Nsp12
		6M71	Nsp7, Nsp12
		7BV1	Nsp7, Nsp12
		7BV2	Nsp7, Nsp12, Remdesivir+Nsp7, Nsp12
		7BW4	Nsp7, Nsp12
		6YYT	Nsp7
		6M51	Nsp7
		6WIQ	Nsp7
		6WQD	Nsp7
		6WTC	Nsp7
		6XIP	Nsp7, Nsp12
		6X2G	Nsp7, Nsp12
		7C2K	Nsp7, Nsp12
		7BZF	Nsp7, Nsp12
		6YHU	Nsp7
		6XEZ	Nsp7, Nsp12, Nsp13
		6XQB	Nsp7, Nsp12
Nsp 9	RNA replicase	6W9Q	Main protease-derived peptide
		6W4B	-
		6WXD	-

Table 4. (Continued)

Protein Name	Function	PDB id(s)	Ligand
Nsp 10	Modulates Nsp14 and Nsp16 activities Nsp10-Nsp11-Nsp14-Nsp16-complex	7C2I	Nsp16, SAM
		7C2J	Nsp16, SAM
		6W61	Nsp16, SAM
		6XKM	Nsp16, SAM
		6WJT	Nsp16, S-Adenosyl-L-Homocysteine
		6WVN	Nsp16, S-Adenosylmethionine
		6WQ3	Nsp16, S-Adenosylmethionine, 7-methyl-GpppA
		6WKS	Nsp16, S-Adenosylmethionine, 7-methyl-GpppA
		6WRZ	Nsp16, S-Adenosylmethionine, 7-methyl-GpppA
		6ZCT	Nsp16, Sinefungin
		6YZ1	Nsp16, Sinefungin
		6WKQ	Nsp16
		7BQ7	Nsp16, S-adenosylmethionine
		6W75	Nsp16
		6W4H	Nsp16, S-adenosylmethionine
Nsp 11	-	-	-
Nsp 12	Part of the replication-transcription complex, involved in virus replication	7BTF	N5p7, Nsp8
		6M71	Nsp7,Nsp8
		7BV1	Nsp7,Nsp8
		7BV2	Nsp7,Nsp8
		7BW4	Nsp7,Nsp8
		6YYT	Nsp7,Nsp8
		6X2G	Nsp7,Nsp8
		7C2K	Nsp7,Nsp8
		7BZF	Nsp7,Nsp8
		6XEZ	Nsp7, Nsp8, Nsp13
		6XQB	Nsp7, Nsp8
Nsp 13	Helicase, part of the RNA polymerase complex, involved in virus replication	6XEZ	Nsp7, Nsp8, Nsp12
Nsp 14	Exonuclease activity and N7-MTase activity. Nsp10-Nsp11-Nsp14-Nsp16 complex	-	-
Nsp 15	RNA endonuclease	6VWW	-
		6WLC	Uridine-5'-
		6W01	Monophosphate Citrate
		6WXC	Tipiracil
		6XDH	-
		6X41	3'-uridinemonophosphate
		6X1B	GpU

Protein Name	Function	PDB id(s)	Ligand
Nsp 16	SAM-dependent, 2'-0 Methyltransferase/Complex with Nsp10. Nsp10-Nsp11-Nsp14-Nsp16 complex	7C21	Nsp10, SAM
		7C2J	Nsp10, SAM
		6W61	Nsp10, SAM
		6XKM	Nsp10, SAM
		6WJT	Nsp10, S-Adenosyl-L-Homocysteine
		6NNVN	Nsp10, S-Adenosylmethionine
		6WQ3	Nsp10, S-Adenosylmethionine, 7-methyl-GpppA
		6WKS	Nsp10, S-
		6WRZ	Adenosylmethionine, 7-methyl-GpppA
		6ZCT	Nsp10, Sinefungin
		6YZ1	Nsp10, Sinefungin
		6WKQ	Nsp10
		7BQ7	Nsp10, S-
		6W75	adenosylmethionine Nsp10,
		6W4H	S-adenosylmethionine
Spike (S) glycoprotein	Receptor recognition, viral attachment, and entry into host cells.	6ZGH	-
		6ZER	EY6A Fab
		6VVV1	ACE2
		6WPS	S309 Fab
		6WPT	S309 Fab
		6VSB	-
		6VXX	-
		6VYB	-
		6X6P	-
		7C01	CB6 antibody
		7BWJ	P2B-2F6 Fab
		7BYR	BD23-Fab
		6M17	ACE2-BOAT1
		6MOJ	ACE2
		6LZG	ACE2
		6LVN	-
		6M1V	-
		6ZGG	-
		6LXT	-
		6YM0	CR3022 Fab
		6YOR	CR3022 Fab
		6YLA	CR3022 Fab
		6W41	CR3022 Fab
		6X2C	CC12.1
		6X2A	-
		6X2C	-
		6X2B	-
		6X29	-

Table 4. (Continued)

Protein Name	Function	PDB id(s)	Ligand
		6YZ5	-
		6ZGE	-
		6ZGI	CC12.1 Fab
		6XC4	CC12.3/CR3022 Fab
		6XC3	CC12.1 Fab
		6XC2	C105 Fab
		6XCN	C105 Fab
		6XCM	Cv30 Fab
		6XE1	-
		6XKL 4A8 Fab	-
		7C2L	-
		6XR8	-
		6XRA	Neutralizing antibody Fab fragment C105, state2
		6XCN	Neutralizing antibody Fab fragment C105, state1
		6XCM	Neutralizing antibody Fab fragment C105
		6XCA	2 Neutralizing antibodies
		6XDG	-
Orf3a	NLRP3 inflammasome activation	6XDC	-
Orf3b	-	-	-
Envelope (E) protein	Envelope protein, involved in virus packaging	-	-
Membrane (M) protein	Membrane protein, involved in virus packaging	-	-
Orf6	Interferon antagonist	-	-
Orf7a	Induces apoptosis	-	6W37
Orf7b	-	-	-
Orf8	-	-	-
Orf9b	Interferon antagonist	-	6Z4U
Orf9c	-	-	-
Orf10	-	-	-
Nucleocapsid (N) protein	RNA binding and packaging	6WZO	-
		6WZQ	-
		6VY0	-
		6ZCO	-
		6YUN	-
		6YI3	-
		7C22	-
		6M3M	-
		6WJI	-
		6WKP	-

Classification of the SARS-CoV-2 Target Proteins

Table 5 describes the SARS-CoV-2 target proteins' protein data bank (PDB) ID, the method used to retrieve the crystal structure, source of organism, macromolécule and their function (Berman et al., 2003).

Table 5. Classification of the SARS-CoV-2 target proteins

PDB ID	Released	Methods	Organism	Macromolecule	Function
6XRZ	2020-08-19	ELECTRON MICROSCOPY 6.9 Å	SARS-CoV-2	Frameshift stimulation element from the SARS-CoV-2 RNA Genome (nucleic acid)	28-kDa frameshift stimulation element
6YVF	2020-05-20	X-RAY DIFFRACTION 1.6 Å	SARS-CoV-2	3C-like proteinase (protein)	Peptide binding protein
6YVA	2020-05-13	X-RAY DIFFRACTION 3.18 Å	SARS-CoV-2	Replicase polyprotein 1a (protein)	Cell invasion
6YWM	2020-05-06	X-RAY DIFFRACTION 2.16 Å	SARS-CoV-2	NSP3 macrodomain (protein)	NSP3 macrodomain in complex with MES
6YWL	2020-05-06	X-RAY DIFFRACTION 2.5 Å	SARS-CoV-2	NSP3 macrodomain (protein)	NSP3 macrodomain in complex with ADP-ribose
6YWK	2020-05-06	X-RAY DIFFRACTION 2.2 Å	SARS-CoV-2	NSP3 macrodomain (protein)	NSP3 macrodomain in complex with HEPES
6Z97	2020-07-01	ELECTRON MICROSCOPY 3.4 Å	SARS-CoV-2	Spike glycoprotein, Fibritin (protein)	Perfusion SARS-CoV-2 spike glycoprotein
6YZ1	2020-05-13	X-RAY DIFFRACTION 2.4 Å	SARS-CoV-2	nsp10 (protein)	Nsp10-nsp16 methyltransferase complex with Sinefungin
6YYT	2020-05-13	ELECTRON MICROSCOPY 2.9 Å	SARS-CoV-2	nsp12 (protein) nsp7 (protein) nsp8 (protein)	Replicating SARS-CoV-2 polymerase
6YZ5	2020-06-03	X-RAY DIFFRACTION 1.8 Å	SARS-CoV-2	Nanobody H11-D4 (protein) Spike glycoprotein (protein)	H11-D4 complex with SARS-CoV-2 RBD
6YZ7	2020-06-03	X-RAY DIFFRACTION 3.3 Å	SARS-CoV-2	Antibody Cr3022 (protein) Antibody light chain (protein)	SARS-CoV-2 RBD, CR3022 ternary complex

Table 5. (Continued)

PDB ID	Released	Methods	Organism	Macromolecule	Function
6ZB5	2020-09-30	ELECTRON MICROSCOPY 2.85 Å	SARS-CoV-2	Spike glycoprotein (protein)	SARS CoV-2 Spike protein, Closed conformation
6ZB4	2020-09-30	ELECTRON MICROSCOPY 3.03 Å	SARS-CoV-2	Spike glycoprotein (protein)	Closed conformation, C1 symmetry
6ZBP	2020-07-29	X-RAY DIFFRACTION 1.85 Å	SARS-CoV-2	H11-H4 (protein) Spike glycoprotein (protein)	H11-H4 complex with SARS-CoV-2
6ZCZ	2020-06-24	X-RAY DIFFRACTION 2.65 Å	SARS-CoV-2	Nanobody (protein) Spike glycoprotein (protein)	Receptor binding domain of SARS-CoV-2
6ZCO	2020-07-01	X-RAY DIFFRACTION 1.361 Å	SARS-CoV-2	Nucleoprotein (protein)	C-terminal Dimerization Domain of Nucleocapsid Phosphoprotein
6ZCT	2020-06-24	X-RAY DIFFRACTION 2.55 Å	SARS-CoV-2	nsp10 (protein)	Nonstructural protein 10 (nsp10) from SARS CoV-2
6ZDH	2020-07-01	ELECTRON MICROSCOPY 3.7 Å	SARS-CoV-2	EY6A heavy chain (protein) EY6A light chain (protein)	Spike glycoprotein in complex with a neutralizing antibody

Medicinal Plants and Their Biological Activities

The COVID-19 pandemic is posing numerous serious issues on a global scale, and its duration is unknown. Our only chance of solving the issue is to manufacture enough vaccinations and deliver them effectively. However, the current shortage of immunizations means that many individuals will continue to experience sickness for some years. We also don't know how long the vaccines will provide immunity, therefore we can't say how long the COVID-19 infections will last worldwide. As a result, there will always be a need for therapies for its symptoms. The use of conventional medicinal herbs is a different approach to provide such care. Table 6 shows the medicinal plants and their biological activities (Babich et al., 2020; Bachar et al., 2021; Ali et al., 2021; Perera et al., 2021; Liang et al., 2023).

Table 6. Medicinal plants and their biological activities

No	Plant Name	Scientific Name	Family	Part of the plant	Phytochemicals	Treatment	Biological activity
1.	Coriander	*Coriandrum Sativum*	Apiaceae	Seeds, whole plant	Flavonoids, alkaloids, tannins, saponin, terpenoids, sterol, carbohydrate, anthraquinones	Fever, nausea, headache, coughs, asthma	• High antimicrobial activity • Antioxidant activity
2.	Wenivel/Ceylon calumba root	*Coscinium Fenestratum*	Menispermaceae	Root, stem of the creeper	Jatrorrhizine, palmatine, berberine, jatrorrhizine	Pains, fever, colds	• Anti-inflammatory
3.	Black Pepper	*Piper nigrum*	Piperaceae	Seeds, roots, leaves	Piperine, tannins and alkaloids	Helps with lung problems, asthma, cough	• Abortifacients • Carminative • Diuretic
4.	Long pepper	*Piper longum*	Piperaceae	Seeds, roots, leaves	(2E, 4E, 14Z)-6-hydroxyl-N-isobutyleicosa-2,4,14-trienamide, Coumaperine, N-5-(4-hydroxy-3-methoxyphenyl)-2E-pentenoyl piperidine, piperolactam A	Tonsils, coughs, vomiting, colds, fevers, vomiting	• Stomachic • Emmenagogue • Abortifacient • Aphrodisiac and digestive • Tonic to the liver
5.	Katuwelbatu/ wild eggplant	*Solanum virginianum*	Solanaceae	Seeds/roots	Terpenoids, glycosides, flavonoids, saponins, coumarins, tannin, proteins and amino acids	Cough, chest pain, vomiting, leprosy, itching scabies, cardiac diseases	• Diuretic • Aphrodisiac • Helps to keep the heart healthy • Have a hypotensive effect • Antiviral activity

Table 6. (Continued)

No	Plant Name	Scientific Name	Family	Part of the plant	Phytochemicals	Treatment	Biological activity
6.	Asamodagam/ Bishop's weed	*Trachyspermum involucratum*	Apiaceae	seeds	Carbohydrates, glycosides, saponins, phenolic compounds, volatile oil (thymol, γ-terpinene, para-cymene, and α- and β-pinene), protein, fat, fiber and mineral matter containing calcium, phosphorous, iron and nicotinic acid	Coughs, asthma, flatulence, atonic dyspepsia, diarrhea, gastro-intestinal ailments, bronchial problems	• Antifungal • Antioxidant • Antimicrobial • Antinociceptive • Cytotoxic • Hypolipidemic • Antihypertensive • Antispasmodic • Broncho-dilating actions • Antilithiasis • Diuretic • Abortifacient • Antitussive • Nematicidal • Anthelmintic • Antifilarial
7.	Yakinaran	*Atalantia ceylanica*	Rutaceae	Leaves, bark, flowers	Bergapten, xanthotoxin, heraclenin, oxypeucedanin and imperatorin	Fever, cough, digestive ailments, bronchitis, healing respiratory ailments, atherosclerosis, cancer, stroke, asthma, arthritis	• Stimulates liver functions • Antioxidant • Hepatoprotective activity

No	Plant Name	Scientific Name	Family	Part of the plant	Phytochemicals	Treatment	Biological activity
8.	Welmee	*Glycyrrhiza glabra*	Fabaceae	Root	Alkaloids, glycosides, carbohydrates, starches, phenolic compounds, flavonoids, proteins, pectin, mucilage, saponins, lipids, tannins, sterols and steroids	Pulmonary tuberculosis, bronchitis, conjunctivitis, pharyngitis	• Laxative • Aphrodisiac
9.	Pathpadagam/ oldenlandia Diamond flower	*Oldenlandia corymbosa*	Rubiaceae	Whole plant	Proteins, polysaccharides, polyphenols, tannins, flavonoids, saponins, steroids, triterpene and glycosides, urosilic acid, oleanolic acid and gamma sitosterol, 0.12% of alkaloids-bifloron, biflorin	Appendicitis, hepatitis, pneumonia, cholecystitis, cellulitis, cough, cold, bronchitis, gynaecologic infections, pelvic inflammatory diseases, cold, cough, bronchitis, necrosis, nervous depression	• Anthelmintic • Remove toxic materials, • Activate blood circulation • Promotes diuresis • Effective curative agent • Cooling properties
10.	Adathoda/ Malabar nut	*Justicia adhathoda*	Acanthaceae	Roots, leaves, flowers, bark, stem	Alkaloids, anthraquinones, flavonoids, saponins, phytosterols, triterpenoids and poly-phenols	Coughs, asthma, tuberculosis, fevers, nausea, swellings, issues related to the respiratory system	• Improves breathing in patients affected by coryza and dyspnea • Effective in reducing cough and clearing congestion in the chest
11.	Saneewani	*Actinopteris dichotoma*	Pterdaceae	Whole plant	Alkaloids, Saponins, Phenols and Tannins, Carbohydrates, Sterols, Glycosides, Proteins	Fever, dysentery, bronchitis, Tuberculosis	• Anthelmintic - Antileprotic • Febrifuge • Cooling agent • Control of blood pressure

Table 6. (Continued)

No	Plant Name	Scientific Name	Family	Part of the plant	Phytochemicals	Treatment	Biological activity
12.	Thebu/canereed	*Costus speciosus*	Zingiberaceae	Roots, rhizome, leaves	Tannin, phenol, flavonoid, alkaloid, saponin, steroid and terpenoids	Catarrhal fever, coughs	• Depurative • Aphrodisiac • Purgative
13.	Wadakaha/ variegated sweet flag	*Acorus calamus*	Acoraceae	Roots	Glycosides, carbohydrates, phenolic compounds, alkaloids, flavonoids and tannins, saponins, steroids and triterpenoids	Cough, rheumatoid arthritis and stroke, colds	• Antimicrobial • Allelopathic • Anticellular • Immunosuppressive
14.	Heen araththa/ Snap ginger	*Alpinia calcarata*	Zingiberaceae	Rhizome	Polyphenols, tannins, flavonoids, steroid glycosides and alkaloids	cough, respiratory ailments, bronchitis, asthma, arthritis and diabetics	• Aphrodisiac • Impurities of blood • Throat inflammation
15.	Karabu nati/ clove	*Syzygiam aromaticum*	Myrataceae	Clove buds	Magnesium, calcium, iron, potassium, sodium, crude fat, crude protein, crude fiber and carbohydrate	Coughs, bronchitis	• Fungicide • Stimulants • Tonic
16.	Kalanduru/ nut grass	*Cyperus rotundus*	Cyperaceae	Tubers	α-cyperone, cyperene α-selinene, flavonoids, terpenoids, sesquiterpenes, sitosterol, cyperene, cyperol, nootkatone and valencene	Fever, asthma, coughs, dysentery	• Antioxidant • DNA damage protective • Anti-cytotoxic activities • Antibacterial activity

No	Plant Name	Scientific Name	Family	Part of the plant	Phytochemicals	Treatment	Biological activity
17.	Kurundu/ Ceylon cinnamon	*Cinnamomun verum*	Lauraceae	Leaves, seeds	pinene, limonene cinnamic acid, cinnamaldehyde, cinnamate, polyphenols, trans-2-hydroxycinnamic, chlorogenic, Rosmarinus, vanillic, gallic, amentoflavone, luteolin, apigenin, quercetin, coumarin, and rutin.	Rheumatism, vomiting, bronchitis, cough	• Carminative • Antibacterial • Antiviral
18.	Wel penela/ winter cherry	*Cardiospermun halicacabum*	Sapindacease	Whole plant, leaves, flowers	Alkaloids, Flavonoids, Saponins, Tannins and Glycosides	Dysentery, rheumatoid arthritis	• Aphrodisiac • Antibacterial activity
19.	Uguressa/ governor and acutes plum	*Flacourtia indica*	Flacourtiaceae	Bark, leaves	Protein, lipids, sugars and polyphenols, calcium, phosphorus, magnesium, sodium, zinc, iron, saponins, tannins, triterpenes	Rheumatic ailments, asthma, bronchitis, diarrhoea, nausea and vomiting, headache	• Carminative • Expectorant
20.	Thibiri/gaub	*Diospyros malabarica*	Ebenaceae	Fruit, whole plant	β-sitosterol and its glucoside, Betulin, monohydroxy triterpene ketone, Betulin, oleanolic acid, myricyl alcohol, Saponin, Betulinic acid	Sore throat, diarrhea, dysentery	• Anti-inflammation • Antimicrobial effect
21.	Shalaparni/ sal leaved desmodium	*Desmodium gangeticum*	Fabaceae	Roots	Allic, protocatechuic, salicylic, chlorogenic, caffeic acids, rutin, quercetin and kaempferol	Fever, bronchitis, pneumonia, diarrhea, liver disease, typhoid, laryngitis	• Cholagogue • Expectorant • Diuretic • Antioxidant activity

Table 6. (Continued)

No	Plant Name	Scientific Name	Family	Part of the plant	Phytochemicals	Treatment	Biological activity
22.	Thotila/Broken Bones tree	*Oroxylum indicum*	Bignoniaceae	Stem bark, root	Terpenoids, phlobatannins, flavonoids, reducing sugars, phenols and tannins, carbohydrates, alkaloids, glycosides, phlobatannins, flavonoids, phenols and tannins and glycosides.	Diarrhea, dysentery, rheumatism, piles	• Astringent
23.	Siriwedi babila/tiny flower hibiscus	*Hibiscus micranthus*	Malvaceae	Leaves, roots	Flavonoids, phenols, steroids, carbohydrates, tannins	Pulmonary disease, stomach disease	• Diuretic • Healing effect • Febrifuge
24.	Saldiri/celery	*Apium graveolens*	Apiaceae	Roots, seeds, whole plants	Caffeic acid, chlorogenic acid, apiin, apigenin, rutaretin, ocimene, bergapten, and isopimpinellin	Rheumatism, inflammation, obesity	• Carminative • Depurative • Diuretic
25.	Rath goraka	*Garcinia quaesita*	Clusiaceae	Fruits, seeds, leaves, bark, bark of the root	Crude protein, Nitrogen, Potassium, Magnesium, Phosphorous, crude fat	Fevers, nausea, vomiting, infections, asthma, abdominal pains,	• Appetizer
26.	Pani dodam/sweet orange	*Citrus sinensis*	Rutaceae	Fruit, leaves	Carbohydrates, flavonoids, glycosides, coumarin glycosides, volatile oils, organic acids, fats and fixed oils	Coughs, bronchitis, Rheumatism, malaria	• Blood purifier • Anti-scorbutic • Diuretic • Purgative

No	Plant Name	Scientific Name	Family	Part of the plant	Phytochemicals	Treatment	Biological activity
27.	Olu/water lily	*Nymphaea pubescens*	Nymphaeaceae	Root, rhizome, fruit, flower, seeds	Oleanolic acid	Diarrhea, fevers, vomiting, coughs	• Purifies the blood • Aphrodisiac
28.	Nil vishnukranthi/ dwarf morning glory	*Evolvulus alsinoides*	Convolvulacease	Leaves, roots, whole plant	alkaloids, tannins, glycosides, steroids, flavonoids, saponins, volatile oil and resins	Dysentery, bronchitis, asthma	• Vermifuge • Alterative • Febrifuge
29.	Navahandi	*Rhipsalis baccifera*	Cactaceae	Root, milky sap	D-limonene, Tannins, alkaloids, saponins, aldehydes, alcohols, ketones, monoterpenes, acids and alkanes	Rheumatism, importance	• Emetic • Rubefacient

Conclusion

In this review, the SARS-CoV-2 virus's structural components; diagnostics tests, clinical manifestations; SARS-CoV-2 target proteins; medicinal plants and their biological activities were discussed. This information will help to choose the best plant for laboratory experimental work.

References

Ali SI, Sheikh WM, Rather MA, Venkatesalu V, Muzamil Bashir S, Nabi SU. Medicinal plants: Treasure for antiviral drug discovery. *Phytotherapy Research* (2021) 35(7):3447-3483.

Babich O, Sukhikh S, Prosekov A, Asyakina L, Ivanova S. Medicinal plants to strengthen immunity during a pandemic. *Pharmaceuticals* (2020) 13(10):313.

Bachar SC, Mazumder K, Bachar R, Aktar A, Al Mahtab M. A Review of medicinal plants with antiviral activity available in Bangladesh and mechanistic insight into their bioactive metabolites on SARS-CoV-2, HIV and HBV. *Frontiers in Pharmacology* (2021) 12:732891.

Berman H, Henrick K, Nakamura H. Announcing the worldwide Protein Data Bank. *Nature Structural & Molecular Biology* (2003) 10(12):980.

COVID-19 Coronavirus pandemic. Worldometer. (2022). Accessed 15 March 2022. https://www.worldometers.info/coronavirus.

Elengoe A. Indian spices boost the immune system against COVID-19. *The Annals of the University Dunarea de Jos of Galati. Fascicle VI-Food Technology* (2020) 44(2):189-206.

Elengoe A. COVID-19 outbreak in Malaysia. *Osong Public Health and Research Perspectives* (2020) 11(3):93-100.

Hossain MF, Hasana S, Mamun AA, Uddin MS, Wahed MI, Sarker S, Behl T, Ullah I, Begum Y, Bulbul IJ, Amran MS. COVID-19 outbreak: pathogenesis, current therapies, and potentials for future management. *Frontiers in Pharmacology* (2020) 11:563478.

Khailany RA, Safdar M, Ozaslan M. Genomic characterization of a novel SARS-CoV-2. *Gene reports* (2020) 19:100682.

Kumar S, Nyodu R, Maurya VK, Saxena SK. Morphology, genome organization, replication, and pathogenesis of severe acute respiratory syndrome coronavirus 2 (SARS-CoV-2). In *Coronavirus Disease* 2019 (COVID-19) 2020 (pp. 23-31). Springer, Singapore.

Liang J, Zheng Y, Tong X, Yang N, Dai S. *In silico* identification of anti-SARS-COV-2 medicinal plants using cheminformatics and machine learning. *Molecules* (2023) 28(1):208.

Mason RJ. Pathogenesis of COVID-19 from a cell biology perspective. *European Respiratory Journal* (2020) 55(4).

Perera WP, Liyanage JA, Dissanayake KG, Gunathilaka H, Weerakoon WM, Wanigasekara DN, Fernando WS, Rajapaksha RM, Liyanage RP, Perera BT. Antiviral potential of selected medicinal herbs and their isolated natural products. *BioMed Research International* (2021) 2021.

Singhal T. A review of coronavirus disease-2019 (COVID-19). *Indian Journal of Pediatrics* (2020) 87(4):281-286.

Viner RM, Ward JL, Hudson LD, Ashe M, Patel SV, Hargreaves D, Whittaker E. Systematic review of reviews of symptoms and signs of COVID-19 in children and adolescents. *Archives of Disease in Childhood* (2021) 106(8):802-807.

Wang MY, Zhao R, Gao LJ, Gao XF, Wang DP, Cao JM. SARS-CoV-2: structure, biology, and structure-based therapeutics development. *Frontiers in Cellular and Infection Microbiology* (2020) 10:587269.

Wu F, Zhao S, Yu B, Chen YM, Wang W, Song ZG, Hu Y, Tao ZW, Tian JH, Pei YY, Yuan ML. A new coronavirus associated with human respiratory disease in China. *Nature* (2020) 579(7798):265-269.

Wu C, Liu Y, Yang Y, Zhang P, Zhong W, Wang Y, Wang Q, Xu Y, Li M, Li X, Zheng M. Analysis of therapeutic targets for SARS-CoV-2 and discovery of potential drugs by computational methods. *Acta Pharmaceutica Sinica B* (2020) 10(5):766-788.

Chapter 9

Bioinformatic Tools to Study SARS-CoV-2 Proteins

Asita Elengoe*, PhD

Department of Biotechnology, Faculty of Applied Science, Lincoln University College, Selangor, Malaysia

Abstract

A major health threat emerged in early 2020. In December 2019, a novel severe acute respiratory syndrome coronavirus 2 (SARS-CoV-2) epidemic outbreak occurred in Wuhan, China and spread rapidly across the world. World Health Organization (WHO) named it coronavirus disease (COVID-19) on 11th February 2020. Vaccines were discovered to slow down the rapid spread of the virus throughout the world. However, there is no specific anti-viral treatment against SARS-CoV-2 until now. Thus, there is an urgency to develop an effective drug for COVID-19. Modern drug discovery practices include a wide range of theoretical and computational methodologies, collectively known as computer-aided drug design (CADD). CADD techniques have played a significant role in the development of some medications that are currently being tested in clinical settings. These techniques have developed in tandem with experimental strategies employed in drug development. Computational chemistry, molecular modeling, molecular design, and rational drug design are the components of computer-aided drug design. In this review, homology modeling, prediction of the secondary structure, protein models validation tools and molecular docking approaches were discussed which help in understanding the SARS-CoV-2 protein to design a safe drug for humans.

---

* Corresponding Author's Email: asitaelengoe@yahoo.com.

In: Medicinal Plants and COVID-19
Editor: Asita Elengoe
ISBN: 979-8-88697-958-9

Keywords: computer-aided drug design (CADD), COVID-19, molecular docking, molecular modeling

Introduction

Recent developments in high-throughput biotechnologies for genomes, proteomes and transcriptomes require the advancement of novel computational methods for a better understanding of the mechanisms of SARS-CoV-2. Structure-based drug design (SBDD) is a technique that involves the acquisition of information on the three-dimensional (3-D) structure of the target protein through methods such as x-ray crystallography, NMR spectroscopy or homology modeling followed by the designing of suitable drug candidates based on the binding affinity and selectivity of the target molecules (Suhaibun et al., 2020). For example, a novel thienopyrimidinone analog was discovered as a potent TAK1 inhibitor by targeting unique amino acid residues, especially the combination of SER111 and ASN114 as well as decreasing the interaction with TYR106 at the hinge position in TAK1. Another example is the development of Ras inhibitors by targeting a novel conformation of Ras and GTP complex called state 1 that possesses 'druggable' surface pockets resulting from the conformational dynamics. Besides, G-protein coupled receptors (GPCRs) are important pharmaceutical targets for the treatment of many diseases including cancer. However, the implementation of valid SBDD considerations in drug design projects targeting GPCRs remains difficult due to limited insights into how ligands interact with and modulate their activity (Zhang et al., 2022).

Currently, with a deeper understanding of the structure and function of phytochemicals, a variety of novel technologies and approaches have been used to further the development of medicinal plants (Yi, 2016). The plant compounds can be obtained from the PubChem database in two or three-dimensional (2-D or 3-D) SDF formats. The pharmacokinetics, drug-likeness and medicinal chemistry friendliness of natural compounds can be determined using the SwissADME server which is a free web tool. The effectiveness of assessing the chemical activities of medicinal plants will be greatly enhanced if a quick and convenient method can be established for accurately predicting a large number of chemical compounds first, followed by performing *in vivo* and *in vitro* pharmacological experiments for verification based on these findings (Yi, 2016). The most well-known antimalarial medication, Artemisinin, was first discovered through extensive herb screening in the

1950s. Following that, numerous *in silico* methods were used to create FDA-approved medications from natural herbs or animal sources. In the 1970s, the angiotensin-converting enzyme inhibitor (Captopril) was the first drug developed by studying the molecular interaction between the enzyme and inhibitor (Tu, 2011). Saquinavir, the first HIV-1 enzyme inhibitor approved by the FDA, was created using computational biology tools such as molecular dynamics simulation and docking (Zaman et al., 2018).

The method of molecular docking consists mainly of three stages (Elengoe and Hamdan, 2018). Firstly, it is important to design the structures of small molecules and target proteins in advance. In this step, in the open-access PDB database, that can be used to explain many physiological processes dependent on crystal structures, and also for homologous template models if docking structures are of interest, abundant experimentally solved structures are available. Second, in the ligand-binding site, it can serve as a motor for predicting conformations, orientations, and positional spaces. Conformational search algorithms work out this mission by using techniques of systematic and stochastic search to determine the conformation of binary complexes (Rallabandi et al., 2020). The *in silico* study of the interaction between SARS-CoV-2 protein and ligand through computational biology tools will aid in developing a new and effective drug for COVID-19 treatment.

In this review, we discussed the homology modeling, prediction of the secondary structure, protein models validation tools and molecular docking approaches for studying the SARS-CoV-2 protein. The world is still looking for more COVID-19 solutions. As a result, this study may provide some insight into how to design an effective drug against SARS-CoV-2 using computational biology tools.

Bioinformatics

Bioinformatics is an interdisciplinary field that involves molecular biology, genetics, computer science, mathematics as well as statistics (Can, 2014). It is also known as computational biology as different computational techniques are used to solve biological problems. The most common problems are modeling biological processes at the molecular level and making inferences from the data collected (Can, 2014). It involves the development and implementation of computational algorithms and software tools that help in the understanding of biological processes. In recent years, the various applications of bioinformatics have changed the pattern of both research and the exploitation of knowledge. However, in bioinformatics, there is also a disadvantage of intruders stealing data which leads to high demand for information security (Sethi and Behera, 2012).

Computer Technology-Aided Drug Design

Computational biology is a prominent and promising bioinformatics branch that involves analyzing biological data using a computer system. Due to it being cheap and more time-intensive than experimental alternatives; this method is currently gaining prominence. This information system aids in the discovery of novel therapeutic targets of protein bioactive constituents in the development of COVID-19 medications.

Substrate Based-Drug Design

COVID-19 continues to be a significant and severe public health issue today. To establish the interactive effects among phytochemicals and the subsequent targets, the structure-based strategy depends on established structural details. Great strides have been made in this area with the advancement of biomolecular spectroscopic technology such as X-ray crystallography and nuclear magnetic resonance (NMR), contributing to substantial advances in our structural understanding of the drug goal. Unique ligands could be logically designed to provoke medicinal benefits, getting the benefit of the 3-D structure of the proteins. By identifying and improving the initial lead molecules, structure-based design (SBD) may also offer crucial research into potential drug design and production. To control particular cellular behaviors, the high-affinity ligand selectively controls approved drug targets, finally producing the desired pharmacological and therapeutic results (Cui et al., 2020).

Since this research, the creation of structure-based drugs has started to serve as a modern and efficient algorithm and methodology to facilitate the development of drugs quicker, cheaper, and more reliable. Extensive developments were made over the past years to encourage the SBD technique, with more and more popular technologies playing a significant role in modern medical studies.

Homology Modeling

Homology modeling is commonly employed due to it generates a high-accuracy 3-D model of the protein of interest (Sanchez et al., 2000; Al-Lazikani et al., 2001). Based on the alignment of the target sequence to one or more template sequences, it predicts the 3-D model of the target sequence. It entails numerous processes, including aligning the target with the template, building the model, refining the model, and evaluating the model. The target sequence is first obtained in FASTA format from the National Center for Biotechnology Information (NCBI). The Protein Basic Local Alignment Search Tool (BLASTP) is then used to search the RCSB Protein Databank for a suitable template that matches the existing 3-D model of the template protein sequence. Through pairwise alignment, the target and template sequences are aligned. MODELLER, also known as the Swiss model program, is a program that predicts a 3-D protein model based on a sequence alignment and a template. Some of the examples of modelers are described in Table 1.

Prediction of Secondary Structure

The secondary structure of amino acid determines whether it is in a helix, strand, or coil. Local hydrogen bonding between C=O and N-H groups near together in the polypeptide chain produces alpha-helices in proteins. The helix is important because it generates dipole moments, which aid in the binding of tiny charged molecules to proteins. Individual peptide dipoles in helices combine to form a macrodipole, with the amino-terminal end polarised positively and the carboxy-terminal end polarised negatively. Positively charged species and the end of the helix dipole form favorable electrostatic contacts in helices, whereas negatively charged side chains and cations interact with the carboxy-terminal ends (Petkso and Ringe, 2004). Examples of secondary structure prediction servers are shown in Table 2.

Protein Model Validation Tools

A variety of web applications and resources can aid in the evaluation of structural model reliability (crystallographic, NMR, or homology-based models). Table 3 explains the protein model validation tools.

Table 1. Examples of modelers for homology modeling

No.	Modelers	Function
1.	Modeller	Uses spatial restrictions to accomplish comparative protein structure modeling
2.	SWISS-MODEL	It is fully automated. Accessible via the web via the DeepView application. With DeepView, interactive model optimization is possible
3.	EasyModeller	A standalone application that can be used on a typical personal computer running Microsoft Windows as its operating system
4.	ModWeb	It is fully automated. The modeling software MODELLER is used to create the final 3-D structure
5.	Pudge	Pipeline for predicting protein structure that incorporates multiple sources of data
6.	Phyre and Phyre2	Remote template detection, alignment, 3-D modeling, multi-templates, *ab initio*
7.	3-D-JIGSAW	It is a fully automated server
8.	CPHmodels	For sequence alignment and distant homology discovery, a sequence-profile and profile-profile comparison technique are used
9.	ESyPred3-D	Template detection, alignment, 3-D modeling
10.	Rosetta	Rosetta homology modeling and *ab initio* fragment assembly with Ginzu domain prediction
11.	EVfold	Evolutionary couplings were discovered through correlated mutations in a protein family and were utilized to predict the 3-D structure and functional residues from sequences alone.
12.	RaptorX	Remote homology detection, protein 3-D modeling, binding site prediction
13.	HHpred	The HH-suite is a free software tool for finding sensitive protein sequences. It includes programs that scan protein sequence databases for related protein sequences
14.	MOE (Molecular Operating Environment)	It's a software platform for drug discovery that combines visualization, modeling, and simulations
15.	IntFOLD	A unified interface for tertiary structure prediction/3-D modeling, 3-D model quality assessment, intrinsic disorder prediction, domain prediction, prediction of protein-ligand binding residues
16.	TIP-STRUCTFAST	Automated comparative modeling
17.	Yasara	A computer program for molecular visualizing, modeling and dynamics
18.	Selvita Protein Modeling Platform	A set of tools for modeling proteins. Users will benefit from a new protein-protein docking module, improved usability, and improved discovery outcomes
19.	SPARKS-X	Fold recognition based on sequence profiles and structural profiles is used to model 3-D structures
20.	I-TASSER	*Ab initio* folding and threading methods combined

Table 2. Secondary structure prediction servers

	Function	Advantage	Disadvantage
SOPMA	Self-Adjusted Prediction Based on the query, the method from many alignments predicts the secondary structure of a protein (primary sequence of a protein)	High accuracy increase the generalization performance of individual prediction method	Two control alignments groups are obtained for each SSSD-aligned pair. The first sequence in the first control alignments group is left alone, while the second is changed
SPIDER2	The most thorough and precise prediction of protein structural features, including secondary structure, local backbone angles, and accessible surface area, by an iterative Deep Neural Network (DNN) (ASA)	Ready availability Multi-core crawling	Each spider kernel has a memory cost of roughly 200 MB, and the accuracy is about 90%; the remaining 10% can be corrected by tweaking Score Rules or Selectors
s2D	In one coherent framework, predicts chaos and secondary structure. Trained on solution-based NMR data	Increasing the amount of storage capacity available and doing so at a lesser cost than other hardware solutions	More expensive
NetSurfP	Profile-based neural network. Surface accessibility and secondary structure of amino acids in a sequence are predicted	Improved accuracy processing time has been optimized	
Meta-PP	Consensus prediction of other servers. By offering a standard and consistent web-based interface, it makes it easier to access a variety of public protein structure and function prediction websites.	Provides access to a restricted group of high-quality servers for comparative modeling, threading/fold recognition, secondary structure prediction, and other specialized disciplines like contact and function prediction	
PREDATOR	Knowledge-based database comparison. Input a single protein sequence to be predicted, and it can best predict the query sequence using a set of unaligned sequences as supplementary information		Wide range of image and speech recognition problems

Table 2. Secondary structure prediction servers

	Function	Advantage	Disadvantage
SymPred	An enhanced dictionary-based method for capturing local sequence similarities among proteins	Benefit from the expanding PDB	
YASSPP	PSI-BLAST profiles are used in a cascaded SVM-based predictor	High accuracy	Similar to many existing methods
PSSpred	PSI-BLAST profiles have several backpropagation neural network predictors		
Frag1D	Using profile-based fragment matching, secondary structure and Shape Strings (discrete states of dihedral angles) can be predicted		

Table 3. Protein model validation tools

No.	Structure validation tools	Function
1.	Procheck	Generates a series of PostScript plots that analyze the overall and residue-by-residue geometry to assess the stereochemical quality of a protein structure
2.	ERRAT	Validates protein structures that have been determined using crystallography. A sliding 9-residue window's position is plotted as a function of error values. The error function is based on the statistics of non-bonded atom-atom interactions in the given structure
3.	Verify 3-D	By assigning a structural class based on its location and environment (alpha, beta, loop, polar, nonpolar, and so on) and comparing the findings to excellent structures, this program determines whether an atomic model (3-D) is consistent with its amino acid sequence (1D)
4.	ProSa	Checks for errors in 3-D models of protein structures. Error detection in experimentally determined structures, theoretical models, and protein engineering are some of its applications
5.	Whatcheck	A protein structural validation system based on the WHAT IF program
6.	MolProbity	Structure validation employing all-atom contact analysis techniques and updated geometrical criteria for phi/psi, sidechain rotamer, and C beta deviation on client-uploaded or PDB ID-specified data

No.	Structure validation tools	Function
7.	WHAT IF	A program for analyzing protein structures, including mutation prediction, structure verification, and molecular graphics
8.	NQ-Flipper	In protein models acquired from X-ray crystallography, NMR, or modeling investigations, NQ-Flipper detects unfavorable rotamers of Asn and Gln residues
9.	Check MyMetal	Metal-binding site and validity are checked
10.	PSVS	PSVS runs quality checks for applications like PROCHECK, Verify3-D, MolProbity ProSA, and others
11.	Eval123-D	Eval123-D runs the applications Verify3-D, Eval23-D ProSA, and EvTree to do quality tests
12.	JCSG structure validation central	Determine any or all of the quality checks provided by the programs PROCHECK, SFCHECK, PROVE, ERRAT, WASP, DDQ, WHAT CHECK and PSQS
13.	MetaMQAPII	Assess the quality of protein models that have been tuned for theoretical models
14.	PARVATI	PARVATI is a technique for analyzing macromolecular structure refinement with anisotropic displacement parameters
15.	STAN	The protein model's quality checks offered by the programs WASP, CISPEP, and MOLEMAN2 are performed by the Structure Analysis server

Table 4. List of molecular docking tools

Program	Description	Reference
FlexX	Incremental construction method	Rarey et al., 1996
DOCK	Shape-based algorithms	Kuntz et al., 1982
GOLD	Genetic algorithms	Jones et al., 1997
Glide, Schrödinger, Portland, OR 97201	Systematic search approaches	Venkatachalam et al., 2003
Monte Carlo simulations (LigandFit)	Systematic search approaches	Venkatachalam et al., 2003
AutoDock	Automated docking of ligand to macromolecule by Lamarckian Genetic Algorithm and Empirical Free Energy Scoring Function	Österberg et al., 2002
AutoDock Vina	Automated docking of ligand to macromolecule by Lamarckian Genetic Algorithm and Empirical Free Energy Scoring Function	Trott and Olson, 2010
rDock	Virtual screening of small molecules against proteins and nucleic acids, binding mode prediction	Ruiz-Carmona et al., 2014
MOE (Molecular Operating Environment)-Dock	Docking application within MOE; choice of placement methods (including alpha sphere methods) and scoring functions (including London dG)	Corbeil et al., 2012

Molecular Docking

Extensive developments were made over the past years to encourage the SBD technique, with more and more popular technologies playing a significant role in modern medical studies. Molecular docking is a traditional system in rational drug design that involves detecting and anticipating the binding patterns and interaction affinities between the ligand and receptor proteins (Tao et al., 2020). It can be classified as stiff docking or fluid docking due to the adaptability of the ligands used in the numerical procedure. To save time, rapid and highly effective rigid docking is extended to high-performance virtual screening using many numbers of small-molecule databases. While the compact approach of docking recognizes knowledge that is more detailed and precise. While the versatile technique of docking finds relevant data that is quite precise and complex. Versatile docking techniques have continually evolved and become more readily available with the increasing development of computational capacity and performance. For docking, various kinds of applications are available, such as Glide, FlexX, DOCK, AutoDock, Discovery Studio and Sybyl (Cui et al., 2020).

Three phases make up the molecular docking procedure. To begin, the structures of tiny molecules and target proteins must be planned ahead of time. In this stage, the open-access PDB database (http://www.rcsb.org) contains a large number of experimentally solved structures that can be utilized to explain various physiological processes that are dependent on crystal structures, as well as homologous template models if docking structures are of relevance. Second, in the ligand-binding site, it can act as a motor, predicting conformations, orientations, and positional spaces. Conformational search methods carry out this task by determining the conformation of binary compounds using systematic and stochastic search strategies (Cui et al., 2020). Table 4 shows some of the examples of docking programs.

Conclusion

Bioinformatics is one of the best techniques to understand better about drug-receptor interaction. This method aids in the development of novel and effective drugs against the SARS-CoV-2 protein.

References

Al-Lazikani B, Jung J, Xiang Z, Honig B. Protein structure prediction. *Current Opinion in Chemical Biology* (2001) 5(1):51-56.

Can T. Introduction to bioinformatics. *InmiRNomics: MicroRNA biology and computational analysis 2014* (pp. 51-71). Humana Press, Totowa, NJ.

Corbeil CR, Williams CI, Labute P. Variability in docking success rates due to dataset preparation. *Journal of Computer-Aided Molecular Design* (2012) 26:775-786.

Cui W, Aouidate A, Wang S, Yu Q, Li Y, Yuan S. Discovering anti-cancer drugs via computational methods. *Frontiers in Pharmacology* (2020) 11:733.

Elengoe A, Hamdan S. Computational approach towards exploring interaction of target protein-phytocompounds in drug development for breast cancer. In *Anticancer plants: mechanisms and molecular interactions 2018* (pp. 131-141). Springer, Singapore.

Jones G, Willett P, Glen RC, Leach AR, Taylor R. Development and validation of a genetic algorithm for flexible docking. *Journal of Molecular Biology* (1997) 267:727-748.

Kuntz ID, Blaney JM, Oatley SJ, Langridge R, Ferrin TE. A geometric approach to macromolecule–ligand interactions. *Journal of Molecular Biology* (1982) 61:269-288.

Österberg F, Morris GM, Sanner MF, Olson AJ, Goodsell DS. Automated docking to multiple target structures: incorporation of protein mobility and structural water heterogeneity in AutoDock. *Proteins* (2002) 46:34-40.

Petsko GA, Ringe D. *Protein structure and function*. 2004 New Science Press.

Rallabandi HR, Mekapogu M, Natesan K, Saindane M, Dhupal M, Swamy MK, Vasamsetti BM. Computational methods used in phytocompound-based drug discovery. In *Plant-derived Bioactives* 2020 (pp. 549-573). Springer, Singapore.

Rarey M, Kramer B, Lengauer T, Klebe G. A fast flexible docking method using an incremental construction algorithm. *Journal of Molecular Biology* (1996) 261:470-489.

Ruiz-Carmona S, Alvarez-Garcia D, Foloppe N, Garmendia-Doval AB, Juhos S, Schmidtke P, Barril X, Hubbard RE, Morley SD. rDock: a fast, versatile and open source program for docking ligands to proteins and nucleic acids. *PLoS Computational Biology* (2014) 10.

Sánchez R, Pieper U, Melo F, Eswar N, Martí-Renom MA, Madhusudhan MS, Mirković N, Šali A. Protein structure modeling for structural genomics. *Nature Structural Biology* (2000) 7(11):986-990.

Sethi PC, Behera PK. Bioinformatics: Applications and issues. *CSI Communications* (2012).

Suhaibun SR, Elengoe A, Poddar R. Technology advance in drug design using computational biology tool. *Malaysian Journal of Medicine Health Sciences* (2020) 16(110):2636-9346.

Tao X, Huang Y, Wang C, Chen F, Yang L, Ling L, Che Z, Chen X. Recent developments in molecular docking technology applied in food science: a review. *International Journal of Food Science & Technology* (2020) 55(1):33-45.

Trott O, Olson AJ. AutoDock Vina: improving the speed and accuracy of docking with a new scoring function, efficient optimization, and multithreading. *Journal of Computational Chemistry* (2010) 31:455-461.

Tu Y. The discovery of artemisinin (qinghaosu) and gifts from Chinese medicine. *Nature Medicine* (2011) 17(10):1217-1220.

Venkatachalam CM, Jiang X, Oldfield T, Waldman M. LigandFit: a novel method for the shape-directed rapid docking of ligands to protein active sites. *Journal of Molecular Graphics and Modelling* (2003) 21:289-307.

Yi F. *In silico* profiling for secondary metabolites from *Lepidium meyenii* (maca) by the pharmacophore and ligand-shape-based joint approach. *Chinese Medicine* (2016) 11(1):42.

Yi F. *In silico* approach for anti-thrombosis drug discovery: P2Y1R structure-based TCMs screening. *Frontiers in Pharmacology* (2016) 7:531.

Zaman MA, Oparil S, Calhoun DA. Drugs targeting the renin-angiotensin-aldosterone system. *Nature Reviews Drug Discovery* (2002) 1(8):621-636.

Zhang Y, Luo M, Wu P, Wu S, Lee TY, Bai C. Application of computational biology and artificial intelligence in drug design. *International Journal of Molecular Sciences* (2022) 23(21):13568.

Chapter 10

Plant-Derived Anti-SARS-CoV-2 Viral Compounds

Matthew Aaron Lopez[1], BSc
and Asita Elengoe[2,*], PhD

[1]School of Biotechnology, Manipal International University (MIU), Negeri Sembilan, Malaysia
[2]Department of Biotechnology, Faculty of Applied Science, Lincoln University College, Selangor, Malaysia

Abstract

A pandemic of acute respiratory disease known as "coronavirus disease 2019" (COVID-19) has been caused by the severe acute respiratory syndrome coronavirus 2 (SARS-CoV-2) virus. It is a highly transmissible and pathogenic virus that poses a hazard to public health and safety. Many parts of the world have been affected by this and mitigation towards the virus management is carried out such as lockdowns, quarantine and introduction of standard operating procedures (SOPs). Nevertheless, Malaysia has practiced the SOPs together with public measures from the government since the virus outbreak in 2020. Since then, cases were controlled with the fast-forward vaccination program, which results in the containment of the virus. Vaccines have been developed to counteract the severity of this disease. However, the approach developed is to minimize the severity and not completely eradicate the virus. Pfizer and Moderna are a few vaccines that have been developed to inoculate the world population from the virus although there are a few more vaccines besides these two. In this review, the timeline of COVID-19 pandemic development; morphology; genomic

* Corresponding Author's Email: asitaelengoe@yahoo.com.

In: Medicinal Plants and COVID-19
Editor: Asita Elengoe
ISBN: 979-8-88697-958-9

characterization; pathogenic mechanism; mode of transmission; similarities and differences between SARS-CoV, MERS- SARS and SARS-CoV-2; sign and symptoms; different types of variants; medicinal plants and their mode of actions were described. This information will aid in designing phytochemical-based drugs for COVID-19.

Keywords: COVID-19, phytochemicals, SARS-CoV-2

Introduction

An unknown acute respiratory tract infectious disease was reported in Wuhan City, China on 12th December 2019 (Guo et al., 2020; Alabed et al., 2020). World Health Organization declared that SARS-CoV2 was the cause of the infectious disease and they named the disease as coronavirus disease 2019 (COVID-19). Chinese scientists discovered that the virus had a high similarity sequence with the bat coronavirus (Singhal, 2020). On 12th March 2020, WHO announced that COVID-19 was a pandemic outbreak (Elengoe, 2020). It has been spreading rapidly from China to 219 countries (Mehmet et al., 2020). It is a deadly virus. It transmits highly from person to person.

The mutation of the COVID-19 strain has resulted in the identification of a number of variations so far. The P.1, N501Y, 501Y.V2, and B.1.617.2 (delta) variants are these variants. While P.1 originated in Brazil and is less concerning, N501Y originated in the United Kingdom and has increased virulence, and 501Y.V2 is a variant from South Africa linked to greater viral load and worst fatality. Among these variants, the delta variant poses a serious threat such as a fatal acute respiratory system that causes an increasing number of cases in India. Since the change of the spike glycoproteins alters the interaction with receptor human angiotensin-converting enzyme 2 (hACE2), these variants diverge from the original Wuhan variant, which influences the function of spike glycoprotein (Mahase et al., 2021; Darby and Hiscox, 2021).

The coronavirus disease (COVID-19) pandemic has become a disastrous and stressful event for every mankind in this world. It has affected many terms of the economy and social structures which has led to changes in economic structure plans for the nations as well as for the people. From social structure, it has made very huge changes that every mankind in this world has to work from home with a virtual platform such as Zoom aside from Microsoft Teams as preference. A certain percentage of people are allowed to work face-to-face in their respective places due to the dangerous outcome presented by the virus which can cause death. Looking back, life has changed instantly in the blink of an eye and COVID-19 has taught us that at this time, encouragement should be given to everyone regardless of race, religion and so on. Therefore, the question is what is COVID-19, how and when did the COVID-19 pandemic starts, and what are the measures taken to manage this situation? All the answers to these questions will be answered in a systematic manner.

Standard operating procedures (SOP) were taken by countries worldwide to overcome the pandemic. Before vaccines were made available, face masks, hand sanitizers and social distancing were among the measures taken to reduce the spread of the virus. Countries have implemented lockdowns and Malaysia is one of the examples for lockdown implementation. The Malaysian government has introduced the Movement Control Order (MCO) as part of the lockdown measures where the movement of people was restricted and limited (Ganasegeran et al., 2020). Despite the lockdown implementation, care lines and virtual counseling were made available to help Malaysian citizens to overcome the stress, depression and anxiety from the introduction of the lockdown measures (Ganasegeran et al., 2020).

Timeline of COVID-19 Pandemic Development

Table 1 shows the timeline of the COVID-19 pandemic development (Coronavirus timeline, 2022).

Structure of the SARS-CoV-2

Table 2 shows the structural components of SARS-CoV-2 (Barupal et al., 2020).

Genomic Characterization

Table 3 shows the genomic characterization of SARS-CoV-2 (Khailany et al., 2020; Wang et al., 2020).

Table 1. COVID-19 pandemic development timeline

Year	Event
December 2019	The first outbreak of coronavirus disease (COVID-19) in Wuhan, China.
30th January 2020	World Health Organization (WHO) declared the coronavirus disease a global health outbreak emergency.
11th February 2020	WHO changed the name of the disease caused by SARS-CoV-2 as COVID-19.
11th March 2020	WHO declared the SARS-CoV-2 outbreak as a pandemic.
16th March 2020	Jennifer Haller was the first recipient to be administered with coronavirus vaccine in the trial stage in the United States.
28th March 2020	US Food and Drug Administration approved the use of a rapid coronavirus test kit which could give patients results in less than 15 minutes.
24th April 2020	The food and Drug Administration (FDA) warned against the use of hydroxychloroquine in clinical trials.
1st May 2020	FDA permitted the use of remdesivir, a drug for emergency use purposes for people with severe COVID-19.
15th May 2020	French researchers found evidence that the virus was in the country as early November and it was spreading in Europe two months before it was officially identified in China.
18th May 2020	The COVID-19 vaccine developed by Moderna has shown an immune response in the human body found to be safe in a small number of patients.
22nd May 2020	An experimental coronavirus vaccine developed by the University of Oxford progressed to advanced stages of human trials.
22nd May 2020	A study conducted by The Lancet shows that hydroxychloroquine increases the death number and does not help with COVID-19 patients.
11th June 2020	Moderna confirmed that start-up trial of 30 000 volunteers for the coronavirus vaccine in July as the company enters the final testing stage.
7th July 2020	The WHO agreed that the virus was an airborne disease.
11th August 2020	The world reached 20 million cases.
18th September 2020	The world reached 30 million cases.
20th October 2020	The world reached 40 million cases.

Year	Event
9th November 2020	The world reached 50 million cases.
16th November 2020	The COVID-19 vaccine developed by Moderna is 94.5% effective.
17th November 2020	FDA authorized first home COVID-19 test kit which gives users results quickly.
18th November 2020	The vaccine developed by Pfizer was submitted for FDA approval within days.
23rd November 2020	The AstraZeneca Vaccine has an efficacy of around 90 percent.
1st December 2020	The vaccine developed by Moderna is submitted to the FDA for approval.
2nd December 2020	The United Kingdom became the first country to approve Pfizer and BioNTech COVID-19 vaccine.
2nd December 2020	The quarantine period was shortened to 10 days with no symptoms and 7 days if the person tested negative and had no symptoms.
11th December 2020	The FDA approved the use of Pfizer's COVID Vaccine for emergency use.
14th December 2020	The vaccine rollout began across the United States.
20th December 2020	Moderna vaccine began its distribution.
27th December 2020	Europe began its Vaccination Campaign.

Table 2. Structural components of the SARS-CoV-2

Structural protein	Functions
Nucleocapsid protein (N)	Bind to RNA genome to make nucleocapsid
Spike glycoprotein (S)	Important for binding of host cell receptors to facilitate entry of host cell
Envelope small membrane protein (E)	Interacts with membrane protein (M) to make the viral envelope
Membrane protein (M)	The organizer of the CoV assembly and determines the shape of the viral envelope
Hemagglutinin-Esterase (HE)	Create a distinct inner fringe of short peplomer

Table 3. Opening reading forms of the SARS-CoV-2 and their function

Open reading frame (ORF)	Function
ORF1	Replicase enzyme into two polypeptides namely ORF1a and ORF1ab
ORF10	Responsible to differentiate infection effectively than the PCR-based method
ORF8	Responsible for the splitting of ORF8 into orf8a and orf8b
ORF3a	Responsible for changes in the viral cycle
ORF3b	Responsible for viral replication *in vitro* and viral pathogenicity
ORF7a	Provides interferon responses
Nsp1	Responsible for RNA processing and replication
Nsp2	Responsible for signaling pathways in host cells
Nsp3	Separate the translated protein
Nsp4	Consist of transmembrane domain 2 and assist in the modification of ER membranes
Nsp5	Participates in the polyprotein process during replication
Nsp6	Presumptive transmembrane domain
Nsp7 and Nsp8	Increase the combination of Nsp12 and template-primer RNA
Nsp9	Acts as an ssRNA-binding protein
Nsp10	Responsible for cap methylation of viral mRNAs
Nsp 12	Consist of RNA-dependent RNA polymerase important for the composition of coronavirus replication or transcription
Nsp 13	Binds with ATP and involve in the replication process
Nsp 14	Proofreading exoribonuclease domain
Nsp 15	Has ($Mn^{2+)}$- dependent endoribonuclease activity
Nsp 16	Suppress the mRNA splicing

Pathogenic Mechanism

In the early stage, the virus infects humans when a person breathes droplets through the respiratory tract (Parasher, 2021). At this stage, the person becomes asymptomatic and thus allows the host receptors to be highly expressive for the virus in the nasal delivery. According to Parasher (2021), the virus replicates and propagates where at this stage, it lasts for a couple of days with a low viral load. However, at a certain stage, the virus travels to the upper respiratory tract (Hossain et al., 2020). The virus established a higher immune response than in the early stage as symptoms such as fever as well as dry cough exist given the release of chemokine ligand 10 and interferons such as interferon beta and lambda from the virus infecting cells (Parasher, 2021). In the last stage, patients become severely ill as the virus causes the release of multiple cytokines and interleukins such as IL-1, IL-6, IL-8, IL-120 and IL-12 affecting the lungs with inflammation and corrosion (Parasher, 2021).

Mode of Transmission

The mode of transmission occurs when the virus spreads from human to human through respiratory droplets when a person has contact with another person who coughs and sneezes (Parasher, 2021). Various sizes of droplet particles cause the COVID-19 pandemic (Hossain et al., 2020). There is evidence that the virus spread through equipment used by an infected person such as the use of bedsheets, blankets, kitchen utensils, thermometers and stethoscopes (Parasher, 2021). There is also proof that the virus spread between front liners such as the use of infected personal protective equipment for instance gloves, gowns or facial masks (Hossain et al., 2020).

Similarities and Differences between SARS-CoV, MERS-SARS and SARS-CoV-2

Table 4 shows the similarities and differences between SARS-CoV, MERS-SARS and SARS-CoV-2 (Rabaan et al., 2020).

Table 4. Similarities and differences between SARS-CoV, MERS- SARS and SARS-CoV-2

Differences/ Virus Name	SARS-CoV	MERS- SARS	SARS-CoV-2
Type of variant	Beta-Coronavirus	Beta - Coronavirus	Divided into alpha, beta, gamma and delta
First case location	Guangdong, China	Saudi Arabia	Wuhan, China
Time is taken to infect the first 1000 people (Days)	130	903	48
Percentage of mortality	10%	34.4%	27.324%
First Outbreak	November 2002	April 2012	December 2019
Incubation period	2-7	5-6	7-14
Cases number	8096	2519	595800 (March 27, 2020)

Types of Variants

Table 5 describes the different types of SARS-CoV-2 variants (Duong, 2021).

Signs and Symptoms

The signs of COVID-19 differ according to the severity level as symptoms presented at each stage have different signs (Table 6). The period of exposure to the virus for symptoms to appear usually takes 5 to 6 days and up to 14 days (Parasher, 2021). Common symptoms are fever, nausea, fatigue, cough, muscle ache, confusion, headache, chest pain, vomiting, chills, sputum production and loss of taste. However, there are three levels of severity presented by COVID-19 with each symptom consisting of mild, severe and critical (Wang et al., 2020).

Medicinal Plants and Their Mode of Actions

The vast majority of people in the world have been using traditional medicine for thousands of years (Seifu, 2004). According to the World Health Organization (WHO) (2007), 80% of people in underdeveloped nations rely on traditionally used medicinal herbs for their main medical care.

Table 5. Different types of SARS-CoV-2 variants

Differences/type of variants	Alpha	Beta	Delta	Gamma
Place of occurrence	United Kingdom	South Africa	India	Brazil
Scientific name	B.1.1.7	B.1.351	B.1.617.2	P.1
First occurrence	September 2020	May 2020	October 2020	November 2020
Mutation	N501Y	E484K	E484Q and L452R	E484K and K417T
Transmission rate	50%	Unknown	60%	1.7-2.4 transmissible however not near transmissible as Alpha or Beta
Threat	Unknown	Unknown	Dangerous	Unknown
Mechanism	It results in changes in spike protein sequence thus changing the cleavage sites where proline amino acid is changed to histidine	More amino acid substitution	It acts as the host for multiple mutations in the S1 subunit. It also includes three subunits in RBD to bind to ACE2 thus changing the cleavage sites where proline is changed to arginine	Unknown

Table 6. Signs and symptoms according to the stages

Stages	Symptoms
Mild	Fever, cough, fatigue and mild pneumonia
Severe	Dyspnea and blood oxygen saturation
Critical	Acute respiratory distress symptoms, respiratory failure, septic shock and multiple organ failure

Table 7. Medicinal plants, their phytochemicals and biological activities

No.	Plant	Scientific Name	Family	Part of the plant	Phytochemicals	Biological activity	Reference
1.	Green chiretta	*Andrographis paniculata*	Acanthaceae	Leaves	Andrographolides	Acts as an antiviral, anti-inflammatory and antioxidant	Cheepsattayakorn et al., (2021)
2.	Ginger	*Zingiber officinale*	Zingiberaceae	Rhizome	6-gingerol	Acts as an antibacterial, antiviral, antioxidant and antihypertensive	Rathinavel et al., (2020)
3.	Garlic	*Allium Sativum*	Amaryllidaceae	Roots and leaves	Allicin, alliin, diallyl sulphide and diallyl disulphide	Has a role in anti-inflammatory and anti-cancer	Alami et al., (2020); Khan et al., (2021)
4.	Black seed	*Nigella sativa*	Ranunculaceae	Seed	Steroids, tannins, flavanoids, coumarins, cardiac glycosides, saponins and diterpenes	Has a role in antioxidant activity	Alami et al., (2020); Khan et al., (2021)
5.	Oregano	*Origanum vulgare*	Lamiaceae	Essential oil	Carvacrol	Effective against respiratory disease	Khan et al., (2021)
6.	Olive	*Olea europaea*	Oleaceae	Fruits and seeds	Glycosides, secoiridoid, flavonoids, several biophenolic compounds and polyunsaturated fatty acids	Has a role in anti-oxidant, antiviral and antimicrobial	Alami et al., (2020)
7.	Onion	*Allium cepa*	Liliaceae	Bulb	Flavonoids, organosulphur and saponins	Has anti-inflammatory and anti-proliferative	Alami et al., (2020)
8.	Thyme	*Thymus maroccanus*	Lamiaceae	Stems and leaves	Carvacrol, p-Cymene, α-Pinene, γ-Terpinene, β-Caryophyllene, Limonene, Linalool and Myrcene	Has anti-microbial and anti-oxidant properties	Alami et al., (2020)

No.	Plant	Scientific Name	Family	Part of the plant	Phytochemicals	Biological activity	Reference
9.	Tasmanian blue gum	*Eucalyptus globulus*	Myrtaceae	Leaves and flowers	Eucalyptol, α-Pinene, α-Terpineol acetate, Alloaromadendrene, β-Pinene, Sabinene, Limonene, Isoledene, α-Gurjunene, Aromadendrene	Has anti-microbial and anti-viral properties	Alami et al., (2020)
10.	Fennel	*Foeniculum vulgare*	Apiaceae	Leaves and flowers	Saponins, flavonoids, cardiac glycosides, sterols, triterpenes, coumarins, proteins, volatile oils, trace elements and vitamins	Has a role in anti-cancer, anti-microbial and immunological responses	Alami et al., (2020)
11.	Javanese turmeric	*Curcuma zanthorrhiza*	Zingiberaceae	Rhizome	Xanthorrhizol, α -Curcumene, Germacrone, Curcumin and Zederone	Has anti-microbial, antioxidant, antiviral and anti-inflammatory	Alami et al., (2020)
12.	Date palm	*Phoenix dactylifera*	Arecaceae	Seed	Polyphenols compounds including phenolic acids, flavonoids, lignans, and carotenoids	Has anti-tumor, antioxidant and anti-inflammatory	Alami et al., (2020)
13.	Rosemary	*Rosmarinus officinalis*	Lamiaceae	Leaves and flowers	Cineole, Camphor, α-Pinene, Camphene, β-Pineno, Borneol, Bornyl acetate, Caryophyllene	Has antioxidant activity	Alami et al., (2020)
14.	Thyme	*Thymus satureioides*	Lamiaceae	Leaves	Borneol and Thymol	Has antibacterial and anti-oxidant	Alami et al., (2020)

Table 7. (Continued)

No.	Plant	Scientific Name	Family	Part of the plant	Phytochemicals	Biological activity	Reference
15.	Pennyroyal	*Mentha pulegium*	Lamiaceae	Leaves	Pulegone, Piperitone, p-Menthane-1,2,3-triol, Elemenene, Guaiene, Carvacrol acetate and Phenyl ethyl alcohol	Has antioxidant and anti-microbial	Alami et al., (2020)
16.	Anise	*Pimpinella anisum*	Apiaceae	Leaves and seeds	cis-dihydrocarvone, Methyl chavicol, α-Himachalene, γ-Himachalene, β-Himachalene and trans-Pseudoisoeugenyl 2-Methylbutyrate	Has antifungal, anti-microbial and antioxidant	Alami et al., (2020)
17.	White wormwood	*Artemisia herbaalba*	Asteraceae	Leaves	Sesquiterpene lactones, flavonoids, phenolic compounds and waxes; essential oils with the major compounds are α-and ß-Thujones, Camphor, Sabinyl acetate, Germacrene D, α-Eudesmol, Caryophyllene acetate, 1,8-Cineole, p-Cymene, Davanone, Camphene, Borneol, Davana ether and Chrysanthenone	Acts as an anticancer	Alami et al., (2020)
18.	French lavender	*Lavandula dentata*	Lamiaceae	Flowers	Essential oils with the major compounds are Eucalyptol, Fenchone, Linalool, Mirtenol, Citronelol, Camphor, Sabinene, ßCaryophyllene, α –bisabolene, α -selinene, ßCaryophyllene oxide and p-Cymene	Has a role in radical scavenging properties, antimicrobial and antioxidant	Alami et al., (2020)

No.	Plant	Scientific Name	Family	Part of the plant	Phytochemicals	Biological activity	Reference
19.	Apple mint	*Mentha suaveolens*	Lamiaceae	Leaves	Piperitenone oxyde, trans-Caryophyllene, Germacrene D, terpinen-4-ol, Nepetalactone, p-Cymen-8-ol and Ehydrate Sabinen	Has cytotoxicity, antimicrobial and anti-viral properties	Alami et al., (2020)
20.	Spanish lavender	*Lavandula stoechas*	Lamiaceae	Flowers	Linalool, Linalyl acetate, Lavandulyl acetate, α - Terpineol, Terpinene-4-ol, Lavandulol, ß-Ocimen	Has anti-bacterial, anti-microbial and cytotoxic activities	Alami et al., (2020)
21.	Lesser galangal	*Alpinia officinarum*	Zingiberaceae	Rhizome and leaves	Eucalyptol, α-Terpineol, γ-Muurolene, α-Farnesene, Caryophyllene, α-Bergamotene and γ-Gurjunene	Has antioxidant and anti-microbial activities	Alami et al., (2020)
22.	Garden cress	*Lepidium sativum*	Brassicaceae	Leaves and stems	Glucotropaeoline, Sinapine, Kaempferol di-hexose rhamnose, Sinapoyl di-glucose, Quercetin di-hexose rhamnose, Sinapoyl malate, and many other substances	Used to treat airway disorders	Alami et al., (2020)
23.	Evergreen coniferous trees	*Tetraclinis articulata*	Cupressoideae	Seeds	Bornyl acetate, Camphor, a-Pinene, Tricyclene, Limonene, Camphene hydrate, Myrtenol, Verbenone, Carvone	Has anti-inflammatory activities	Alami et al., (2020)
24.	Arar	*Juniperus phoenicea*	Cupressaceae	Bark, leaves and woods	α-Pinene,3-Carene, Myrcene, Fenchone, Camphene, E-β-Caryophelene, Germacrene D, β-Cadinene, βPinene, Limonene, Terpinolene	Has anti-oxidant properties	Alami et al., (2020)
25.	Cade	*Juniperus oxycedrus*	Cupressaceae	Seeds and leaves	α-Pinene, Limonene, α-Phellandrene, 14-hyd rox y-9- epi-E-Caryophyllene and Germacrene D	Has antioxidant and anti-inflammatory functions	Alami et al., (2020)

Dehydrozingerone's antiproliferative activities on HT-29 human colon cancer cells were examined by Yogosawa and his colleagues (2012), and it was discovered that it dose-dependently reduced growth at the G2/M phase via up-regulating p21. Zhou and Raffoul's (2012) study has been demonstrated that the use of grape antioxidants is promising against a variety of cancer cells by reducing overexpression of COX-2 and prostaglandin E2 receptors, targeting EGFR and its downstream pathways or changing estrogen receptor pathways, causing cell cycle arrest and apoptosis. Therefore, a significant public interest in medicinal plants against SARS-CoV-2. Understanding the biological activities of medicinal plants will help in designing plant-based drug which is safe for humans. Table 7 describes medicinal plants, their family, phytochemicals and biological activities.

Conclusion

In this review, the COVID-19 pandemic development timeline; the SARS-CoV-2 virus's structural components; genomic characterization; signs and symptoms; different types of variants; and medicinal plants and their biological activities were discussed. This information will help in designing an effective drug against SARS-CoV-2.

References

Alabed A A, Elengoe A, Anandan E S, Almahdi A Y. Recent perspectives and awareness on transmission, clinical manifestation, quarantine measures, prevention and treatment of COVID-19 among people living in Malaysia in 2020. *Journal of Public Health* (2020):1-0.

Barupal T, Tak P K, Meena M. COVID-19: Morphology, characteristics, symptoms, prevention, clinical diagnosis and current scenario. *Coronaviruses* (2020) 1(1):82-89.

Cheepsattayakorn A, Chawapon R C. *Andrographis Paniculata*, A promising medicinal plant fighting against SARS-CoV-2. *Medical and Research Publications* (2021) 6(2).

Coronavirus timeline: Tracking the critical moments of COVID-19. Accessed 12 November 2022. https://www.nbcnews.com/health/health-news/coronavirus-timeline-tracking-critical-moments-covid-19-n1154341.

Darby A C, Hiscox J A. *COVID-19: variants and vaccination.* (2021).

Duong D. Alpha, Beta, Delta, Gamma: What's important to know about SARS-CoV-2 variants of concern? *Canadian Medical Association Journal* (2021).

El Alami A, Fattah A, Chait A. Medicinal plants used for the prevention purposes during the COVID-19 pandemic in Morocco. *Journal of Analytical Sciences and Applied Biotechnology* (2020) 2(1):2-1.

Elengoe A. COVID-19 Outbreak in Malaysia. *Osong Public Health and Research Perspectives* (2020) 11(3):93.

Ganasegeran K, Swee H C A, Looi I. Coronavirus (COVID-19): evidence relevant to critical care. *Cochrane Database of Systematic Review* (2020) 10(2):1-4.

Guo Y R, Cao Q D, Hong Z S, Tan Y Y, Chen S D, Jin H J, Tan K S, Wang D Y, Yan Y. The origin, transmission and clinical therapies on coronavirus disease 2019 (COVID-19) outbreak-an update on the status. *Military Medical Research* (2020) 7(1): 1-0.

Hossain M F, Hasana S, Mamun A A, Uddin M S, Wahed M I, Sarker S, Behl T, Ullah I, Begum Y, Bulbul I J, Amran M S. COVID-19 outbreak: pathogenesis, current therapies, and potentials for future management. *Frontiers in Pharmacology* (2020) 11:563478.

Khailany R A, Safdar M, Ozaslan M. Genomic characterization of a novel SARS-CoV-2. *Gene Reports* (2020) 19:100682.

Khan T, Khan M A, Ullah N, Nadhman A. Therapeutic potential of medicinal plants against COVID-19: The role of antiviral medicinal metabolites. *Biocatalysis and Agricultural Biotechnology* (2021) 31:101890.

Mahase E. COVID-19: What new variants are emerging and how are they being investigated? *British Medical Journal (Clinical Research Ed.)* (2021) 372:n158.

Mehmet N, Al-Abed A A, Gökler M E, Elengoe A, Ünal E, Mollahaliloğlu S. Knowledge, attitudes and practices regarding COVID-19 among the Turkish and Malaysian general populations during lockdown: A cross-sectional online survey. *The Ethiopian Journal of Health Development* (2020) 34(4).

Parasher A. COVID-19: Current understanding of its pathophysiology, clinical presentation and treatment. *Postgraduate Medical Journal* (2021) 97(1147):312-320.

Rabaan A A, Al-Ahmed S H, Haque S, Sah R, Tiwari R, Malik Y S, Dhama K, Yatoo M I, Bonilla-Aldana D K, Rodriguez-Morales A J. SARS-CoV-2, SARS-CoV, and MERS-COV: A comparative overview. *Infezioni in Medicina* (2020) 28(2):174-184.

Rathinavel T, Palanisamy M, Palanisamy S, Subramanian A, Thangaswamy S. Phytochemical 6-Gingerol-A promising Drug of choice for COVID-19. *International Journal on Advanced Science Engineering* (2020) 6(4):1482-1489.

Seifu T. *Ethnobotanical and ethnopharmaceutical studies on medicinal plants of Chifra district, Afar region, North Eastern Ethiopia*. PhD Thesis. Addis Ababa University: School of Graduate Studies; 2004.

Singhal T. A review of coronavirus disease-2019 (COVID-19). *The Indian Journal of Pediatrics* (2020) 87(4):1-6.

Wang M Y, Zhao R, Gao L J, Gao X F, Wang D P, Cao J M. SARS-CoV-2: structure, biology, and structure-based therapeutics development. *Frontiers in Cellular and Infection Microbiology* (2020) 10:587269.

World Health Organisation. *Health of indigenous peoples*. Geneva, Switzerland: Factsheets N° 326; 2007.

Yogosawa S, Yamada Y, Yasuda S, Sun Q, Takizawa K, Sakai T. Dehydrozingerone, a structural analogue of curcumin, induces cell-cycle arrest at the G2/M phase and accumulates intracellular ROS in HT-29 human colon cancer cells. *Journal of Natural Products* (2012) 75(12):2088-2093.

Zhou K, Raffoul J J. Potential anticancer properties of grape antioxidants. *Journal of Oncology* (2012) 2012:803294.

Chapter 11

The Mechanistic Aspects of Using Phytochemicals against SARS-CoV-2

Dharithri Hansani[1], BSc
and Asita Elengoe[2,*], PhD

[1]School of Biotechnology, Manipal International University (MIU), Negeri Sembilan, Malaysia
[2]Department of Biotechnology, Faculty of Applied Science, Lincoln University College, Selangor, Malaysia

Abstract

The coronavirus disease 2019 (COVID-19) has emerged and is spreading quickly, posing a threat to public health around the world. It was caused by the severe acute respiratory syndrome-coronavirus-2 (SARS-CoV-2). Since COVID-19 does not have a known specific and efficient antiviral treatment, new or modified antivirals are urgently required. Many studies are being carried out on phytochemicals because they provide some promise for human health in the midst of the pandemic. Phytochemicals have been utilized as antiviral medicines against a variety of viruses such as human immunodeficiency virus (HIV), hepatitis virus, human papillomavirus, influenza virus, etc. Since, they can block them directly at the viral entry point or throughout the replication phases, as well as via immunomodulation potentials. Medicinal plants and their extracts, as well as herbs, have demonstrated promising effects in combating SARS-CoV-2 infection and may be beneficial in treating COVID-19 patients as alternatives to treatment under phytotherapy techniques. In this review, the timeline of the COVID-19 pandemic development; morphology of the SARS-CoV-2; structural proteins and their functions; transmission;

* Corresponding Author's Email: asitaelengoe@yahoo.com.

In: Medicinal Plants and COVID-19
Editor: Asita Elengoe
ISBN: 979-8-88697-958-9

signs and symptoms; diagnostics test; therapeutic developments; types of variants; and SARS-CoV-2 target proteins. Moreover, 25 medicinal plants were tabulated in this review.

Keywords: COVID-19, mechanism, phytochemicals, SARS-CoV-2

Introduction

On December 31, 2019, a number of pneumonia cases with an unknown source were reported to the World Health Organization (WHO) in the Chinese city of Wuhan, Hubei province. The 2019 novel coronavirus, which was discovered in January 2020, was the cause of the outbreak, according to samples taken from cases and virus genetic analysis. WHO referred to this novel coronavirus disease 2019 as COVID-19 in February 2020. SARS-CoV-2 is the name of the virus, while COVID-19 is the name of the illness (Fong et al., 2020).

A type of virus called coronavirus is responsible for ailments like respiratory and gastrointestinal issues. Disorders of the respiratory system can range from the common cold to more serious illnesses. Beta coronavirus, Gammacoronavirus, Deltacoronavirus and Alphacoronavirus are the four main types of coronaviruses that are known to cause gastrointestinal and respiratory tract illnesses. While the last two primarily infect mammals, the first two primarily affect birds. A new coronavirus strain that has never been identified in humans is known as a novel coronavirus (nCoV). The way that coronaviruses look under a microscope is how they got their name (Fong et al., 2020). The virus was composed of a genetically engineered core that was encased in a protein-filled envelope. This gives it the impression of being a crown. In Latin, the name Corona means "crown."

Coronaviruses had a diameter of around 120 nanometers and were single-stranded RNA viruses. Because of their propensity for mutation and recombination, they were very diverse. There are over 40 different kinds, and most of them affect both human and non-human mammals as well as birds. They can spread to other animals, including humans, from wild birds and bats (Aronson et al., 2020). It was thought that the virus that causes COVID-19 began in bats and traveled widely to snakes and pangolins before eventually infecting humans, probably through contamination of wild animal meat sold in China's meat markets.

Spike glycoproteins, or peplomers, give coronaviruses their corona-like appearance and are necessary for the viruses to enter host cells.

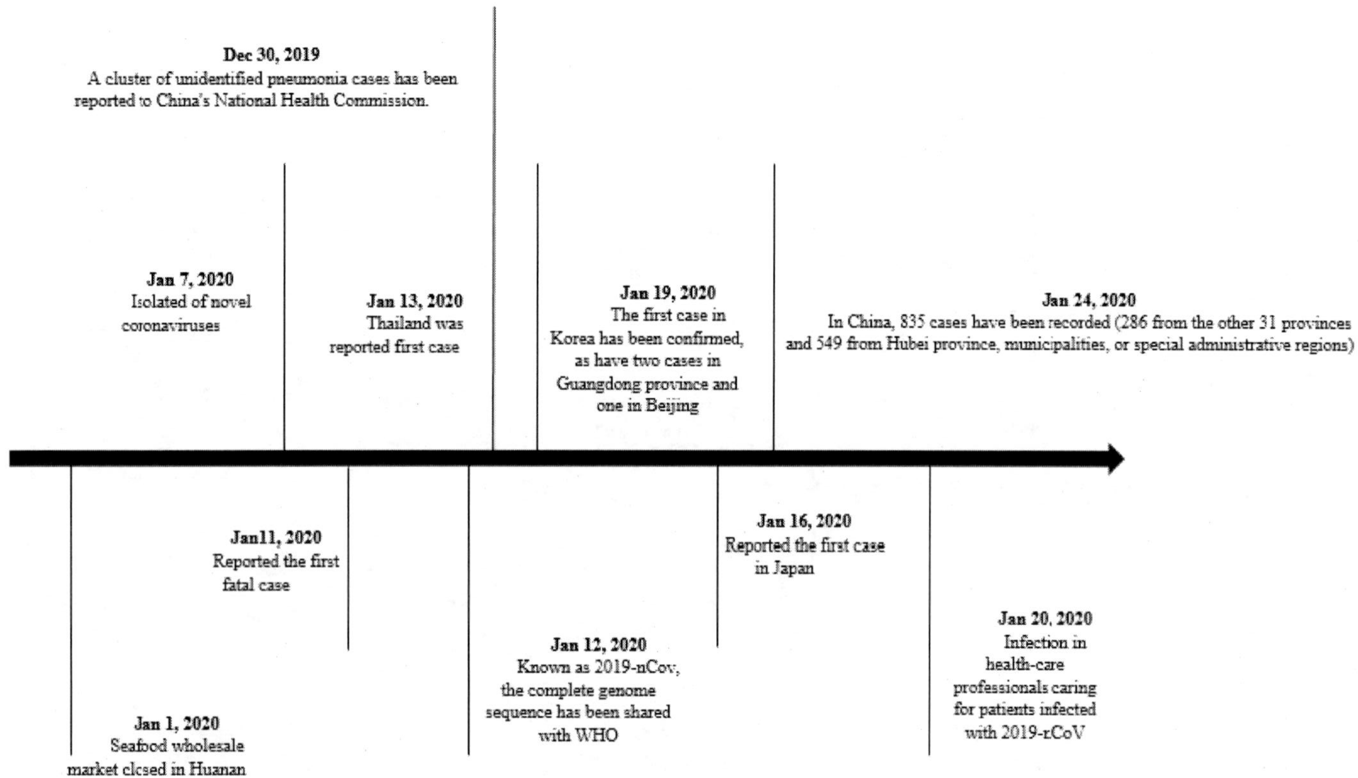

Figure 1. (Continued).

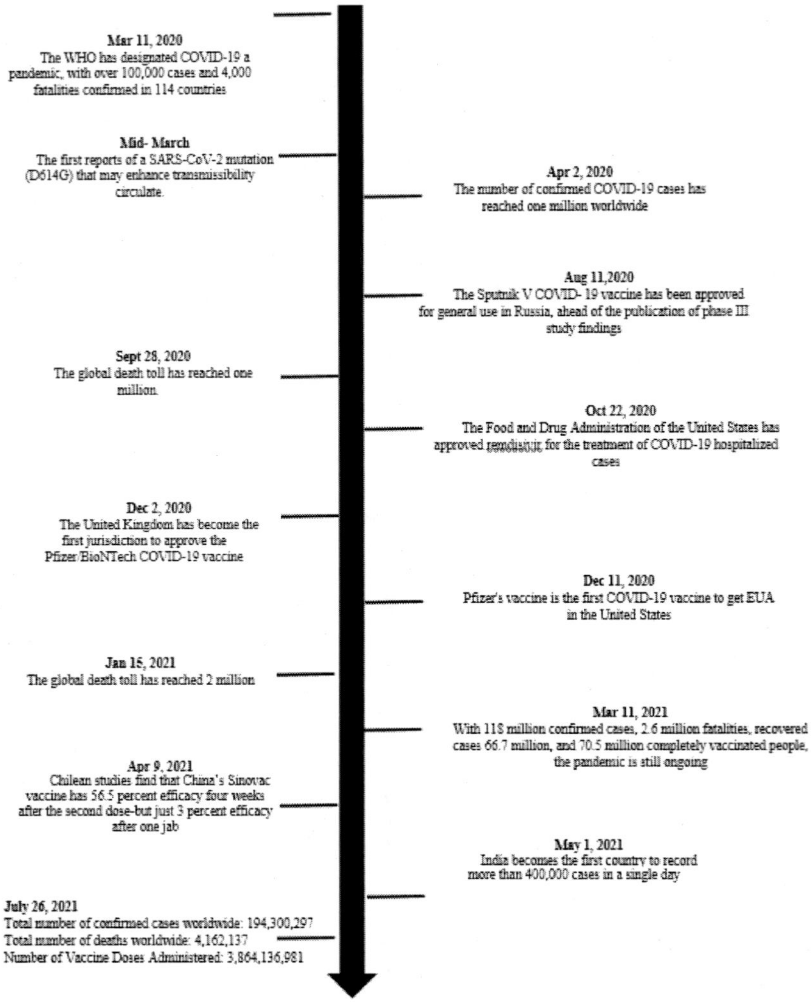

Figure 1. Timeline of COVID-19 pandemic development-chronological list of 2019-nCoV global outbreak.

The spike consists of two subunits, S1 and S2, which fuse with the cell membrane and connect to a receptor on the surface of the host cell. Angiotensin-converting enzyme-2 (ACE-2) would be the specific cell membrane receptor for both SARS-CoV-1 and SARS-CoV-2. Standard ACE-1 medications like enalapril and ramipril would inhibit this enzyme (Aronson et al., 2020; Elengoe et al., 2022). The ACE-2 enzyme interacts with the S1 subunit of the spike on the cell membrane. A host transmembrane serine protease called TMPRSS2 then activated and cleaved the spike. The S2 subunit and TMPRSS2 also interact, enabling the virus to attach to the cell membrane. A cell is invaded by the virus. The cell's cysteine protease cathepsin or acidification both helped the virus escape from endosomes (Elengoe, 2020).

This virus spreads from one individual to another by droplet transmission. The virus thus spreads rapidly in densely populated places. A high body temperature and respiratory symptoms including a sore throat, cough, and headache are the only symptoms that the majority of people experience. Some patients may experience severe symptoms like pneumonia and acute respiratory distress syndrome (Khailany et al., 2020).

Morphology of the Virus

Coronaviruses were divided into four genera; alpha, beta, gamma, and delta coronaviruses. There are six human coronaviruses that might be present (α-HCoV-NL63, α-HCoV-229E, β -SARS-CoV, β -HCoV-OC43, β -HCoV-HKU1, and β -MERS-CoV). β -SARS-CoV and β -MERS-CoV were viruses that cause severe respiratory tract illnesses (Elengoe and Selvam, 2022; Li et al., 2014). Because of their low pathogenicity, the other four human coronaviruses cause mild respiratory symptoms similar to a common cold. Some viral structural protein characteristics shared by -SARS-CoV and -MERS-CoV include spike glycoprotein (S), membrane (M), envelope (E), and nucleocapsid (N). The SARS-CoV-2 beta coronavirus is a member of the Coronaviridae family, Coronavirinae suborder, which is part of the Nidovirales order. It is an enclosed, positive-sense, non-segmented, single-stranded RNA virus. It has a big RNA genome of around 30 kb. The viral particle (virion) was spherical, 60-140 nm in diameter, with 9-12 nm spikes. The mRNA transcript has a 5' terminal cap structure and a 3' poly A-tail structure.

Genomic Characterization

Structural Proteins and Their Function

- ✓ Spike glycoprotein (S): Involved in the binding of the virion to the host cell receptor. It enables viral particles to infiltrate the host cell.

- ✓ Membrane glycoprotein (M): Maintain the virion's form. It increases the curvature of the membrane and aids in nucleocapsid binding.
- ✓ Envelope (E): Involved in virus assembly and release.
- ✓ Nucleocapsid protein (N): This was required for RNA synthesis and the packaging of the genome into virions.

Pathogenic Mechanism

The viral infection begins when SARS-CoV-2 virus particles bind to the human angiotensin-converting enzyme 2 (hACE2) via the spike protein. SARS-CoV and SARS-CoV-2 use hACE2 as an entrance for the host receptor. It is a transmembrane glycoprotein that contains a zinc-binding carboxypeptidase. It mostly affects the lungs and the host immune system. The spike protein was made up of two domains, S1 and S2. The spike protein and N protein help to keep the viral particle stable. The N protein is a structural protein that is involved in vision assembly and is critical in viral transcription and assembly efficiency. S protein has the ability to attach to the cellular receptors of susceptible cells and infect them (Elengoe et al., 2022).

The S2 domain then initiates the contact between the viral particle and the host cell membrane. Finally, the virus infiltrates the host cell via the endosomal route. At low pH in the cellular environment, the cysteine protease cathepsin L1 (CTSL1) can also aid in membrane fusion between the virion and the host cell. During SARS-CoV-2 infection, this host protease enhances endosomal viral entrance into the cell membrane by proteolytic cleavages that activate the S protein (Wang et al., 2020).

Human airway proteases include transmembrane protease serine 2 (TMPRSS2) and trypsin-like protease (TMPRSS11D). They were found on the host cell surface membrane and are responsible for cleaving the S1/S2 domain in order to activate the S protein and allow the virus to enter the host cells via the non-endosomal route. S protein's receptor-binding motif can make direct contact with hACE2. hAEC2 has been identified as a major binding residue and functional receptor for SARS-CoV, as well as a protective factor for alveolar cells. During SARS, spike protein binding to hACE2 and subsequent downregulation of this receptor lead to severe alveolar harm (Elengoe et al., 2022). Down-regulation of hACE2 causes an increase in angiotensin II synthesis by the related enzyme hCE, and activation of type 1A angiotensin II receptor can lead to an increase in pulmonary vascular permeability, which may explain the increased lung pathology when hACE2 expression was reduced (Hu et al., 2021).

The viral RNA was once released into the cytoplasm. Orf 1a and Orf 1b were translated into polypeptides 1a (pp1a) and 1b (pp1b) via the ribosomal frameshifting process (pp1b). The polypeptides were broken by both the host and viral proteases, resulting in non-structural proteins (NSPs) (Xu et al., 2020). They then were put together to create the RNA replicase-transcriptase complex. This complex was primarily involved in viral replication. To construct structural and auxiliary proteins, sub-genomic RNAs are transcribed and translated. The viral progeny particles are packed and constructed in the cytoplasm from genomic RNA incorporating nucleocapsid and other structural proteins such as the S, E, and M proteins. Then, by exocytosis, they bud into the lumen of the endoplasmic Reticulum-Golgi intermediate compartment (ERGIC) and emerge as a smooth-walled vesicle on the plasma membrane. Finally, developed virions were liberated from infected cells and spread to other host cells. The viral infectious cycle will restart (Elengoe et al., 2022).

Transmission Mode

SARS-CoV-2 is a very dangerous virus. It spreads readily from person to person. This was spread by respiratory secretions such as fluid droplets when an infected person sneezes or coughs in the early stage, causing dry cough, fever and fatigue; later on, respiratory clinical symptoms such as difficulty in breathing and lung infection occur (Elengoe et al., 2022; Zhou et al., 2021). It can lead to life-threatening pneumonia, bronchiolitis, and organ failure in extreme instances. Aches and pains, headache, nasal congestion, sore throat, conjunctivitis, diarrhea, loss of taste or fragrance, skin rash and finger or toe discoloration have all been identified as novel COVID-19 symptoms, according to the Centres for Disease Control and Prevention (Elengoe et al., 2022). These less common symptoms may occur in some patients. The SARS-CoV-2 infection may alter the immune system by reducing innate immune responses and causing delayed activation of pro-inflammatory signals from cytokines and chemokines. Elderly persons, pregnant women, children, immune-compromised patients and those with chronic illnesses such as hypertension, diabetes, lung, kidney, and heart issues are more vulnerable to SARS-CoV-2 infection.

Sign and Symptoms

Cough, fever, myalgia, dyspnea and diarrhea were the early symptoms of SARS patients, and the prognosis of patients was related to host characteristics. During hospitalization, 90.8% of SARS patients experienced respiratory difficulty. The average time from illness start to severe respiratory distress was 9.8-3.0 days. Some patients developed leukopenia, lymphopenia, and thrombocytopenia throughout the disease, which was associated with an increase in aspartate transaminase (AST), alanine aminotransferase (ALT), lactic dehydrogenase (LDH), and C-reactive protein (CRP). In comparison, COVID-19 revealed comparable patterns in SARS patients. The most common symptoms of the patients are fever, tiredness, and a dry cough, with nasal congestion, runny nose, and other upper respiratory tract symptoms being uncommon (Guo et al., 2020).

Diagnostics Test for SARS-CoV-2

Table 1 describes the different types of diagnostic tests such as next gene sequencing (NGS), polymerase chain reaction (PCR), clustered regularly interspaced short palindromic repeats (CRISPRs), serological antigen tests and viral antigen tests.

Therapeutic Developments

Table 2 shows the variety of drugs, their group, mechanism and common adverse events; and Table 3 describes the vaccines produced against SARS-CoV-2.

Types of SARS-CoV-2 Variants

Table 4 shows the different types of variants of SARS-CoV-2 such as alpha, beta, gamma and delta.

Table 1. List of diagnostics tests and their functional roles

Diagnostic Tests	Purpose
Viral genome (NGS)	• Deep-sequencing techniques, such as next-generation sequencing (NGS), have transformed virology and the discovery of new viruses. • The use of NGS to sequence new viruses in clinical samples, such as SARS-CoV-2, will be essential in our knowledge of virus evolution and the discovery of novel targets for the development of diagnostic tests and antiviral treatments.
Viral RNA test (PCR)	• This COVID-19 test identifies viral genetic material using a laboratory technique known as polymerase chain reaction (PCR). • A fluid sample was obtained by inserting a long nasal swab into a nostril and taking fluid from the back of the nose, or by using a shorter nasal swab. • The main benefits of the COVID-19 PCR test are its accuracy and dependability. This is the most accurate COVID-19 detection test available.
Viral RNA test (CRISPR, Iso.Amp)	• The gold-standard diagnostic for direct detection of SARS-CoV-2 is now based on the PCR technique. • To obtain the required high rate of sensitivity and specificity in diagnostic testing, PCR is employed to amplify minute quantities of viral genetic material in clinical samples.
Serological antibody tests	• By identifying SARS-CoV-2 specific antibodies IgG and IgM in host blood, researchers can track the host immune response to COVID-19 infection. • This was crucial to note that antibody seroconversion and the detection of SARS-CoV-2-specific IgG/IgM antibodies in host blood require 7–14 days following COVID-19 infection.
Viral antigen tests	• To identify the presence of a specific virus in fluid samples, take use of the high binding affinity between an antibody and a distinct viral antigen. • The S protein and other structural proteins on cell surfaces are major targets of the SARS-CoV-2 virus for the development of antigen-based assays.

Elengoe et al., 2022.

Table 2. List of anti-SARS-CoV-2 drugs and their mechanisms

Name Of Drugs	Group	Mechanism	Common Adverse Events
Remdesivir	Nucleoside analog	• The active form of remdesivir acts as a nucleoside analog and inhibits the RNA replication from the RNA template of coronaviruses including SARS-CoV-2.	• Hepatic adverse events
Chloroquine and hydroxychloroquine	Antiparasitic	• An important mode of action of chloroquine and hydroxychloroquine is the interference of lysosomal activity and autophagy. • Inhibits endocytosis by increasing the acidity in endosomes and prevents the entry of the virus into the host cell.	• Pale stools • Bitter taste in the mouth • Nausea
Azithromycin	Antibiotic	• Prevents bacteria from growing by interfering with their protein synthesis.	• Vomiting • Headache • Fever • Loss of energy
Lopinavir and Ritonavir	HIV protease inhibitor	• Inhibits the HIV protease enzyme by forming an inhibitor-enzyme complex thereby preventing cleavage of the gag-pol polyproteins	• Diarrhea • Nausea • Vomiting
Camostatmesylate	Serine protease inhibitor	• Inhibits the enzymatic activity of type II transmembrane serine protease which is important for infectious viral entry into the host cell.	• Skin rashes • Diarrhoeas • Itching
Darunavir	Protease inhibitor	• Blocks viral replication. • Inhibits the HIV protease enzyme by forming an inhibitor-enzyme complex thereby preventing cleavage of the gag-pol polyproteins.	• Nausea • Headache
Ribavarin	Nucleoside Analog	• Inhibits viral RNA synthesis and mRNA capping.	• Cough • Insomnia • Rash • Nausea
Umifenovir	Broad spectrum antiviral	• Inhibits fusion of virions with the endosomal membrane.	• Nausea • Headache

Name Of Drugs	Group	Mechanism	Common Adverse Events
Oseltamivir	Neuraminidase inhibitor	• Inhibits RNA translation process.	• Abdominal pain • Insomnia • Headaches
Interferon- beta	Cytokines	• Triggers the activation of innate antiviral immunity.	• Vomiting • Diarrhoeas
Favipiravir	Nucleoside analog	• Inhibits the RNA-dependent RNA polymerase of RNA viruses. • Shows anti-viral activities against other RNA viruses.	• Nausea • Vomiting • Increase of serum in uric acid

Khan et al., 2021; Freidel and Armen, 2021.

Table 3. List of vaccines

Vaccine	Country of origin	Date of origin	Function
Oxford–AstraZeneca COVID-19 vaccine University of Oxford, AstraZeneca, CEPI	United Kingdom, Sweden	April 19, 2021	• Safe and effective. • Adults of all ages showed a strong immune response.
Pfizer–BioNTech COVID-19 vaccine. BioNTech, Pfizer	Germany, United States	December 11, 2020	• Instructing cells to produce the spike protein. • Triggers an immunological response, which includes the production of antibodies specific to the SARS-CoV-2 spike protein. • Strong immune response in all groups tested.
Moderna COVID-19 vaccine. Moderna, NIAID, BARDA, CEPI	United States	December 18, 2020	• mRNA vaccines are a new type of vaccination that protects against infectious illnesses. • mRNA vaccines instruct our cells on how to produce a protein or simply a portion of a protein that causes an immunological response inside our bodies.
Janssen COVID-19 vaccine. Janssen Vaccines (Johnson & Johnson), BIDMC	United States, Netherlands	April 23, 2021	• The most effective way to prevent infectious disease. • They prepared the immune system to recognize and defend itself against a specific virus.

Table 3. (Continued)

Vaccine	Country of origin	Date of origin	Function
BBIBP-CorV Sinopharm: Beijing Institute of Biological Products	China	December 31, 2020	• The Sinopharm BBIBP-CorV vaccine encourages the immune system to produce antibodies against the SARS-CoV-2 which is the beta coronavirus. • Inactivated viral vaccinations, such as rabies and hepatitis A vaccines, have been used for decades.
Sputnik V COVID-19 vaccine Gamaleya Research Institute of Epidemiology and Microbiology	Russia	August 11, 2020	• The Sputnik V COVID-19 vaccine employs two harmless viruses that transmit genetic information to human cells to produce a protein from the new coronavirus. • This prepares our immune systems to resist future infections caused by the new coronavirus.
CoronaVac Sinovac	China	February, 2021	• CoronaVac is a vaccination that has been inactivated. • It employs a dead form of the SARS-CoV-2 virus to prevent replication, but it leaves the surface spike protein intact to stimulate the body's immune system to produce antibodies to defend against the live virus if it invades.
Covaxin Bharat Biotech, Indian Council of Medical Research	India	December 6, 2020	• Covaxin is included in the vaccination along with immune potentiators, also known as vaccine adjuvants, which are added to raise and boost immunogenicity.
Novavax	United Kingdom, United States	February 9, 2021	• It is a protein subunit vaccine, which means it is manufactured in a lab from a lab-made version of the SARS-CoV-2 spike protein. • This spike protein cannot make someone ill on its own. To ensure that the body continues to produce protective antibodies against it, Novavax has been included as an adjuvant. • It works as a hypeman for the immune system, signaling it to spring into action.

Elengoe et al., 2022.

Table 4. Different types of SARS-CoV-2 variants

Name of variants	Alpha	Beta	Gamma	Delta
Scientific Name	B.1.1.7	B.1.3251	P.1	B.1.617.2
First Documentation	UK Sep 2020	South Africa May 2020	Brazil Nov 2020	India Oct 2020
Mutations	N501Y,P681H	K417N, E484K, N501Y	K417T, E484K, N501Y	L452R, T478K, P681R
Transmissibility	24- 33% (+29%)	20-30% (+25%)	29-48% (+38%)	76-117% (+97%)
Hospitalizations	47-57% (+52%)	Under investigations	Possibly increased	39- 147% (85%) Relative to Alpha
Mortality	44-74% (+59%)	Possible increased	20- 90% (+50%)	50-230% (+137%)
From natural infection	Minimal reduction	Reduced, the D614G virus's T-cell reaction remains effective	Reduced	Evidence of reinfection
From vaccination	Minimal reduction	Efficacy; decreased against symptomatic diseases, but preserved against severe illnesses	Retained by many	Minimal decrease in efficacy
Transmission medium	- Animal to Human - Human to Human	- Animal to Human - Human to Human	- Animal to Human - Human to Human	- Animal to Human - Human to Human
Transmission mode	- Coughing - Sneezing from infected person	- Coughing - Sneezing from infected person	- Coughing - Sneezing from infected person	- Coughing - Sneezing from infected person
Pathogenicity	High	High	High	High
Symptoms	- Cough - Fever - Tiredness	- Cough - Fever - Breathing problems	- Cough - Fever - Tiredness	- Sore throat - Headache - Sneezing - Fever - Runny nose
Reported in	173 countries	122 countries	74 countries	104 countries
Effectiveness vaccine	Oxford -AstraZeneca Pfizer Novavax	Oxford -AstraZeneca Pfizer Janssen Sinopharm	Oxford –AstraZeneca Pfizer Moderna Coronavax	Oxford -AstraZeneca Pfizer Covaxin

Roberts, 2021.

Table 5. SARS-CoV-2 target proteins

Proteins	PDB ID	Function	References
Nsp 3	6YI3, 6YMN, 6YWL, 6XAA	• Binding of RNA • SARS-CoV-2 Replication-Transcription Complex Assembly (RTC) • Replicase protein maturation • Replication of the viral RNA genome • The activity of an RNA-directed 5'-3' RNA polymerase	(Freidel et al., 2021)
Nsp 1	6ZLW, 6ZMO, 6ZP4	• The inflammatory response is influenced by the control of respiratory bursts • Binding of poly(U) RNA • Translation at the presynapse is regulated, modulating synaptic transmission	(Khan et al., 2021)
Nsp 5	6W63, 6XCN, 6XBO, 7BRO	• Maturation of replicase proteins in the SARS-CoV-2 Replication-Transcription Complex (RTC) • The activity of the CMP-N-acetylneuraminate transmembrane transporter • Spike protein maturation • Structural protein translation • Assembly and release of virion	(Freidel et al., 2021)
Nsp9	6W9Q 6W4B	• Binding of RNA	(Khan et al., 2021)
Nsp 7 – Nsp 8	7BV1, 7BV2, 6W10, 6M71, 6M51	• Synthesis of RNA for SARS-CoV-2 • Assembly of the SARS-CoV-2 Replication-Transcription Complex (RTC) • Maturation of replicase proteins	(Khan et al., 2021)

Table 6. Medicinal plants and their biological activities

No.	Plant	Scientific Name	Family	Part of the plant	Phytochemicals	Biological activity	References
1.	Liquorice	*Glycyrrhiza glabra*	Fabaceae	Centuries, dried roots and rhizomes	Glycyrrhizin	• Antiviral • Antimicrobial • Antitumor activity	(Metin et al., 2020)
2.	Ceylon Cinnamon	*Cinnamomum verum*	Lauraceae	Bark	Cinnamic acid Apigenin Rosmarinic	• Antiviral • Anti-inflammatory • Inhibits attachment of the human respiratory syncytial virus	(Sytar et al., 2021)
3.	Amla fruits / Indian gooseberry	*Phyllanthus emblica*	Phyllanthaceae	Fruits Seed	Chebulagic acid Corilagin Mucic acid	• Antioxidant • Anti-inflammatory	(Singh et al., 2021)
4.	Garlic	*Allium sativum*	Amaryllidaceae	Bulb	Organosulfur Flavonoid	• Immunomodulatory • Antibacterial • Anti-inflammatory • Antimutagenic • Infection by boosting immune system cells and suppressing the synthesis and release of proinflammatory cytokines as well as the proinflammatory adipose tissue-derived hormone leptin	(Metin et al., 2020)

Table 6. (Continued)

No.	Plant	Scientific Name	Family	Part of the plant	Phytochemicals	Biological activity	References
5.	Drumstick tree	*Moringa Oleifera*	Moringaceae	Roots, flowers, leaves and pod	Zeatin Kaempferol	• Has the potential to suppress the SARS-CoV-2 virus • Antioxidant • Antispasmodic • Antiviral, • Anti-inflammatory • Stimulant • Cardiac circulatory tonic properties	(Gupta et al., 2021)
6.	Turmeric	*Curcuma longa*	Zingiberaceae	Rhizomes	Curcumin Demethoxycurcumin Turmerin Zingiberene	• Antiviral • Suppressing inflammatory cytokine production • Role in preventing COVID-related lung injury	(Gupta et al., 2021)
7.	Black Cumin	*Nigella sativa*	Ranunculaceae	Seeds	Thymoquinone Nigellidine	• Antiviral • Antioxidant • Anti-inflammatory • Anticoagulant • Immunomodulatory • Bronchodilator • Antitussive • Analgesic properties	(Khan et al., 2020)

No.	Plant	Scientific Name	Family	Part of the plant	Phytochemicals	Biological activity	References
7.	Black Cumin (continued)					• To be identified as a possible SARS CoV-2 inhibitor	
8.	Ginger	*Zingiber officinale*	Zingiberaceae	Rhizome	6- gingerol 8- gingerol 10- gingerol	• Anti-inflammatory qualities that help in toxin removal from the respiratory tract	(Bandyopadhyay, 2020)
9.	Black pepper	*Piper nigrum*	Piperaceae	Fruits	Piperine	• To prevent RNA packing in the nucleocapsid • To prevent viral replication • To aid in the direct battle of SARS-CoV-2 via potential antiviral actions and immunomodulatory activities	(Choudhary et al., 2020)
10.	Neem tree	*Azadirachta indica*	Meliaceae	Leaves, roots, twigs and seeds	Flavonoids Tannins Glycosides	• Antiviral • Antimicrobial • Anti-inflammatory	(Anand et al., 2021)
11.	Cinnamon	*Cinnamomum verum*	Lauraceae	Roots	Cinnamic acid Gallic Vanillic	• To extract as potential immunomodulatory for server COVID-19	(Anand et al., 2021)
12.	Ashwagandha	*Withania somnifera*	Solanaceae	Roots	Phenolics Alkaloids	• To used treat stress, immunomodulatory, inflammatory, and nervous disorder	(Anand et al., 2021)

Table 6. (Continued)

No.	Plant	Scientific Name	Family	Part of the plant	Phytochemicals	Biological activity	References
13.	Common sage	*Salvia officinalis*	Lamiaceae	Whole plant	1,8- cineole Camphor	• High binding to COVID-19 proteases • Inhibits SARS-CoV and HSV-1 replication	(Sytar et al., 2021)
14.	Woad	*Isatis tinctoria*	Brassicaceae	Root	Tryptanthrin Indirubin	• Inhibits cleavage activity of SARS-3CLpro enzyme • High antioxidant potential • Anti-inflammatory effects	(Sytar et al., 2021)
15.	Coriander	*Coriandrum sativum*	Apiaceae	Leaves and seeds	Flavanoids Alkaloids	• Anti-inflammatory • To inhibit the activation of NF-kB	(Choudhary et al., 2020)
16.	Green chireta	*Andrographis paniculata*	Acanthaceae	Leaves and roots	Flavonoids Xanthones	• Different formulations against various diseases sore throat, fever, viral hepatitis, and upper respiratory infections	(Anand et al., 2021)
17.	Fenugreek	*Trigonella foenum-graecum*	Fabaceae	Seeds	Alkaloids Phenol Steroids	• To potent antioxidant • Anti-inflammatory properties.	(Choudhary et al., 2020)
18.	Bay leaf	*Laurus nobilis*	Lauraceae	Leaf	Flavonoids Tannins Citric acid	• Anti-inflammatory	(Gupta et al., 2020)

No.	Plant	Scientific Name	Family	Part of the plant	Phytochemicals	Biological activity	References
19.	Curry leaf	*Murraya koenigii*	Rutaceae	Leaf	Alkaloids Steroids	• To boost immunity to fight COVID-19 • To against foreign bodies and pathogens like the new coronavirus as a safeguard	(Gupta et al., 2022)
20.	Mint	*Mentha spicata*	Lamiaceae	Leaf	Tannin Phenols Flavonoids	• To inhibit inflammatory factors such as ROS, NF-κB, 8-hydroxy-2′-deoxyguanosine	(Serlahwaty and Giovani, 2021)
21.	Chilli	*Capsicum annuum*	Solanaceae	Fruit	Capsaicinoids Carotenoids Vitamins	• Anti-inflammatory	(Illian et al., 2021)
22.	Oregano	*Origanum vulgare*	Lamiaceae	Leaves, Szddtems	Carvacrol Thymol	• Respiratory and antiviral activity	(Anand et al., 2021)
23.	Onion	*Allium cepa*	Amaryllidaceae	Bulb	Flavonoids Fructans	• Anti-inflammatory • Anti-oxidant	(Gupta et al., 2020)
24.	Fennel seed	*Foeniculum vulgare*	Apiaceae	Seeds	Anethole Flavonoids Limonene	• To stop the virus entering from and causing a lung infection	(Gupta et al., 2020)
25.	Common nettle	*Urtica dioica*	Urticaceae	Leaves	Serotonin Histamine Lectin	• Inhibition of SARS coronavirus replication	(Choudhary, 2020)

SARS-CoV-2 Target Proteins

Table 5 describes the non-structural proteins (nsps), their Protein Data Bank ID (PDB ID) and functional roles.

Medicinal Plants and Their Mechanisms

The search for therapies and vaccines for the new coronavirus illness (COVID-19) is underway. Desperation in the community, particularly among middle- and low-income groups who are particularly hard hit by the economic consequences of forced lockdowns, has sparked a rise in interest in alternative medicinal plant-based therapies. Although there appears to be a significant public interest in using the chosen medicinal plants for COVID-19, there are concerns about their effectiveness and safety (Lim et al., 2021). Exploring the possible functions of medicinal plants, lobbying, and extrapolating from established scientific data on safety and efficacy might be beneficial as COVID-19 therapy research advances. Table 6 shows the medicinal plants and their biological activities.

Conclusion

This review describes the SARS-CoV-2 virus's morphology; genomic characterization; signs and symptoms; diagnostics tests; anti-viral drugs; vaccines; different types of variants; and medicinal plants and their mechanisms against SARS-CoV-2. The relationship between the SARS-CoV-2 target protein's (three-dimensional, 3-D) structure and phytochemicals, which is essential to the development of medications that are effective against lethal viruses, has not been adequately studied. Therefore, additional research should be required to elucidate the molecular interaction between the target proteins of SARS-CoV-2 and bioactive compounds to develop a plant-based drug for COVID-19 treatment.

References

Anand AV, Balamuralikrishnan B, Kaviya M, Bharathi K, Parithathvi A, Arun M, Senthilkumar N, Velayuthaprabhu S, Saradhadevi M, Al-Dhabi NA, Arasu MV. Medicinal plants, phytochemicals, and herbs to combat viral pathogens including SARS-CoV-2. *Molecules* (2021) 26(6):1775.

Aronson JK, Brassey J, Mahtani KR. When will it be over?: An introduction to viral reproduction numbers, R0 and Re. *The Centre for Evidence-Based Medicine* (2020) 14.

Bandyopadhyay P. Role of functional foods in COVID-19 situation. *International Research Journal of Modernization in Engineering Technology and Science* (2021) 3:2582-5208.

Choudhary P, Chakdar H, Singh D, Selvaraj C, Singh SK, Kumar S, Saxena AK. Computational studies reveal piperine, the predominant oleoresin of black pepper (*Piper nigrum*) as a potential inhibitor of SARS-CoV-2 (COVID-19). *Current Science* (2020) 119(8):1333-1342.

Duong D. Alpha, Beta, Delta, Gamma: What's important to know about SARS-CoV-2 variants of concern? *The Canadian Medical Association Journal* (2021).

Elengoe A. Indian spices boost the immune system against COVID-19. *The Annals of the University Dunarea de Jos of Galati. Fascicle VI-Food Technology* (2020) 44(2):189-206.

Elengoe A, Selvam S, Selvarajah M, Manikam H, Vijaya R. Sequence analysis and structure prediction of Malaysia SARS-CoV-2 strain's structural and accessory proteins. *Biointerface Research in Applied Chemistry* (2022) 12(1):3259-3304.

Elengoe A, Selvam S. Overview on origin, transmission, treatment and clinical researches of SARS-COV-2. *Journal of Health Sciences & Surveillance System* (2022) 10(1):2-12.

Fong SJ, Dey N, Chaki J. An introduction to COVID-19. In *Artificial intelligence for coronavirus outbreak* 2021 (pp. 1-22). Springer, Singapore.

Freidel MR, Armen RS. Mapping major SARS-CoV-2 drug targets and assessment of druggability using computational fragment screening: Identification of an allosteric small-molecule binding site on the Nsp13 helicase. *PloS One* (2021) 16(2):e0246181.

Guo YR, Cao QD, Hong ZS, Tan YY, Chen SD, Jin HJ, Tan KS, Wang DY, Yan Y. The origin, transmission and clinical therapies on coronavirus disease 2019 (COVID-19) outbreak-an update on the status. *Military Medical Research* (2020) 7(1):1-0.

Gupta H, Gupta M, Bhargava S. Potential use of turmeric in COVID-19. *Clinical and Experimental Dermatology* (2020) 45(7):902-903.

Gupta S, Singh V, Varadwaj PK, Chakravartty N, Katta AK, Lekkala SP, Thomas G, Narasimhan S, Reddy AR, Reddy Lachagari VB. Secondary metabolites from spice and herbs as potential multitarget inhibitors of SARS-CoV-2 proteins. *Journal of Biomolecular Structure and Dynamics* (2022) 40(5):2264-2283.

Henderson R, Edwards RJ, Mansouri K, Janowska K, Stalls V, Gobeil S, Kopp M, Li D, Parks R, Hsu AL, Borgnia MJ. Controlling the SARS-CoV-2 spike glycoprotein conformation. *Nature Structural & Molecular Biology* (2020) 27(10):925-933.

Hu B, Guo H, Zhou P, Shi ZL. Characteristics of SARS-CoV-2 and COVID-19. *Nature Reviews Microbiology* (2021) 19(3):141-154.

Huang Y, Yang C, Xu X F, Xu W, Liu S W. Structural and functional properties of SARS-CoV-2 spike protein: potential antivirus drug development for COVID-19. *Acta Pharmacologica Sinica* (2020) 41(9):1141-1149.

Illian DN, Siregar ES, Sumaiyah S, Utomo AR, Nuryawan A, Basyuni M. Potential compounds from several Indonesian plants to prevent SARS-CoV-2 infection: a mini-review of SARS-CoV-2 therapeutic targets. *Heliyon* (2021) 7(1):e06001.

Khailany RA, Safdar M, Ozaslan M. Genomic characterization of a novel SARS-CoV-2. *Gene Reports* (2020) 19:100682.

Khan MT, Irfan M, Ahsan H, Ahmed A, Kaushik AC, Khan AS, Chinnasamy S, Ali A, Wei DQ. Structures of SARS-CoV-2 RNA-binding proteins and therapeutic targets. *Intervirology* (2021) 64(2):55-68.

Khan MT, Ali A, Wei X, Nadeem T, Muhammad S, Al-Sehemi AG, Wei D. Inhibitory effect of thymoquinone from *Nigella sativa* against SARS-CoV-2 main protease. An *in-silico* study. *Brazilian Journal of Biology* (2022):84.

Lim XY, Teh BP, Tan TY. Medicinal plants in COVID-19: potential and limitations. *Frontiers in Pharmacology* (2021) 12:611408.

Liu DX, Fung TS, Chong KK, Shukla A, Hilgenfeld R. Accessory proteins of SARS-CoV and other coronaviruses. *Antiviral Research* (2014) 109:97-109.

Metin S, Didinen BI, Kara N, Diler Ö. Effect of rosemary (*Rosmarinus officinalis*) and sage (*Salvia officinalis*) essential oils on disease resistance against *Aeromonas sobria* in goldfish (*Carassius auratus*). *The Israeli Journal of Aquaculture-Bamidgeh* (2020):72.

Serlahwaty D, Giovani C. *In silico* screening of mint leaves compound (*Mentha piperita L.*) as a potential inhibitor of SARS-CoV-2. *Pharmacy Education* (2021) 21:81-86.

Singh NA, Kumar P, Kumar N. Spices and herbs: potential antiviral preventives and immunity boosters during COVID-19. *Phytotherapy Research* (2021) 35(5):2745-57.

Sytar O, Brestic M, Hajihashemi S, Skalicky M, Kubeš J, Lamilla-Tamayo L, Ibrahimova U, Ibadullayeva S, Landi M. COVID-19 prophylaxis efforts based on natural antiviral plant extracts and their compounds. *Molecules* (2021) 26(3):727.

Wang C, Horby PW, Hayden FG, Gao GF. A novel coronavirus outbreak of global health concern. *The Lancet* (2020) 395(10223):470-473.

Zhou L, Ayeh SK, Chidambaram V, Karakousis PC. Modes of transmission of SARS-CoV-2 and evidence for preventive behavioral interventions. *BMC Infectious Diseases* (2021) 21(1):1-9.

Chapter 12

An *In Silico* Study-Exploring the Interaction between *Cinnamomum verum* Plant Compounds and SARS-CoV-2 Receptor Binding Domain (RBD)

Dharithri Hansani[1], BSc
and Asita Elengoe[2,*], PhD

[1]School of Biotechnology, Manipal International University (MIU), Negeri Sembilan, Malaysia

[2]Department of Biotechnology, Faculty of Applied Science, Lincoln University College, Selangor, Malaysia

Abstract

Currently, the coronavirus disease 2019 (COVID-19) pandemic, which is caused by the severe acute respiratory syndrome corona virus-2 (SARS-CoV-2), is the world's most serious health problem. New or repurposed antivirals are urgently needed because there is no specific and effective antiviral treatment for COVID-19. Phytochemicals provide a glimmer of hope for human health in the midst of the pandemic. In this research, the molecular interaction between the SARS-CoV-2 receptor binding domain (RBD) and natural compounds from *Cinnamomum verum* plant was studied using a molecular docking approach. Twenty-four phytochemicals from cinnamon and target protein (PDB ID: 6YLA) were obtained from the PubChem database and RCSB Protein Data Bank (PDB) respectively. They were docked using the SwissDock server. After that, the drug-likeness of the plant compounds that demonstrated interaction was evaluated. The results show that 6YLA had the lowest

* Corresponding Author's Email: asitaelengoe@yahoo.com.

In: Medicinal Plants and COVID-19
Editor: Asita Elengoe
ISBN: 979-8-88697-958-9

binding affinity with the alpha-copaene (PubChem ID: 19725) at the value of -6.6 kcal/mol. 6YLA protein had the most stable interaction with the alpha-copaene. The toxicity prediction analysis revealed that alpha-copaene did not show any properties that may cause harmful effects to humans. Therefore, it was chosen to be used as a safe drug. The findings of this study should aid pharmaceutical researchers in identifying cinnamon-based medications.

Keywords: *Cinnamomum verum*, docking, receptor binding domain (RBD), SARS-CoV-2, toxicity

Introduction

On December 31, 2019, the World Health Organization (WHO) received a report of a cluster of pneumonia patients from an unknown cause in Wuhan, Hubei province. In January 2020, it was discovered that a novel coronavirus was the source of the outbreak. In February 2020, WHO called this new coronavirus Disease 2019 as COVID-19. The virus is known as severe acute respiratory syndrome coronavirus 2 (SARS-CoV-2) (Fong et al., 2021). Till 15th March 2022, a sum of 460,187,168 positive cases, 6,067,672 death cases and 393,287,895 recovered cases has been reported throughout the world (Worldometer, 2022).

The search for therapies and vaccines for the new coronavirus illness (COVID-19) is underway. Desperation in the community, particularly among middle- and low-income groups who are particularly hard hit by the economic consequences of forced lockdowns, has sparked a rise in interest in alternative medicinal plant-based therapies. Although there appears to be a significant public interest in using the chosen medicinal plants for COVID-19, there are concerns about their effectiveness and safety (Xin et al., 2021). Exploring the possible functions of medicinal plants, lobbying, and extrapolating from established scientific data on safety and efficacy might be beneficial as COVID-19 therapy research advances.

Cinnamomum verum belongs to Lauraceae family. The most common name for *Cinnamomum verum* is cinnamon. Cinnamon is native to Sri Lanka. It is a small evergreen tree that is about 30-50 feet tall. The main bioactive constituents of cinnamon are eugenol, cinnamaldehyde, cinnamyl acetate, copane, and camphor. Cinnamon has great medical and pharmacological activities such as anti-microbial, antioxidant, anti-inflammatory, anticancer, anti-diabetic, wound healing activity, anti-HIV, antianxiety, anti-depressant, anti-Parkinson, spasmolytic and cardiovascular and anti-obesity (Pathak and Sharma, 2021).

Computational biology tools include molecular modeling, docking, pharmacokinetics and toxicity approach helping in developing substrate-based drugs (SBD) and understanding better the molecular interaction between macromolecule (target protein) and phytocompound (ligand). The structure-based technique relies on existing structural features to determine the interacting effects among phytochemicals and future targets. Unique ligands could be logically constructed to elicit medical effects, taking advantage of the proteins' three-dimensional (3-D) structure. SBD may provide vital research into prospective drug design and manufacture by finding and refining the initial lead compounds. The high-affinity ligand selectively modulates approved drug targets in order to control certain cellular activities, resulting in the intended pharmacological and therapeutic outcomes (Suhaibun et al., 2020; Yu and MacKerell, 2017). According to Singh and Bast (2015) study, it has been demonstrated that the molecular interaction between IR (PDB; 3ETA), IGF1R (PDB ID: 1K3A), EGFR (PDB ID: 1M17), VEGFIR (PDB ID: 3HNG), and VEGFIIR (PDB ID: 2OH4) and 20 plant compounds were performed through molecular docking approach. The results show that epigallocatechin gallate (EGCG) (plant compound) had the best G_{score} with IGF1R (PDB ID: 1K3A) and VEGFIIR (PDB ID: 2OH4). Based on Elengoe and Loganathan (2021) study, it has been reported that breast cancer cell line proteins (Casp3, BRCA1 and Rb) were created through a Swiss model and docking with natural compounds (ferulic acid and quercetin) were analyzed via Swissdock server. They discovered that the Rb-quercetin complex had the best binding affinity (-7.8 kcal/mol).

In this study, the molecular interaction between the SARS-CoV-2 receptor binding domain (PDB ID: 6YLA) and the twenty-four bioactive constituents from the cinnamon plant was analyzed via docking, pharmacokinetics and toxicity prediction tools.

Methodology

Plant Selection

The cinnamon (*Cinnamomum verum*) plant was chosen due to its high concentration of phytochemical components and related therapeutic applications such as anti-viral, anti-inflammation, anti-bacterial and ability to treat bronchitis, cough, flu and lung infection.

Phytochemical Preparation

Natural compounds were used as ligands. Twenty-four plant compounds were chosen, and the three-dimensional (3-D) conformers of the selected plant compounds were downloaded in SDF format from the PubChem database (Kim et al., 2016; Elengoe and Selvam, 2022).

Preparation of Ligand

The obtained 24 plant compounds (alpha-pinene, camphene, benzaldehyde, beta-pinene, beta-myrcene, 1-phellandrene, delta-3-careen, o-cymene, limonene, beta-phellandrene, 1,8-cineole, benzyl alcohol, linalool, phenylethyl alcohol, trans-cinnamaldehyde, eugenol, alpha-copaene, trans-caryophyllene, cinnamaldehyde-p-methoxy, caryophyllene oxide, benzyl benzoate, cinnamyl acetate, coumarin, alpha-humulene) in SDF format from the PubChem database; were prepared using the DS 4.0 'Prepare 'ligand' technique (Brooks et al., 2009). The retrieved ligands are screened using Lipinski's Rule of Five, which establishes a set of defined requirements that a ligand must meet in order to be appropriate for drug development. It develops drug-like quality parameters and concentrates on medicine bioavailability (Lipinski, 2004; Veber et al., 2002; Jagtap et al., 2020). These ligands are screened to prepare for molecular docking with the SARS-CoV-2 receptor binding domain (PDB ID: 6YLA).

Protein Preparation

SARS-CoV-2 receptor binding domain (RBD) is the virus's small immunogenic segment that binds to a specific endogenous receptor sequence in order to gain access into host cells and leads to infection. This refers to the S-domain of the 'spike' glycoprotein, which is required for membrane fusion and transport to the cytoplasm by interacting with endogenous receptors. The S-domain is frequently the target of neutralizing antibodies. RBD is the key target in the prevention and treatment of SARS-CoV-2 viral infection. The 3D model of the target protein (PDB ID: 6YLA) was obtained from RCSB Protein Data Bank (PDB) and downloaded in PDB format (Berman et al., 2003). The 3D target protein model was publicly available.

Molecular Docking

The docking of the target protein (PDB ID: 6YLA) with its relevant plant compound was carried out using SwissDock (Grosdidier et al., 2011). The 3-D model of the RBD-plant compound-complex was visualized using DS 4.0. The binding energy (kcal/mol), the number of hydrogen bonds and the distance of hydrogen bond between the RBD and plant compound were determined (Parmar et al., 2021).

Prediction of Toxicity

The toxicity of the chosen phytocomponents was assessed using the web-based SwissADME server (Tian et al., 2015).

Results and Discussion

Phytocomponents Screening and Pharmacokinetics (PK) Analysis

Lipinski's Rule of 5 was used to screen phytocomponents (ligands) retrieved from the PubChem database. Lipinski's Rule of 5 can be used to predict the ability and strength of absorption and penetration in medication development. When there are more than 5 hydrogen bond donors (HBD$\geq$ 5), 10 hydrogen bond acceptors (HBA$\geq$ 10), 10 rotatable bonds (RB$\geq$10), the molecular weight is larger than 500 (MW$\geq$ 500 daltons), and the computed Log P is greater than 5 (LogP$\geq$5), poor absorption and penetration are more likely. The pharmacokinetics of the specified ligands/phytocompounds was evaluated using this set of standards. All the 24 phytocomponents (ligands) ((alpha-pinene, camphene, benzaldehyde, beta-pinene, beta-myrcene, 1-phellandrene, delta-3-careen, o-cymene, limonene, beta-phellandrene, 1,8-cineole, benzyl alcohol, linalool, phenylethyl alcohol, trans-cinnamaldehyde, eugenol, alpha-copaene, trans-caryophyllene, cinnamaldehyde-p-methoxy, caryophyllene oxide, benzyl benzoate, cinnamyl acetate, coumarin, alpha-humulene) pass the evaluation test. All the selected ligands have less than 5 hydrogen bond donors, less than 10 hydrogen bond acceptors, a molecular weight of less than 500, and a calculated Log P of less than 5, indicating that these compounds have good absorption and permeation, possess the chemical and physical properties to be orally active drugs, and can proceed to dock. Table 1 shows a list of the pharmacokinetic parameters of the selected ligands.

Docking Analysis

SARS-CoV-2 attaches to host cells via a trimeric spike whose receptor binding domain (RBD) targets angiotensin-converting enzyme 2, causing conformational changes that lead to membrane fusion. RBD is the major immunogenic target for virus neutralization and the present focus of many vaccine development efforts. The intensity of the binding relationship between a single biomolecule and its ligand/binding partner is referred to as binding affinity (e.g., drug or inhibitor). Non-covalent intermolecular interactions such as electrostatic interactions, hydrogen bonding, hydrophobic and Van der Waals forces between the two molecules impact binding affinity. Furthermore, the presence of additional molecules might influence the binding affinity of a ligand and its target molecule. Mahnashi and her colleagues (2021) discovered that 4-methyl-5-oxo-tetrahydrofuran-3-yl acetate and methyl 4-hydroxy-3-methoxybenzoate (natural compounds) had the lowest binding affinities with EGFR and HER2 receptors (target proteins). Both the bioactive constituents had a strong and stable molecular interaction with VEGFR at the binding residues of LYS 868, VAL 916 and ASP 1046. The docking activity was carried out using the molecular operating environment (MOE) software. 4-methyl-5-oxo-tetrahydrofuran-3-yl acetate and methyl 4-hydroxy-3-methoxybenzoate were the bioactive compounds derived from the crude extract of *Polygonum hydropiper* leaves. Furthermore, these two phytocompounds exhibited anti-cancer activity against breast cancer cells (MCF-7), cervical cancer cells (HeLA) and NIH/3T3 fibroblasts cells. The anticancer activity was performed using the MTT assay. IC_{50} was calculated. These plant compounds can further proceed for *in vivo* experiments. In this study, the molecular interaction between RBD (PDB ID: 6YLA) and phyto-component (ligand) was determined via the SwissDock server. The best binding affinity was chosen based on the lowest negative value for each RBD-plant compound-complex because they had the most stable intermolecular interaction between them. 6YLA had the lowest binding energy with the alpha-copaene (PubChem ID: 19725) at the value of -6.6 kcal/mol (Table 2). 6YLA interacted with alpha-copaene at the residues LEU 368, VAL 367, PHE 342 PHE 374, and TRP 436 through Alkyl, Pi-Alkyl, and Pi-sigma bonds (Table 3).

Prediction of Toxicity

Using the admetSAR 2.0 web server, hERG toxicity, AMES toxicity, carcinogenicity, *tetrahymena pyriformis* (TP) toxicity, and honeybee (HB) toxicity, plasma protein binding, fish toxicity, rat acute toxicity (LD_{50} in mol/kg) with hepatotoxicity activity of all compounds were evaluated (Table 4). Based on the result obtained from *in silico* toxicity test of the selected plant compounds using admetSAR 2.0 web server, beta-myrcene, trans-cinnamaldehyde, cinnamaldehyde-p-methoxy, benzyl benzoate and coumarin were discovered to possess properties that may cause harmful effects to humans. Trans-cinnamaldehyde, cinnamaldehyde-p-methoxy and benzyl benzoate were identified to have hepatotoxicity which may cause harm to the liver. However, they could still be used for drug design and by low dosage administration to reduce the harmful effects they may cause on the liver. Beta-myrcene was identified to have carcinogenicity properties and therefore this compound is not suitable for drug design since the compound has the ability and tendency to produce cancer. On the other hand, trans-caryophyllene was identified to possess hERG (human ether-à-go-go-related gene) toxicity. According to Garrido et al., 2020 study, hERG toxicity is one of the most frequent adverse side effects and may cause cardiac side effects to humans. Furthermore, trans-cinnamaldehyde and coumarin were identified to have AMES toxicity which is similar to carcinogens where it may cause cancer.

Conclusion

In this study, alpha-copaene (PubChem ID: 19725) was chosen to use as a safe drug because it did not show any properties that may cause harmful effects to humans based on the toxicity prediction analysis. Moreover, it had the lowest binding affinity with 6YLA protein at the value of -6.6 kcal/mol. The findings of this study should aid pharmaceutical researchers in identifying cinnamon-based medications.

Table 1. List of pharmacokinetics properties, molecular weight (MW), hydrogen bond donor (HBD), hydrogen bond acceptor (HBA), partitioning coefficient (LogP), number of rotatable bond (RB), number of heavy atoms, number of aromatic heavy atoms, synthetic accessibility (SA), gastrointestinal (GI) absorption, and Lipinski's Rule of 5 of all plant compounds

No.	Compound Name	PubChem CID	MW g/m (<500)	HBD (≤5)	HBA (≤10)	RB (≤10)	Log P (≤5)	Log S (ESOL)	SA	GI	LR	Num. heavy atoms	Num. arom. Heavy atoms
1.	Alpha-pinene	6654	136.23 g/mol	0	0	0	2.63	-3.51	4.44 Moderate	Low	Yes; 1 violation: MLOGP>4.15	10	0
2.	Camphene	6616	136.23 g/mol	0	0	0	2.58	-3.34	3.50 Moderate	Low	Yes; 1 violation: MLOGP>4.15	10	0
3.	Benzaldehyde	240	106.12 g/mol	0	1	1	1.36	-1.92	1.00 Easy	High	Yes; 0 violation	8	6
4.	Beta-pinene	14896	136.23 g/mol	0	0	0	2.59	-3.31	3.73 Moderate	Low	Yes; 1 violation: MLOGP>4.15	10	0
5.	Beta-myrcene	31253	136.23 g/mol	0	0	4	2.89	-3.05	2.85 Easy	Low	Yes; 0 violation	10	0
6.	1-Phellandrene	7460	136.23 g/mol	0	0	1	2.64	-2.64	4.15 Moderate	Low	Yes; 0 violation	10	0
7.	Delta-3-careen	26049	136.23 g/mol	0	0	0	2.63	-3.44	3.84 Moderate	Low	Yes; 1 violation: MLOGP>4.15	10	0
8.	*o*-cymene	10703	134.22 g/mol	0	0	1	2.43	-3.81	1.00 Easy	Low	Yes; 1 violation: MLOGP>4.15	10	6
9.	Limonene	22311	136.23 g/mol	0	0	1	2.72	-3.50	3.46 Easy	Low	Yes; 0 violation	10	0
10.	Beta-phellandrene	11142	136.23 g/mol	0	0	1	2.65	-2.79	3.73 Moderate	Low	Yes; 0 violation	10	0

No.	Compound Name	PubChem CID	MW g/m (<500)	HBD (≤5)	HBA (≤10)	RB (≤10)	Log P (≤5)	Log S (ESOL)	SA	GI	LR	Num. heavy atoms	Num. arom. Heavy atoms
11.	1,8-cineole	2758	154.25 g/mol	0	1	0	2.58	-2.52	3.65 Moderate	High	Yes; 0 violation	11	0
12.	Benzyl alcohol	244	108.14 g/mol	1	1	1	1.66	-1.69	1.00 Easy	High	Yes; 0 violation	8	6
13.	Linalool	6549	154.25 g/mol	1	1	4	2.70	-2.40	2.74 Easy	High	Yes; 0 violation	11	0
14.	Phenylethyl alcohol	6054	122.16 g/mol	1	1	2	1.70	-1.82	1.00 Easy	High	Yes; 0 violation	9	6
15.	*trans*-cinnamaldehyde	637511	211.60 g/mol	0	3	3	1.47	-2.66	2.42 Easy	High	Yes; 0 violation	14	6
16.	Eugenol	3314	164.20 g/mol	1	2	3	2.37	-2.46	1.58 Easy	High	Yes; 0 violation	12	6
17.	Alpha-copaene	19725	204.35 g/mol	0	0	1	3.40	-3.86	4.62 Moderate	Low	Yes; 1 violation: MLOGP>4.15	15	0
18.	*trans*-caryophyllene	26318	204.35 g/mol	0	0	0	3.29	-3.87	4.51 Moderate	Low	Yes; 1 violation: MLOGP>4.15	15	0
19.	Cinnamaldehyde-*p*-methoxy	641294	162.19 g/mol	0	2	3	1.96	-2.17	1.61 Easy	High	Yes; 0 violation	12	6
20.	Caryophyllene oxide	1742210	220.35 g/mol	0	1	1	3.15	-3.45	4.35 Moderate	High	Yes; 0 violation	16	0
21.	Benzyl benzoate	2345	212.24 g/mol	0	2	4	2.68	-3.95	1.44 Easy	High	Yes; 0 violation	16	12
22.	Cinnamyl acetate	5282110	176.21 g/mol	0	2	4	2.17	-2.43	1.98 Easy	High	Yes; 0 violation	13	6
23.	Coumarin	323	146.14 g/mol	0	2	0	1.75	-229	2.74 Easy	High	Yes; 0 violation	11	10
24.	Alpha-humulene	6508206	204.35 g/mol	0	0	0	3.27	-3.97	3.66 Moderate	Low	Yes; 1 violation: MLOGP>4.15	15	0

Table 2. Binding affinity score between phytocompound and target protein (6YLA)

No.	Compound Name	PubChem CID	Ligand	Binding Affinity
1.	Alpha-pinene	6654	6yla_6654_uff_E=633.64	-5.3 kcal/mol
2.	Camphene	6616	6yla_6616_uff_E=354.78	-5.1 kcal/mol
3.	Benzaldehyde	240	6yla_240_uff_E=52.88	-4.5 kcal/mol
4.	Beta-pinene	14896	6yla_14896_uff_E=713.31	-5.3 kcal/mol
5.	Beta-myrcene	31253	6yla_31253_uff_E=90.99	-4.0 kcal/mol
6.	1-Phellandrene	7460	6yla_7460_uff_E=154.38	-5.0 kcal/mol
7.	Delta-3-careen	26049	6yla_26049_uff_E=1484.95	-5.4 kcal/mol
8.	*o*-cymene	10703	6yla_10703_uff_E=80.24	-5.2 kcal/mol
9.	Limonene	22311	6yla_22311_uff_E=102.50	-4.7 kcal/mol
10.	beta-phellandrene	11142	6yla_11142_uff_E=124.80	-4.9 kcal/mol
11.	1,8-cineole	2758	6yla_2758_uff_E=344.57	-5.1 kcal/mol
12.	Benzyl alcohol	244	6yla_244_uff_E=68.48	-4.6 kcal/mol
13.	Linalool	6549	6yla_6549_uff_E=118.54	-4.7 kcal/mol
14.	Phenylethyl alcohol	6054	6yla_6054_uff_E=95.74	-4.7 kcal/mol
15.	*trans*-cinnamaldehyde	637511	6yla_637511_uff_E=76.45	-5.1 kcal/mol
16.	Eugenol	3314	6yla_3314_uff_E=169.59	-5.3 kcal/mol
17.	Alpha-copaene	19725	6yla_19725_uff_E=770.21	-6.6 kcal/mol
18.	*trans*-caryophyllene	26318	6yla_26318_uff_E=718.31	-6.1 kcal/mol
19.	Cinnamaldehyde-*p*-methoxy	641294	6yla_641294_uff_E=97.53	-5.0 kcal/mol
20.	Caryophyllene oxide	1742210	6yla_1742210_uff_E=2180.22	-6.0 kcal/mol
21.	Benzyl benzoate	2345	6yla_2345_uff_E=139.93	-5.7 kcal/mol
22.	Cinnamyl acetate	5282110	6yla_5282110_uff_E=95.07	-5.4 kcal/mol
23.	Coumarin	323	6yla_323_uff_E=92.32	-5.4 kcal/mol
24.	Alpha-humulene	6508206	6yla_6508206_uff_E=247.53	-5.9 kcal/mol

Table 3. List of hydrogen bond interactions between target protein (6YLA) and ligands (plant compounds)

No.	Compound Name	PubChem CID	Residues	Bond category	Distance
1.	Alpha-pinene	6654	TYR 473	Pi-Alkyl	5.26
			ARG 457	Alkyl	5.42
				Alkyl	4.62
			LYS 458	Alkyl	3.75
				Alkyl	4.32
				Alkyl	4.48
2.	Camphene	6616	LYS 458	Alkyl	4.48
			PRO 491	Alkyl	4.52
			ARG 457	Alkyl	5.21
				Alkyl	5.03
3.	Benzaldehyde	240	GLY 496	Pi-Donor Hydrogen bond	2.45
			TYR 505	Pi-Pi T-shaped	4.80
4.	Beta-pinene	14896	PHE 456	Van der waals	4.47
				Van der waals	4.12
			LYS 458	Alkyl	4.24
				Alkyl	4.48
			ARG 457	Alkyl	5.47
				Alkyl	4.12
				Alkyl	4.12
			TYR 473	Pi-Alkyl	5.29
5.	Beta-myrcene	31253	TYR 473	Pi-Alkyl	4.92
			PRO 491	Alkyl	4.33
			ARG 457	Alkyl	4.58
			LYS 458	Alkyl	4.08
				Alkyl	4.69
6.	1-Phellandrene	7460	LYS 458	Alkyl	4.18

Table 3. (Continued)

No.	Compound Name	PubChem CID	Residues	Bond category	Distance
7.	Delta-3-careen	26049	LYS 458	Alkyl	4.69
				Alkyl	3.92
8.	*o*-cymene	10703	PRO 426	Pi-Alkyl	4.97
			PHE 464	Pi-Pi T-shaped	4.84
				Pi-sigma	3.84
9.	Limonene	22311	LYS 458	Alkyl	5.04
				Alkyl	4.25
			PRO 491	Alkyl	4.26
			TYR 473	Pi-Alkyl	4.89
10.	Beta-phellandrene	11142	LYS 458	Alkyl	4.24
11.	1,8-cineole	2758	LYS 458	Conventional hydrogen bond	2.18
				Alkyl	4.64
			ARG 457	Alkyl	5.47
12.	Benzyl alcohol	244	LYS 458	Pi-Alkyl	4.25
			ARG 457	Carbon	3.56
13.	Linalool	6549	THR 430	Donor-Donor	1.31
			TYR 396	Pi-Alkyl	5.38
				Pi-Alkyl	4.89
			PHE 464	Pi-Alkyl	4.80
				Pi-Sigma	3.95
14.	Phenylethyl alcohol	6054	GLU 516	Pi-Anion	4.84
			PHE 464	Conventional Hydrogen Bond	2.65
			THR 430	Pi-Pi Stacked	4.16
15.	*trans*-cinnamaldehyde	637511	LYS 458	Pi-Alkyl	4.68
			TYR 473	Pi-Pi T-Shaped	5.54
			ARG 457	Pi-Alkyl	4.86
			SER 459	Conventional Hydrogen Bond	2.14

No.	Compound Name	PubChem CID	Residues	Bond category	Distance
16.	Eugenol	3314	TYR 453	Pi-Alkyl	4.69
			GLY 496	Pi-Donor	2.45
			TYR 505	Pi-Pi T-Shaped	4.84
				Pi-Alkyl	4.59
				Conventional Hydrogen Bond	2.51
17.	Alpha-copaene	19725	LEU 368	Alkyl	5.35
			VAL 367	Alkyl	5.35
			PHE 342	Pi-Alkyl	5.25
				Pi-Alkyl	5.09
			PHE 374	Pi-Alkyl	5.50
				Pi-Alkyl	4.87
			TRP 436	Pi-Sigma	3.82
				Pi-Sigma	3.62
18.	*trans*-caryophyllene	26318	PHE 464	Pi-Alkyl	4.90
19.	Cinnamaldehyde-*p*-methoxy	641294	ASP 364	Conventional Hydrogen Bond	2.20
			PHE 338	Amide Pi-Stacked	3.86
			VAL 367	Pi-Sigma	3.69
			ASN 343	Conventional Hydrogen Bond	2.52
20.	Caryophyllene oxide	1742210	PRO 426	Alkyl	5.03
			PHE 464	Pi-Alkyl	5.03
21.	Benzyl benzoate	2345	TYR 505	Pi-Pi T-shaped	4.80
				Pi-Pi stacked	3.96
			GLY 496	Pi-Donor	2.49
22.	Cinnamyl acetate	5282110	LEU 335	Pi-Alkyl	5.25
			PHE 338	Pi-Sigma	3.94
			PHE 342	Pi-Alkyl	5.34
			VAL 367	Pi-Alkyl	4.20
				Alkyl	482

Table 3. (Continued)

No.	Compound Name	PubChem CID	Residues	Bond category	Distance
23.	Coumarin	323	ARG 457	Pi-Alkyl	4.83
				Pi-Alkyl	4.62
			LYS 458	Pi-Alkyl	4.43
				Pi-Alkyl	4.85
24.	Alpha-humulene	6508206	PHE 464	Pi-Sigma	3.56
				Pi-Alkyl	4.89

Table 4. List of the drug-induced hERG inhibition, AMES toxicity, carcinogens, *Tetrahymena pyriformis* (TP) toxicity, honeybee (HB) toxicity, rat acute toxicity (LD_{50} in mol/kg) along with P-glycoprotein inhibitor (PGI) activity of all compounds

Compound Name	hERG inhibition	AMES	Carcinogenicity	Hepatotoxicity	Honeybee toxicity	Fish toxicity	Plasma protein binding	Rat Acute Toxicity	TP Toxicity	PGI
Alpha-pinene	No	No	No	No	Yes	Yes	0.94%	1.527 kg/mol	0.733 µg/L	No
Camphene	No	No	No	No	Yes	Yes	0.616%	2.088 kg/mol	0.376 µg/L	No
Benzaldehyde	No	No	No	No	Yes	Yes	0.393%	1.81 kg/mol	0.076 µg/L	No
Beta-pinene	No	No	No	No	Yes	Yes	0.751%	1.405 kg/mol	0.733 µg/L	No
beta-myrcene	No	No	Yes	No	Yes	Yes	0.433%	1.66 kg/mol	1.193 µg/L	Yes
1-Phellandrene	No	No	No	No	Yes	Yes	0.858%	0.813 kg/mol	0.168 µg/L	No
delta-3-careen	No	No	No	No	Yes	Yes	0.837%	1.5 kg/mol	0.252 µg/L	No
o-cymene	No	No	No	No	Yes	Yes	1.063%	1.669 kg/mol	0.495 µg/L	No
Limonene	No	No	No	No	Yes	Yes	0.373%	1.856 kg/mol	0.113 µg/L	No
Beta-phellandrene	No	No	No	No	Yes	Yes	0.457%	2.045 kg/mol	-0.3 µg/L	No
1,8-cineole	No	No	No	No	Yes	Yes	0.732%	1.59 kg/mol	0.523 µg/L	No

Compound Name	hERG inhibition	AMES	Carcinogenicity	Hepatotoxicity	Honeybee toxicity	Fish toxicity	Plasma protein binding	Rat Acute Toxicity	TP Toxicity	PGI
Benzyl alcohol	No	No	No	No	Yes	No	0.43%	1.998 kg/mol	-0.892 µg/L	No
Linalool	No	No	No	No	Yes	Yes	0.64%	2.338 kg/mol	0.202 µg/L	No
Phenylethyl alcohol	No	No	No	No	Yes	No	0.436%	1.655 kg/mol	-0.773 µg/L	No
trans-cinnamaldehyde	No	Yes	No	Yes	No	No	1.089%	2.598 kg/mol	1.693 µg/L	No
Eugenol	No	No	No	No	Yes	Yes	0.94%	1.645 kg/mol	1.124 µg/L	No
Alpha-copaene	No	No	No	No	Yes	Yes	1.212%	1.937 kg/mol	0.894 µg/L	No
trans-caryophyllene	Yes	No	No	No	Yes	Yes	0.832%	2.366 kg/mol	1.034 µg/L	No
Cinnamaldehyde-*p*-methoxy	No	No	No	Yes	Yes	Yes	0.973%	1.928 kg/mol	1.248 µg/L	No
Caryophyllene oxide	No	No	No	No	Yes	Yes	0.848%	2.236 kg/mol	1.154 µg/L	No
Benzyl benzoate	No	No	No	Yes	Yes	Yes	0.778%	1423 kg/mol	1.299 µg/L	No
Cinnamyl acetate	No	No	No	No	Yes	Yes	0.646%	1.819 kg/mol	0.341 µg/L	No
Coumarin	No	Yes	No	No	Yes	Yes	0.865%	2.514 kg/mol	2.026 µg/L	No
Alpha-humulene	No	No	No	No	Yes	Yes	0.912%	2.013 kg/mol	0.93 µg/L	No

References

Berman H, Henrick K, Nakamura H. Announcing the worldwide Protein Data Bank. *Nature Structural & Molecular Biology* (2003) 10(12):980.

Brooks BR, Brooks III CL, Mackerell Jr AD, Nilsson L, Petrella RJ, Roux B, Won Y, Archontis G, Bartels C, Boresch S, Caflisch A. CHARMM: The biomolecular simulation program. *Journal of Computational Chemistry* (2009) 30(10):1545-1614.

COVID-19 Coronavirus pandemic. Worldometer. (2022). Accessed 15 March 2022. https://www.worldometers.info/coronavirus.

Elengoe A, Loganthan V. Molecular modeling and docking studies on phyto-compounds against Caspase-3, BRCA1, and Rb. *Biointerface Research in Applied Chemistry* (2021) 12(6):7606-7620.

Elengoe A, Selvam S. Overview on origin, transmission, treatment and clinical researches of SARS-COV-2. *Journal of Health Sciences & Surveillance System* (2022) 10(1):2-12.

Fong SJ, Dey N, Chaki J. An Introduction to COVID 19. In *artificial intelligence for coronavirus outbreak* 2021 (pp. 1 22).

Garrido A, Lepailleur A, Mignani SM, Dallemagne P, Rochais C. hERG Toxicity assessment: Useful guidelines for drug design. *European Journal of Medicinal Chemistry* (2020) 195:112290.

Grosdidier A, Zoete V, Michielin O. Swissdock, A protein-small molecule docking web service based on Eadock DSS. *Nucleic Acids Research* (2011) 39(suppl_2):W270-277.

Jagtap N, Yadav A, Mohite S. Synthesis, Molecular docking studies and anticancer activity of 1, 3, 4-oxadiazole-3 (2h)-thione derivatives. *Journal of University of Shanghai for Science and Technology* (2020) 22(11):535-550.

Kim S, Thiessen PA, Bolton EE, Chen J, Fu G, Gindulyte A, Han L, He J, He S, Shoemaker BA, Wang J. PubChem substance and compound databases. *Nucleic Acids Research (*2016) 44(D1):D1202-1213.

Lim XY, Teh BP, Tan TY. Medicinal plants in COVID-19: Potential and limitations. *Frontiers in Pharmacology* (2021) 12:611408.

Lipinski CA. Lead-and Drug-Like Compounds: The rule-of-five revolution. *Drug Discovery Today: Technologies* (2004) 1(4):337-341.

Mahnashi MH, Alqahtani YS, Alyami BA, Alqarni AO, Ullah F, Wadood A, Sadiq A, Shareef A, Ayaz M. Cytotoxicity, anti-angiogenic, anti-tumor and molecular docking studies on phytochemicals isolated from *Polygonum Hydropiper* L. *BMC Complementary Medicine and Therapies* (2021) 21(1):1-4.

Parmar DR, Soni JY, Guduru R, Rayani RH, Kusurkar RV, Vala AG, Talukdar SN, Eissa IH, Metwaly AM, Khalil A, Zunjar V. Discovery of new anticancer thiourea-azetidine hybrids: Design, synthesis, *in vitro* antiproliferative, SAR, *in silico* molecular docking against VEGFR-2, ADMET, toxicity, and DFT studies. *Bioorganic Chemistry* (2021) 115:105206.

Pathak R, Sharma H. A Review on Medicinal uses of *Cinnamomum verum* (Cinnamon). *Journal of Drug Delivery and Therapeutics* (2021) 11(6-S):161-166.

Singh P, Bast F. Screening of Multi-targeted natural compounds for receptor tyrosine kinases inhibitors and biological evaluation on cancer cell lines, *in silico* and *in vitro*. *Medical Oncology* (2015) 32(9):1-8.

Suhaibun SR, Elengoe A, Poddar R. Technology advance in drug design using computational biology tool. *Malaysian Journal of Medical Health Sciences* (2020) 6(110):2636-9346.

Tian S, Wang J, Li Y, Li D, Xu L, Hou T. The application of *in silico* drug-likeness predictions in pharmaceutical research. *Advanced Drug Delivery Reviews* (2015) 86:2-10.

Veber DF, Johnson SR, Cheng HY, Smith BR, Ward KW, Kopple KD. Molecular properties that influence the oral bioavailability of drug candidates. *Journal of Medicinal Chemistry* (2002) 45(12):2615-2623.

Yu W, MacKerell AD. Computer-aided drug design methods. In *antibiotics* 2017 (pp. 85-106). Humana Press, New York, NY.

Chapter 13

A Molecular Docking Study on Welmee Plant (*Glycyrrhiza glabra.* L.) Compounds against SARS-CoV-2 Spike Proteins

Hitheshi Saluwadana[1], BSc
and Asita Elengoe[2,*], PhD
[1]School of Biotechnology, Manipal International University (MIU), Putra Nilai, Negeri Sembilan, Malaysia
[2]Department of Biotechnology, Faculty of Applied Science, Lincoln University College, Selangor, Malaysia

Abstract

SARS-CoV-2 (severe acute respiratory syndrome coronavirus-2) virus is the main cause of the COVID-19 disease (coronavirus disease 2019) in humans. This mystery virus pandemic has claimed the lives of over 4 million people around the world, and vaccination is presently the only treatment that has been confirmed to be effective against the SARS-CoV-2 virus. As a result, the goal of this research is to find natural therapeutic candidates against SARS-CoV-2 in everyday herbal plants. In order to find phytochemical compounds that are effective against the SARS-CoV-2 spike proteins, *in silico* drug design strategies like molecular docking, pharmacokinetics and toxicity evaluation were used in this study. The molecular interaction between 25 phytochemical compounds from the Welmee plant (*Glycyrrhiza glabra.* L.) and SARS-CoV-2 spike protein (PDB ID: 6M0J) was explored using the SwissDock server. The docking analysis reveals that 6M0J had the lowest binding energy with the alpha-

* Corresponding Author's Email: asitaelengoe@yahoo.com.

In: Medicinal Plants and COVID-19
Editor: Asita Elengoe
ISBN: 979-8-88697-958-9

terpineol (PubChem ID: 17100) at the value of -5.9 kcal/mol. 6M0J had good interaction with the alpha-terpineol. Alpha-terpineol (plant compound) shows no toxicity for humans. Thus, it is suitable to develop as a drug for COVID-19 treatment.

Keywords: docking, *Glycyrrhiza glabra.* L., SARS-CoV-2 spike protein, toxicity

Introduction

Wuhan, a city in China's Hubei province, was the epicenter of the recent outbreak. The first COVID-19 instances were reported in December 2019 (Elengoe, 2020). COVID-19 stands for coronavirus disease in 2019. The letters CO, VI, and D stand for Corona, Virus, and Disease, respectively. SARS-CoV-2 is the virus that causes COVID-19. Coronaviruses acquire their name from the spike proteins that protrude from them, giving them a crown-like appearance (Wu et al., 2020).

Coronaviruses are a huge family that may infect humans as well as a variety of animals, including camels, cattle, cats, and bats. Coronaviruses come in a variety of forms, including some that cause a typical head or chest cold. Other coronavirus diseases, such as severe acute respiratory syndrome (SARS) and Middle East respiratory syndrome (MERS), are far more hazardous than colds and COVID-19, but they are considerably less common (Kumar et al., 2020).

SARS-CoV-2 is transferred from person to person in close quarters. When a person with COVID-19 coughs or exhales, small droplets containing the virus are expelled. These droplets can infect someone who does not have the virus by entering their mouth or nose. Close contact with someone who has the virus is the most likely route for this illness to spread (Baruah et al., 2020). Close touch is defined as being within 6 feet of each other. COVID-19 is still spreading over the globe. Therefore, there is an urgent need for the treatment of COVID-19. Medicinal plants may have the potential to be anti-SARS-CoV-2 agents because plant compounds have been proven to treat a variety of diseases such as diabetes, hypertension, obesity, etc.

Glycyrrhiz glabra belongs to the Fabaceae family. Flavonoids, hydroxybenzoic acids, hydroxycinnamic acids, and hydroxyphenyl propens, as well as other volatile chemicals with antioxidant, antibacterial, antinociceptive, and antiviral properties, are found in the welmee plant

(*Glycyrrhiz glabra*). It has been utilized for therapeutic purposes for millennia such as respiratory problems, hyperdipsia, epilepsy, fever, sexual debility, paralysis, stomach ulcers, rheumatism, skin illnesses, hemorrhagic diseases and jaundice (Sharma et al., 2017).

Bioinformatics AKA *in-silico* screening, docking, and molecular dynamics simulation research can be viewed as a significant benefit that aids in the identification of novel and natural treatments against the SARS-CoV-2 virus, allowing for simple drug design. Mendie and Hemalatha (2022) explored the molecular interaction between fifty plant compounds from and EGFRs (HER-2 and HER-3), VEGFR (VEGFR-2 and VEGFR-3), and FGFR (FGFR-2) using Autodock version 4.2.6. They identified that Withaferin A had the lowest binding energy (-9.5, -11.59, -9.53, -7.27 and -8.63 kcal/mol) with HER2, HER3, VEGFR2, VEGFR3 and FGFR respectively. Withaferin A had one, three, two, three and three hydrogen bonds with HER-2, HER-3, VEGFR-2, VEGFR-3 and FGFR-2 respectively.

In this research, the molecular interaction between the SARS-CoV-2 spike protein (PDB ID: 6M0J) and the twenty-five phytochemicals from the welmee plant was determined using docking, pharmacokinetics and toxicity evaluation analysis.

Methodology

Plant Selection

Glycyrrhiz glabra plant was selected due to its high concentration of phytochemical compounds and related medicinal uses such as anti-inflammation, anti-viral, anti-bacterial, etc.

Phytochemical Preparation

Phytoconstituents were used as ligands. Twenty-one plant constituents (liquiritin, shinpterocarpin, glisoflavone, coumarin, shinflavanone, glucoliquiritin apioside, licoarylcoumarin, licopyranocoumarin, glycyrrhizin, isoangustone A, semilicoisoflavone B, licoriphenone, 1-methoxyficifolinol, kanzonol R, 1-pentanol, 2,3,5,6-tetramethylpyrazine, 1-hexanol, terpinen-4-ol, linalool, oxide, geraniol, alpha-terpineol) were selected and the three-dimensional (3-D) conformers of the chosen phytocompounds were downloaded in SDF format from the PubChem database (Kim et al., 2016; Elengoe and Selvam, 2022).

Preparation of Ligand

The retrieved 3-D models of the 21 plant compounds were prepared using the DS 4.0 'Prepare ligand' technique (Brooks et al., 2009). The obtained ligands are screened using Lipinski's Rule of Five, which establishes a set of defined requirements that a ligand must meet in order to be appropriate for drug development. It develops drug-like quality parameters and concentrates on medicine bioavailability (Lipinski, 2004; Veber et al., 2002; Jagtap et al., 2020). These ligands are screened to prepare for molecular docking with the SARS-CoV-2 spike protein (PDB ID: 6M0J).

Protein Preparation

The SARS-CoV-2 spike (S) protein is made up of two subunits, S1 and S2, and is essential for receptor identification and cell membrane fusion. The host receptor angiotensin-converting enzyme 2 is recognized and bound by the receptor-binding domain of the S1 subunit, while the two-heptad repeat domain of the S2 subunit facilitates the fusing of the viral cell membrane. The SARS-CoV-2 spike (S) protein is the key target in the development of an antiviral drug for COVID-19 disease. The 3-D conformer of SARS-CoV-2 spike protein (PDB ID: 6M0J) in PDB format was downloaded from The RCSB Protein Data Bank (Berman et al., 2003; Suhaibun et al., 2020). It was publicly available.

Molecular Docking

The docking analysis was performed between the target protein (PDB ID: 6M0J) with its relevant phytoconstituent using PyRx software (Dallakyan and Olson, 2015). The 3-D model of the SARS-CoV-2 spike protein and phytocompound-complex was viewed using DS 4.0. The binding affinity (kcal/mol) was determined (Parmar et al., 2021).

Prediction of Pharmacokinetics of Plant Compounds

The phytochemical compounds' pharmacokinetic properties were determined using SwissADME, a free web application provided by the Swiss Institute of Bioinformatics.

Prediction of Toxicity

AdmetSAR, a free web application, was used to determine the toxicity of phytochemical compounds (Tian et al., 2015).

Results and Discussion

COVID-19 is an infectious disease that causes mild to life-threatening respiratory illness. The SARS-CoV-2 virus is a single-stranded RNA virus with structural and non-structural protein gene components wrapped in a single-stranded RNA envelope. Non-structural proteins contribute to replication and release, while structural proteins encode the spike, envelope, membrane, and nucleocapsid proteins that enable viral attachment and entrance (Khailany et al., 2020). The spike protein of the SARS-CoV-2 virus has been identified as a first-line therapeutic target for drug and vaccine development. This is because the spike (S) protein is involved in receptor recognition and cell membrane fusion, both of which are necessary for viral entry, and the virus must be attached to the host cell to replicate and infect. Creating or developing drugs that target the SARS-CoV-2 spike (S) protein is the most straightforward technique.

The plant compounds that meet Lipinski's rule of five [(hydrogen bond donor (HBD) ≤ 5), hydrogen bond acceptor (HBA ≤ 10), rotatable bonds (RB ≤ 10), molecular weight (MW ≤ 500 daltons) and (LogP ≤ 5)] are thought to be readily absorbable into the human body, with good gastrointestinal absorption, facile binding, and permeability. Moreover, it is thought to be compatible with the development of orally active drugs. In this study, only 19 phytochemical compounds (liquiritin, shinpterocarpin, glisoflavone, coumarin, shinflavanone, licoarylcoumarin, licopyranocoumarin, isoangustone A, semilicoisoflavone B, licoriphenone, 1-methoxyficifolinol, kanzonol R, 1-pentanol, 2,3,5,6-tetramethylpyrazine, terpinen-4-ol, linalool oxide, geraniol, alpha-terpineol) were met the Lipinski's rule of five and proceeded for docking analysis. Therefore, these plant compounds could be a potential and effective drug to be designed. Table 1 shows a list of the pharmacokinetic parameters of the selected plant compounds.

Table 1. List of pharmacokinetics properties, molecular weight (MW), hydrogen bond donor (HBD), hydrogen bond acceptor (HBA), partitioning coefficient (LogP), number of rotatable bond (RB), synthetic accessibility (SA), gastrointestinal (GI) absorption, and Lipinski's Rule of 5 of all plant compounds

Compound name	PubChem CID	MW (g/mol)	HBD	HBA	RB	logP	logS	SA	GI	LR
Liquiritin	503737	418.39	5	9	4	2.22	S	4.91	L	Yes
Shinpterocarpin	10336244	322.35	1	4	0	3.09	MS	4.26	H	Yes
Glisoflavone	5487298	368.38	3	6	4	2.99	MS	3.64	H	Yes
Coumarin	323	146.14	0	2	0	1.75	S	2.74	H	Yes
Shinflavanone	197678	390.47	1	4	3	4.17	MS	4.49	H	Yes
Glucoliquiritin apioside	74819335	712.65	10	18	10	3.12	S	6.99	H	NO
Licoarylcoumarin	10090416	368.38	3	6	4	3.04	MS	3.67	H	Yes
Licopyranocoumarin	122851	384.38	3	7	3	2.61	MS	4.10	H	Yes
Glycyrrhizin	3495	822.93	8	16	7	2.15	PS	8.84	L	NO
Isoangustone A	21591148	422.47	4	6	5	3.85	PS	4.02	L	Yes
Semilicoisoflavone B	5481948	352.34	3	6	1	3.05	MS	3.79	H	Yes
Licoriphenone	21591149	372.41	3	6	7	2.44	MS	3.12	H	Yes
1-Methoxyficifolinol	480872	422.51	2	5	5	4.20	PS	4.72	H	Yes
Kanzonol R	131753027	370.44	2	5	5	3.74	MS	4.01	H	Yes
1-Pentanol	6276	88.15	1	1	3	1.82	S	1.00	H	Yes
2,3,5,6-Tetramethylpyrazine	14296	136.19	0	2	0	1.99	S	1.83	H	Yes
1-Hexanol	8103	102.17	1	1	4	2.03	S	1.07	H	Yes
Terpinen-4-ol	11230	154.25	1	1	1	2.58	S	3.28	H	Yes
Linalool, oxide	102611	170.25	1	2	4	2.56	S	3.21	H	Yes
Geraniol	637566	154.25	1	1	4	2.75	S	2.85	H	Yes
alpha-Terpineol	17100	154.25	1	1	1	2.51	S	3.24	H	Yes

*S = Soluble, MS = Moderate soluble, PS = Poorly soluble.

Molecular docking investigation of SARS-CoV-2 spike protein and phytochemical components of the welmee plant yielded 18 phytochemical molecules that met the criteria for medication preparation. The stability of the chemicals varied depending on the spike protein, but overall, all phytochemical compounds had the lowest binding affinity (Table 2).

Table 2. Binding affinity between phytocompound and target protein (6M0J)

No.	Compound name	PubChem CID	Ligand	Binding affinity
1.	Liquiritin	503737	6m0j_503737_uff_E=412.91	-8.6
2.	Shinpterocarpin	10336244	6m0j_10336244_uff_E=375.70	-8.4
3.	Glisoflavone	5487298	6m0j_5487298_uff_E=492.53	-8
4.	Coumarin	323	6m0j_323_uff_E=92.32	-7.1
5.	Shinflavanone	197678	6m0j_197678_uff_E=339.20	-10.2
6.	Licoarylcoumarin	10090416	6m0j_10090416_uff_E=511.76	-7.5
7.	Licopyranocoumarin	122851	6m0j_122851_uff_E=517.22	-7.4
8.	Isoangustone A	21591148	6m0j_21591148_uff_E=450.13	-8.4
9.	Semilicoisoflavone B	5481948	6m0j_5481948_uff_E=381.20	-9
10.	Licoriphenone	21591149	6m0j_21591149_uff_E=351.40	-6.4
11.	1-Methoxyficifolinol	480872	6m0j_480872_uff_E=556.75	-8.5
12.	Kanzonol R	131753027	6m0j_131753027_uff_E=386.36	-7.5
13.	1-Pentanol	6276	6m0j_6276_uff_E=29.98	-3.8
14.	2,3,5,6-Tetramethylpyrazine	14296	6m0j_14296_uff_E=63.51	-5.6
15.	Terpinen-4-ol	11230	6m0j_11230_uff_E=155.64	-5.6
16.	Linalool, oxide	102611	6m0j_102611_uff_E=1523.41	-5.9
17.	Geraniol	637566	6m0j_637566_uff_E=110.10	-5.8
18.	alpha-Terpineol	17100	6m0j_17100_uff_E=147.80	-5.9

Apart from ADMET characteristics, phytochemicals were categorized based on their predicted toxicity. *In silico* toxicity study of phytochemical compounds saves time and money while also eliminating the necessity for animal testing. Toxicity testing of phytochemical substances for hepatotoxicity, carcinogenicity, AMES toxicity, and hERG inhibition is primarily focused on developing medications that are as safe as possible for humans. 18 of the 21 welmee phytochemical compounds with ADMET property predictions can be employed as SARS-CoV-2 spike protein inhibitors.

Based on the toxicity analysis, 1-pentanol, 2,3,5,6-tetramethylpyrazine, 1-hexanol, terpinen-4-ol, geraniol, alpha-terpineol show no human toxicity, indicating that they can be easily employed to make anti-SARS-CoV-2 medicines (Table 3). ADMET profiling of phytochemical substances from plants can provide a solid foundation for pharmaceutical industry researchers and scientists to develop new plant-based medications.

Table 3. List of the drug-induced hERG inhibition, AMES toxicity, carcinogens, *Tetrahymena pyriformis* (TP) toxicity, honeybee (HB) toxicity, rat acute toxicity (LD_{50} in mol/kg) along with P-glycoprotein inhibitor (PGI) activity of all natural compounds

Compound name	hERG inhibition	AMES	Carcino-genicity	Fish toxicity	Honeybee toxicity	Hepato-toxicity	PGI	Rat acute toxicity (kg/mol)	TP toxiciy (ug/L)
Liquiritin	No	No	No	Yes	Yes	Yes	No	2.668	1.016
Shinpterocarpin	No	No	No	Yes	Yes	Yes	No	2.875	1.747
Glisoflavone	No	No	No	Yes	Yes	Yes	No	2.304	0.642
Coumarin	No	Yes	No	Yes	Yes	No	No	2.514	2.026
Shinflavanone	Yes	No	No	Yes	Yes	Yes	Yes	2.907	1.61
Glucoliquiritin apioside	Yes	Yes	No	Yes	Yes	Yes	Yes	3.588	0.98
Licoarylcoumarin	No	No	No	Yes	Yes	Yes	No	2.965	1.591
Licopyranocoumarin	Yes	No	No	Yes	Yes	Yes	No	3.103	1.062
Isoangustone A	No	No	No	Yes	Yes	Yes	Yes	2.295	1.08
Semilicoisoflavone B	No	No	No	Yes	Yes	Yes	No	2.691	1.306
Licoriphenone	No	No	No	Yes	Yes	Yes	No	1.577	0.823
1-Methoxyficifolinol	No	Yes	No	Yes	Yes	Yes	Yes	2.077	0.757
Kanzonol R	No	No	No	Yes	Yes	Yes	Yes	2.244	1.366
1-Pentanol	No	No	No	No	Yes	No	No	2.176	-0.661
2,3,5,6-Tetramethylpyrazine	No	No	No	No	Yes	No	No	0.712	1.349
1-Hexanol	No	No	No	Yes	Yes	No	No	2.235	-0.262
Terpinen-4-ol	No	No	No	Yes	Yes	No	No	2.047	-0.092
Linalool oxide	No	Yes	No	Yes	Yes	No	No	2.04	0.053
Geraniol	No	No	No	Yes	Yes	No	No	2.109	0.376
alpha-terpineol	No	No	No	Yes	Yes	No	No	2.137	-0.468

Conclusion

In this study, alpha-terpeniol (PubChem ID: 17100) was chosen to use as a safe drug because it lacked any characteristics that might have adverse effects on people. Additionally, it had a binding affinity of -5.9 kcal/mol for the 6M0J protein, which was the lowest. The results of this investigation should help pharmaceutical researchers develop welmee-based drugs.

References

Baruah C, Devi P, Sharma DK. Sequence analysis and structure prediction of SARS-CoV-2 accessory proteins 9b and ORF14: evolutionary analysis indicates close relatedness to bat coronavirus. *BioMed Research International* (2020) 2020.

Berman H, Henrick K, Nakamura H. Announcing the worldwide Protein Data Bank. *Nature Structural & Molecular Biology* (2003) 10(12):980.

Brooks BR, Brooks III CL, Mackerell Jr AD, Nilsson L, Petrella RJ, Roux B, Won Y, Archontis G, Bartels C, Boresch S, Caflisch A. CHARMM: The biomolecular simulation program. *Journal of Computational Chemistry* (2009) 30(10):1545-1614.

Dallakyan S, Olson AJ. Small-molecule library screening by docking with PyRx. *Methods in Molecular Biology* (2015).

Elengoe A. Indian spices boost the immune system against COVID-19. *The Annals of the University Dunarea de Jos of Galati. Fascicle VI-Food Technology* (2020) 44(2):189-206.

Elengoe A, Selvam S. Overview on origin, transmission, treatment and clinical researches of SARS-COV-2. *Journal of Health Sciences & Surveillance System* (2022) 10(1):2-12.

Jagtap N, Yadav A, Mohite S. Synthesis, molecular docking studies and anticancer activity of 1, 3, 4-oxadiazole-3 (2h)-thione derivatives. *Journal of University of Shanghai for Science and Technology* (2020) 22(11):535-550.

Khailany RA, Safdar M, Ozaslan M. Genomic characterization of a novel SARS-CoV-2. *Gene reports* (2020) 19:100682.

Kim S, Thiessen PA, Bolton EE, Chen J, Fu G, Gindulyte A, Han L, He J, He S, Shoemaker BA, Wang J. PubChem substance and compound databases. *Nucleic Acids Research* (2016) 44(D1):D1202-1213.

Kumar S, Nyodu R, Maurya VK, Saxena SK. Morphology, genome organization, replication, and pathogenesis of severe acute respiratory syndrome coronavirus 2 (SARS-CoV-2). *Coronavirus Disease 2019 (COVID-19)* (2020) 23-31.

Kumar S, Nyodu R, Maurya VK, Saxena SK. Morphology, genome organization, replication, and pathogenesis of severe acute respiratory syndrome coronavirus 2 (SARS-CoV-2). In *Coronavirus disease 2019 (COVID-19)* 2020 (pp. 23-31). Springer, Singapore.

Lipinski CA. Lead-and Drug-Like Compounds: The rule-of-five revolution. *Drug Discovery Today: Technologies* (2004) 1(4):337-341.

Mendie LE, Hemalatha S. Molecular docking of phytochemicals targeting GFRs as therapeutic sites for cancer: An *In silico* study. *Applied Biochemistry Biotechnology* (2022) 194(1):215-231.

Parmar DR, Soni JY, Guduru R, Rayani RH, Kusurkar RV, Vala AG, Talukdar SN, Eissa IH, Metwaly AM, Khalil A, Zunjar V. Discovery of new anticancer thiourea-azetidine hybrids: Design, synthesis, *in vitro* antiproliferative, SAR, *in silico* molecular docking against VEGFR-2, ADMET, toxicity, and DFT studies. *Bioorganic Chemistry* (2021) 115:105206.

Sharma V, Katiyar A, Agrawal RC. *Glycyrrhiza glabra*: Chemistry and pharmacological activity. *Sweeteners* (2017):87-100.

Suhaibun SR, Elengoe A, Poddar R. Technology advance in drug design using computational biology tool. *Malaysian Journal of Medicine and Health Science* (2020) 16(110):2636-9346.

Tian S, Wang J, Li Y, Li D, Xu L, Hou T. The application of *in silico* drug-likeness predictions in pharmaceutical research. *Advanced Drug Delivery Reviews* (2015) 86:2-10.

Veber DF, Johnson SR, Cheng HY, Smith BR, Ward KW, Kopple KD. Molecular properties that influence the oral bioavailability of drug candidates. *Journal of Medicinal Chemistry* (2002) 45(12):2615-2623.

Wu F, Zhao S, Yu B, Chen YM, Wang W, Song ZG, Hu Y, Tao ZW, Tian JH, Pei YY, Yuan ML, Zhang YL, Dai FH, Liu Y, Wang QM, Zheng JJ, Xu L, Holmes EC, Zhang YZ. A new coronavirus associated with human respiratory disease in China. *Nature* (2020) 579(7798):265-269.

Chapter 14

An *In Silico* Study-Mechanism Action of Tasmanian Blue Gum Plant Compounds against SARS-CoV-2

Asita Elengoe[1,*], PhD
and Matthew Aaron Lopez[2], BSc
[1]Department of Biotechnology, Faculty of Applied Science, Lincoln University College, Selangor, Malaysia
[2]School of Biotechnology, Manipal International University (MIU), Negeri Sembilan, Malaysia

Abstract

Severe acute respiratory syndrome coronavirus-2 (SARS-CoV-2) results from the infection of animals, namely bats, to humans. This disease has affected many in different ways, from social approaches to personal well-being. Vaccines have been developed to counteract the severity of this disease however the approach developed is to minimize the severity but not to completely eradicate the virus. Plant compounds have been proven to treat a variety of chronic diseases. In this study, molecular docking was performed to investigate the binding mechanism between the SARS-CoV-2 main protease protein (M^{pro}), (PDB ID: 6M2N) and seven phytoconstituents (eucalyptol, α-terpineol acetate, alloaroma dendrene, sabinene, limonene, isoledene and aromadendrene) from the Tasmanian Blue Gum plant using SwissDock server. ADME prediction server was used to determine the pharmacokinetics of the seven plant compounds. Moreover, the evaluation of the toxicity of natural compounds was carried out using the admetSAR online database. The docking analysis

* Corresponding Author's Email: asitaelengoe@yahoo.com.

In: Medicinal Plants and COVID-19
Editor: Asita Elengoe
ISBN: 979-8-88697-958-9

revealed that eucalyptol has displayed good binding affinity (-5.4 kcal/mol) in comparison to the other six plant compounds. Eucalyptol exhibited an acceptable drug-like property with no carcinogenicity or mutagenicity. Thus, this potential drug candidate can further be validated through lab experiments for its proper function without causing any harmful effects in the human body.

Keywords: docking, pharmacokinetics, SARS-CoV-2 main protein protease (Mpro), Tasmanian Blue Gum, toxicity

Introduction

COVID-19 is a respiratory disease. It was first detected in December 2019 and originated in Wuhan, China. This disease was named severe acute respiratory syndrome coronavirus-2 (SARS-CoV-2) due to its 80% similarity with SARS-CoV given its high mortality rate of the outbreak between 2002-2003 (Yuki et al., 2020). Currently, there are several variants identified from the mutation of the COVID-19 strain. These variants are namely P.1, N501Y, 501Y.V2 (Mahase, 2021) and B.1.617.2 (delta) variant (Stowe et al., 2021). Among these variants, the delta variant poses a serious threat such as a fatal acute respiratory system which causes an increasing number of cases in India (Stowe et al., 2021) and P.1 originated from Brazil which has less concern whereas N501Yoriginated from the United Kingdom with increased virulence while 501Y.V2 is a variant from South Africa linked to greater viral load thus worst fatality (Mahase, 2021). These variants differ from the original Wuhan variant which influences the function of spike glycoprotein as the alteration of the spike glycoproteins affects the interaction with the human angiotensin-converting enzyme 2 (hACE2) receptor (Darby and Hiscox, 2021).

This outbreak started in the Wuhan seafood market in December 2019 after early cases were discovered there. Cases then spread from this market to other locations in China, particularly Hubei province, where they have infected many people quickly through human-to-human contact (Hua and Shaw, 2020). For this virus to spread throughout entire China, it took just 30 days for the Chinese population to be infected and this virus was detected in as many as 184 countries in the following months (Hua and Shaw, 2020). Malaysia recorded the first wave of this pandemic from 25th January 2020 until 16th February 2020 when 22 cases were reported and mostly these were import cases (Ganasegeran et al., 2020). Despite the cases, Malaysia continues to record a high number of daily cases and some measures have been taken to overcome the pandemic outcome. Thus, there is an urgency to develop vaccines and drugs for the treatment of COVID-19 disease.

Medicinal plants have been proven to treat a variety of diseases and infections such as diabetes mellitus, heart diseases, kidney-related diseases, cancer, inflammation, hypertension, neurological diseases, etc. Therefore, plant compounds may be the potential to be the medicine for COVID-19 disease. *Eucalyptus globulus* belongs to the Myrtle family. Tasmanian Blue Gum is the common name for *Eucalyptus globulus*. Aboriginals have used eucalyptus leaves as a traditional herbal treatment. The leaves contain an essential oil that is a potent antiseptic and is used all over the world to treat sore throats, colds, coughs, and other illnesses. Many over-the-counter cold treatments contain essential oil as a common ingredient. The adult leaves are antiperiodic, antiseptic, fragrant, deodorizing, expectorant, febrifuge, hypoglycemic, and stimulating even without their petioles. The leaves and the essential oil they contain are stimulant, expectorant, febrifuge, antiseptic, and antispasmodic. The leaf extracts have antimicrobial properties. Because ozone is created in it when exposed to air, the essential oil extracted from different eucalyptus species has a very strong antibacterial effect, especially when it has been around for a while. It has a clear disinfecting effect and kills all lesser forms of life. The oil can be administered topically to cuts and other skin infections, breathed to cure blocked nasal passages, gargled for sore throats, and consumed internally for a variety of problems (Hayat et al., 2015). However, some caution is suggested because, like all essential oils, it can be harmful to the body in higher dosages. This species' oil has a slightly unpleasant smell, hence other members of the genus are now more commonly employed for therapeutic purposes as a result. The tree releases an oleo-resin.

Computational biology analysis helps to identify the drug target; screen the drug candidate; refinement of drug candidate; toxicity of the drug candidate; prediction of drug resistance, etc. Lenin and his colleagues (2022) studied the interaction between phytocompounds and breast cancer protein (estrogen receptor alpha) using FlexX molecular docking approach. They determined the binding affinity and orientation of phytoconstituents in the target protein active site. They found that daidzein had the best binding energy (-23.89 kcal/mol) when compared with the other nine plant compounds (emodin, delphinidin, cyanidin, epigenin, cordycepin, ascorbigen, biochanin A, chrysin, demethoxy curcumin) and cytarabine (standard drug). They concluded that daidzein may potential to be a ligand for estrogen receptor alpha.

In this study, the molecular interaction between SARS-CoV-2 main protease protein, M^{pro} (PDB ID: 6M2N) and the seven phytoconstituents from the Tasmanian Blue Gum plant was determined using docking, pharmacokinetics and toxicity evaluation analysis.

Methodology

Plant Selection

The Tasmanian Blue Gum (*Eucalyptus globulus*) plant was selected due to its high concentration of natural components and related therapeutic properties such as asthmatic, decongestant, soothing, expectorant, anticatarrhal, anti-inflammation of the respiratory and digestive tract.

Phytochemical Preparation

PubChem database was used to search the bioactive compounds from the Tasmanian blue gum plant. The three-dimensional (3-D) structures of the seven chosen phytoconstituents were obtained in SDF format (Kim et al., 2016; Elengoe and Selvam, 2022).

Preparation of Ligand

The retrieved 7 phytocompounds (eucalyptol, α-pinene, α-terpineol acetate, alloaroma dendrene, sabinene, limonene, isoledene, aroma dendrene) were prepared using the Discovery Studio (DS) 4.0 'Prepare 'ligand' technique (Brooks et al., 2009). The obtained ligands are screened using Lipinski's Rule of Five, which establishes a set of defined requirements that a ligand must meet in order to be appropriate for drug development. It develops drug-like quality parameters and concentrates on medicine bioavailability (Lipinski, 2004; Veber et al., 2002; Jagtap et al., 2020). These ligands are screened to prepare for molecular docking with the SARS-CoV-2 main protease protein, M^{pro} (PDB ID: 6M2N).

Protein Preparation

The main protease (M^{pro}), also known as 3-chymotrypsin-like protease ($3CL^{pro}$), is a group of cysteine hydrolases found in coronaviruses that are extremely conserved. This protein is important for viral replication and is therefore an important target for the prevention and treatment of infectious illnesses brought on by coronaviruses, such as COVID-19. The 3-D model of the target protein (PDB ID: 6M2N) was obtained from RCSB Protein Data Bank (PDB) and downloaded in PDB format (Berman et al., 2003; Elengoe et al., 2015). It was publicly available.

Molecular Docking

The docking of the M^{pro} (PDB ID: 6M2N) with its relevant plant compound was performed using SwissDock (Grosdidier et al., 2011; Supramaniam and Elengoe, 2020). The 3D model of the M^{pro}-plant compound complex was visualized using DS 4.0. The binding energy (kcal/mol), binding residues, binding distance, bond category and bond type between the M^{pro} and plant constitute were analyzed (Parmar et al., 2021).

Prediction of Pharmacokinetics of Plant Compounds

The phytochemical compounds' pharmacokinetic properties were determined using SwissADME, a free web application provided by the Swiss Institute of Bioinformatics.

Prediction of Toxicity

The toxicity of the chosen phytocomponents was assessed using the web-based SwissADME server (Tian et al., 2015).

Results and Discussion

COVID-19 is a respiratory disease that belongs to a family of single-stranded RNA. It is transmissible from human to human (Aw et al., 2021). There are a few types of variants that arise from the mutation of this virus namely alpha, beta, gamma and delta with the different severity levels imposed by these strains towards individuals (Roy et al., 2021). According to Dousari et al., (2020), the structure of the virus is necessary to enable the development of disease in a person which causes the person to develop different levels of disease severity depending on the person's immune system.

Based on Benarba and Pandiella (2020), the phytochemicals in medicinal plants act as active molecules that are effective in the fight against COVID-19 disease. Hasan and his colleagues (2022) carried out a molecular docking analysis to explore the interaction between quercetin (natural compound) and mycolic acid cyclopropane synthase (PDB ID: 1KPI), plasmepsin II (PDB ID: 1SME), cyclooxygenase-1 (PDB ID: 2OYE), human estrogen receptor alpha (PDB ID: 3ERT), fat mass and obesity-associated (FTO) protein (PDB ID: 3LFM), BACE1 (PDB ID: 4IVT). The results have demonstrated that quercetin had the lower binding energy with 1KPI (-9.00), 1SME (- 6.36), 2OYE (-8.53), 3ERT (-7.28), 3LFM (-7.89) and 4IVT (-6.68) using SP glide of Schrodinger maestro (v11.1). They concluded that Quercetin may potent to be anti-tuberculosis, anti-malarial, anti-inflammatory, anti-cancer, anti-obesity and anti-Alzheimer's drug.

In this study, computational biology tools such as molecular docking, pharmacokinetics and toxicity approaches were used to study the interaction between the M^{pro} and seven selected natural compounds. The absorption, distribution, metabolism, and excretion of phytochemical compounds are referred to as pharmacokinetics. The logarithm of the inorganic and aqueous phase partition coefficient of the target molecule is referred to as the partition coefficient or LogP, which expresses lipophilicity. It has an impact on molecular pharmacokinetics and absorption. The number of hydrogen bond donors and acceptors represents the potential of a phytochemical molecule to traverse the lipid bilayer of a cell membrane, while the logarithm of the solubility, or LogS, indicates the solubility of the selected phytochemical compounds. The oral bioavailability of phytochemical substances is linked to rotatable bonds, and only by taking these factors into account can a safe and effective medicine be designed. Table 1 shows the pharmacokinetics of seven selected plant compounds.

Table 1. List of pharmacokinetics properties, molecular weight (MW), hydrogen bond donor (HBD), hydrogen bond acceptor (HBA), partitioning coefficient (LogP), number of rotatable bond (RB), number of heavy atoms (HA), number of aromatic heavy atoms (AHA), synthetic accessibility (SA), gastrointestinal (GI) absorption, and Lipinski's Rule of 5 of all plant compounds

Compound name	PubChem CID	MW g/mol (< 500)	HBD (≤ 5)	HBA (≤ 10)	RB (≤ 10)	HA	AHA	logP (≤ 5)	Log S (ESOL)	SA	GI	LR
Eucalyptol	2758	154.25	0	1	0	11	0	2.58	S	3.65	H	Yes
α-Terpineol Acetate	91752241	196.29	0	2	3	14	0	2.99	S	3.26	H	Yes
Alloaroma Dendrene	42608158	204.35	0	0	0	15	0	3.26	MS	3.70	L	Yes
Sabinene	18818	136.23	0	0	1	10	0	2.65	S	2.87	L	Yes
Limonene	22311	136.23	0	0	1	10	0	2.72	MS	3.46	L	Yes
Isoledene	530426	204.35	0	0	0	15	0	3.26	S	4.80	L	Yes
Aroma dendrene	91354	204.35	0	0	0	15	0	3.26	MS	3.70	L	Yes

* S=Soluble, MS=Moderate Soluble, H=High, L=Low.

Table 2. Binding affinity score between phytocompound and target protein (6M2N)

No.	Compound name	Pubchem ID	Ligand	Binding Affinity
1.	Eucalyptol	2758	6m2n_2758_uff_E=344.81	-5.4 kcal/mol
2.	α-Terpineol Acetate	91752241	6m2n_91752241_uff_E=161.28	-5.7 kcal/mol
3.	Alloaroma Dendrene	42608158	6m2n_42608158_uff_E=1770.46	-6.5 kcal/mol
4.	Sabinene	18818	6m2n_18818_uff_E=1440.19	-5.0 kcal/mol
5.	Limonene	22311	6m2n_22311_uff_E=102.89	-5.2 kcal/mol
6.	Isoledene	530426	6m2n_530426_uff_E=1658.87	-6.3 kcal/mol
7.	Aroma dendrene	91354	6m2n_91354_uff_E=1697.68	-6.4 kcal/mol

Table 3. List of hydrogen bond interactions between target protein (6M2N) and ligands (plant compounds)

No.	Compound Name	PubChem CID	Residues	Bond category	Bond type	Distances
1.	Eucalyptol	2758	ILE 249	Pi-Alkyl	Hydrophobic	5.46
			VAL 202	Pi-Alkyl	Hydrophobic	4.64
			PRO 293	Pi-Alkyl	Hydrophobic	5.28
			HIS 246	Pi-Alkyl	Hydrophobic	4.77
			GLN 110	Hydrogen	Hydrogen conventional	2.44
			PHE 294	Pi-Sigma	Hydrophobic	3.87
2.	α-Terpineol Acetate	9152241	SER 46	Carbon-Hydrogen	Hydrogen	4.97
3.	Alloaroma Dendrene	42608158	ARG 294	Pi-Alkyl	Hydrophobic	4.43
			PHE 298	Pi-Alkyl	Hydrophobic	4.80
4.	Sabinene	18818	MET 49	Pi-Alkyl	Hydrophobic	3.61
			CYS 44	Pi-Alkyl	Hydrophobic	4.14
5.	Limonene	22311	PRO 108	Pi-Alkyl	Hydrophobic	4.17
			PRO 132	Pi-Alkyl	Hydrophobic	4.64
				Pi-Alkyl	Hydrophobic	5.11
			ILE 200	Pi-Alkyl	Hydrophobic	4.65
			VAL 202	Pi-Alkyl	Hydrophobic	4.72
			HIS 246	Pi-Alkyl	Hydrophobic	4.37
6.	Isoledene	530426	LYS 458	Pi-Alkyl	Hydrophobic	4.34
7.	Aroma dendrene	91354	LYS 458	Pi-Alkyl	Hydrophobic	4.94
				Pi-Alkyl	Hydrophobic	4.52

Table 4. List of the drug-induced hERG inhibition, AMES toxicity, carcinogens, *Tetrahymena pyriformis* (TP) toxicity, fish toxicity, honeybee (HB) toxicity, hepatotoxicity, rat acute toxicity (LD_{50} in mol/kg), plasma protein binding along with P-glycoprotein inhibitor (PGI) activity of all bioactive compounds

Compound Name	hERG inhibition	AMES	Carcinogenicity	Fish toxicity	TP toxicity	HB toxicity	PGI	RAT (LD_{50})	Hepatotoxicity	Plasma protein binding/ %
Eucalyptol	No	No	No	Yes	0.652	Yes	No	0.8147	No	0.716
Alloaroma Dendrene	Yes	No	No	Yes	0.665	Yes	No	0.8388	No	0.914
Sabinene	No	No	No	Yes	1.032	Yes	No	0.7601	No	0.555
Limonene	No	No	No	Yes	0.113	Yes	No	0.9069	No	0.373
Isoledene	No	No	No	Yes	0.773	Yes	No	0.8165	No	0.92
Aroma dendrene	Yes	No	No	Yes	0.665	Yes	No	0.8388	No	0.914
α-Terpineol Acetate	No	No	No	Yes	0.535	Yes	No	0.8303	No	0.87

Eucalyptol and α-terpineol acetate display the highest gastrointestinal absorption value as compared to alloaroma dendrene, sabinene, limonene, isoledene and aromadendrene with the lowest gastrointestinal absorption value. Alloaroma dendrene shows the lowest binding affinity with -6.5 kcal/mol compared to the other six plant compounds (Table 2). It had two hydrophobic bonds (Pi-Alkyl) with the 6M2N at the residue ARG 294 and PHE 298. However, Alloaroma dendrene and Aroma dendrene were identified to possess hERG (human ether-à-go-go-related gene) toxicity (Table 4). Based on Garrido et al., 2020 study, hERG toxicity is one of the most common unfavorable side effects and may have cardiac side effects in humans. Therefore, it was not suitable to be a potent drug for the treatment of COVID-19 disease. According to Dev and Kaur (2020), eucalyptol and α-terpineol acetate act as non-inhibitors to the cytochrome P enzyme found in the liver function for the detoxification of toxins from the liver. 6M2N had the negative value of binding affinity (-5.4 kcal/mol) with the eucalyptol (PubChem ID: 2758) (Table 2). 6M2N interacted with eucalyptol at the residue GLN 110 through a hydrogen bond with a distance of 2.44 nm (Table 3). The target protein had five hydrophobic bonds (Pi-Alkyl and Pi-Sigma) with the Eucalyptol at the residue ILE 249, VAL 202, PRO 293, HIS 246 and PHE 294 (Table 3).

Conclusion

In this study, eucalyptol (PubChem ID: 2758) had a good binding score with the SARS-CoV-2 main protease protein, M^{pro} (PDB ID: 6M2N). The toxicity prediction analysis revealed that eucalyptol exhibited an acceptable drug-like property with no carcinogenicity or mutagenicity. With a safety profile and thorough medication interaction research, this potential therapeutic candidate can be further confirmed in wet lab tests for its proper function in human health.

References

Aw S B, Teh B T, Ling G H T, Leng P C, Chan W H, Ahmad M H. The COVID-19 pandemic situation in Malaysia: Lessons learned from the perspective of population density. *International Journal of Environmental Research and Public Health* (2021).

Benarba B, Pandiella A. Medicinal plants as sources of active molecules against COVID-19. *Frontiers in Pharmacology* (2020).

Berman H, Henrick K, Nakamura H. Announcing the worldwide Protein Data Bank. *Nature Structural & Molecular Biology* (2003) 10(12):980.

Brooks B R, Brooks III C L, Mackerell Jr A D, Nilsson L, Petrella R J, Roux B, Won Y, Archontis G, Bartels C, Boresch S, Caflisch A. CHARMM: The biomolecular simulation program. *Journal of Computational Chemistry* (2009) 30(10):1545-1614.

Darby A C, Hiscox J A. *COVID-19: variants and vaccination.*

Dousari A S, Moghadam M T, Satarzadeh N. COVID-19 (Coronavirus disease 2019): A new coronavirus disease. *Infection and Drug Resistance* (2020).

Dev S, Kaur I. Bioactive molecules from eucalyptus essential oil as potential inhibitors of COVID 19 corona virus infection by molecular docking studies. *Kragujevac Journal of Science* (2020) (42):29-43.

Elengoe A, Naser M A, Hamdan S. A novel protein interaction between nucleotide binding domain of Hsp70 and p53 motif. *International Journal of Genomics* (2015):2015.

Elengoe A, Selvam S. Overview on origin, transmission, treatment and clinical researches of SARS-COV-2. *Journal of Health Sciences & Surveillance System* (2022) 10(1):2-12.

Ganasegeran K, Swee H C A, Looi I. Coronavirus (COVID-19): Evidence relevant to critical care. *Cochrane Database of Systematic Review* (2020) 10(2): 1-4.

Garrido A, Lepailleur A, Mignani S M, Dallemagne P, Rochais C. hERG Toxicity assessment: Useful guidelines for drug design. *European Journal of Medicinal Chemistry* (2020) 195:112290.

Grosdidier A, Zoete V, Michielin O. Swissdock, A protein-small molecule docking web service based on Eadock DSS. *Nucleic Acids Research* (2011) 39(suppl_2):W270-277.

Hasan M M, Khan Z, Chowdhury M S, Khan M A, Moni M A, Rahman M H. *In silico* molecular docking and ADME/T analysis of Quercetin compound with its evaluation of broad-spectrum therapeutic potential against particular diseases. *Informatics in Medicine Unlocked* (2022) 29:100894.

Hayat U, Jilani M I, Rehman R, Nadeem F. A review on *Eucalyptus globulus*: A new perspective in therapeutics. *International Journal of Chemical and Biochemical Science* (2015) 8:85-91.

Hua J, Shaw R. Coronavirus (COVID-19) "infodemic" and emerging issues through a data lens: The case of China. *International Journal of Environmental Research and Public Health* (2020) 17(7).

Jagtap N, Yadav A, Mohite S. Synthesis, Molecular docking studies and anticancer activity of 1, 3, 4-oxadiazole-3 (2h)-thione derivatives. *Journal of University of Shanghai for Science and Technology* (2020) 22(11):535-550.

Kim S, Thiessen P A, Bolton E E, Chen J, Fu G, Gindulyte A, Han L, He J, He S, Shoemaker B A, Wang J. PubChem substance and compound databases. *Nucleic Acids Research (*2016) 44(D1):D1202-1213.

Lenin M, Ramakrishnan E, Sankaran M. Structure-Based molecular docking studies toward exploring phytoestrogen against breast cancer. *Eurasian Journal of Medicine and Oncology* (2022) 6(2):142-149.

Lipinski C A. Lead-and Drug-Like Compounds: The rule-of-five revolution. *Drug Discovery Today: Technologies* (2004) 1(4):337-341.

Mahase E. COVID-19: What new variants are emerging and how are they being investigated? *British Medical Journal (Clinical Research Ed.)* (2020) 372:n158.

Parmar D R, Soni J Y, Guduru R, Rayani R H, Kusurkar R V, Vala A G, Talukdar S N, Eissa I H, Metwaly A M, Khalil A, Zunjar V. Discovery of new anticancer thiourea-azetidine hybrids: Design, synthesis, *in vitro* antiproliferative, SAR, *in silico* molecular docking against VEGFR-2, ADMET, toxicity, and DFT studies. *Bioorganic Chemistry* (2021) 115:105206.

Roy B, Dhillon J, Habib N, Pugazhandhi B. Global variants of COVID-19: Current understanding. *Journal of Biomedical Sciences* (2021).

Stowe J, Andrews N, Gower C, Gallagher E, Utsi L, Simmons R., Lopez Bernal J. Effectiveness of COVID-19 vaccines against hospital admission with the Delta (B.1.617.2) variant. *In Public Health England* (2021).

Supramaniam G, Elengoe A. In silico molecular docking of glycyrrhizin and breast cancer cell line proteins. In *Plant-derived Bioactives* 2020 (pp. 575-589). Springer, Singapore.

Tian S, Wang J, Li Y, Li D, Xu L, Hou T. The application of *in silico* drug-likeness predictions in pharmaceutical research. *Advanced Drug Delivery Reviews* (2015) 86:2-10.

Veber D F, Johnson S R, Cheng H Y, Smith B R, Ward K W, Kopple K D. Molecular properties that influence the oral bioavailability of drug candidates. *Journal of Medicinal Chemistry* (2002) 45(12):2615-2623.

Yuki K, Fujiogi M, Koutsogiannaki S. COVID-19 pathophysiology: A review. *Clinical Immunology* (2020):215.

Chapter 15

A Molecular Interaction Study between *Syzygium aromaticum* Plant Compounds and SARS-CoV-2 Target Proteins

Abdul Haleem Mohamad Wahhaj[1], BSc and Asita Elengoe[2,*], PhD

[1]School of Biotechnology, Manipal International University (MIU), Putra Nilai, Negeri Sembilan, Malaysia

[2]Department of Biotechnology, Faculty of Applied Science, Lincoln University College, Petaling Jaya, Selangor, Malaysia

Abstract

The COVID-19 pandemic is caused by SARS-CoV-2 virus which belongs to SARS or severe acute respiratory syndrome virus family. This unknown outbreak of the virus has resulted in more than 4 million deaths worldwide. Currently, vaccination is the only measure that has been proven to be efficient against the SARS-CoV-2 virus. Therefore, this study has been mainly designed to identify natural drug candidates against SARS-CoV-2 from day-to-day herbal plants. To identify phytochemical compounds against the SARS-CoV-2 spike proteins, *in silico* drug designing strategies including molecular docking, ADMET (Absorption, Distribution, Metabolism, Excretion and Toxicity) and toxicity approaches have been carried out. 15 phytochemical compounds from the clove plant (*Syzygium aromaticum*) were retrieved from the PubChem database and screened against SARS-CoV-2 main protease protein, M^{pro} (PDB ID: 6LU7), RNA-dependent RNA polymerase, RdRp (PDB ID: 7BW4) and spike receptor (PDB ID: 6M0J) using the

* Corresponding Author's Email: asitaelengoe@yahoo.com.

In: Medicinal Plants and COVID-19
Editor: Asita Elengoe
ISBN: 979-8-88697-958-9

SwissADME tool. Only 14 phytochemical compounds successfully passed Lipinski's rule of 5. Then, the molecular docking between the target protein-bioactive compound was performed using PyRx software. The results show that all 14 natural compounds have good binding affinities. However, ellagic acid (CID:5281855) had the best binding energy with 6LU7 (-8.3 kcal/mol), 7BW4 (-7.6 kcal/mol) and 6M0J (-8.1 kcal/mol). Moreover. It did not show any carcinogenicity and mutagenesis in humans. Therefore, it is a suitable and safe drug for COVID-19 disease. The computational biology results will aid in designing active drugs from the phytochemicals of the clove.

Keywords: docking, SARS-CoV-2, *Syzygium aromaticum,* toxicity

Introduction

The occurrence of an unknown acute respiratory tract infection was first reported in Wuhan City, Hubei Province, China which led to an outbreak of a global pandemic. The coronavirus disease that caused the COVID-19 pandemic in recent years was due to a virus that caused severe acute respiratory syndrome (SARS) which developed into unexplained pneumonia in humans and animals (Allaq et al., 2021). Thus, it was named severe acute respiratory syndrome coronavirus 2 (SARS-CoV-2) or most commonly known as COVID or coronavirus, etc. (Elengoe et al., 2022). It affects the respiratory, gastrointestinal, liver, and central nervous systems of humans and other vertebrates. It was first identified in the city of Wuhan, in the southern part of China (Elengoe, 2020). By the end of January 2020, the World Health Organization (WHO) declared it a Public Health Emergency of International Concern due to the unexpected spread of SARS-CoV-2 rapidly to more than 206 countries around the globe on 11th March 2020. Therefore, it was declared a global pandemic threatening humans and other vertebrate animals.

SARS-CoV-2 is a single-stranded RNA virus that is contagious in humans (Khailany et al., 2020). It is known to have originated from the consumption of infected bats from a wet market in Wuhan City, China. This information was devastating as it was released by the National Health Commission of the People's Republic of China. The first identified people with SARS-CoV-2 had links to a live animal and seafood market or a wet market that sold exotic animals and their meat. The Huanan wet market where the virus has been expected to have originated contained a wild animal section where live and slaughtered species like snakes, beavers, porcupines, and baby crocodiles were among other animals that were sold. The SARS-CoV-2 has been identified to have close genetic similarity and behavior patterns to bat coronaviruses depicting that it emerged from bats and current research whether SARS-CoV-2 came directly from bats or indirectly through any intermediate hosts indicating a spillover event is in process. The initial source and pattern of spreading are still a mystery as there is very little conclusive information about the virus's origin (Harrison et al., 2020).

Observation of the SARS-CoV-2 virus outbreak began since its first appearance in late 2019 and scientists have worked earnestly to identify the causes which lead to the rapid spread of the virus globally. The human-to-human transmission was confirmed on 20th January 2020 as the virus primarily spread between people through close contact with infected people. Later, it was identified that respiratory droplets or aerosols exhaled from infected people when talking and breathing and from coughs or sneezes also facilitated the spread of the virus. Studies have suggested that it enters human cells by binding to an angiotensin-converting enzyme 2 (ACE2) which is a membrane protein that regulates the renin-angiotensin system and several other studies stated that the virus is merely airborne, where it has the potential to transmit the virus indirectly through contact of contaminated surfaces (Michel et al., 2020). Pre-research has indicated that the virus can remain viable on plastic and stainless-steel surfaces for up to three days but cannot survive on cardboard or paper surfaces for more than one day and on copper surfaces for more than four hours and effective sanitation eliminates the virus. The exact incubation period of the virus is still uncertain, but research has indicated that peak viral load on the pharynx is attained by the virus after 4 days of infection and shedding duration is estimated to be between 3 - 46 days after symptom appearance (Li et al., 2021).

Most unknown viruses are known to be zoonotic and a deep understanding of viruses, virus diversity and evolution, transmission events, interactions and the factors contributing to pandemics is needed for taking preventive measures and to avoid the spread of a virus. Currently, after vast research and dedication towards the study of SARS-CoV-2 vaccines are being produced and people are being vaccinated to create immunity against the coronavirus. A highly adaptive virus-like SARS-CoV-2 which evolves through natural selection can only be defeated through preparedness or being in a prepared state and should be made a priority to avoid global outbreaks in the future (Shang et al., 2020).

Currently, traditional plants are used to treat chronic health problems such as cardiovascular disease, kidney-related disease, lung-related disease, inflammation, bacterial infections, diabetes mellitus, hypertension, cancer, etc. Traditional herbal remedies were heavily involved in China's 2003 effort to control and treat severe acute respiratory syndrome (SARS), and in Africa, the Africa flower has long been used to treat HIV-related wasting symptoms (De Smet 2005; Tilburt and Kaptchuk 2008). In most industrialized nations, one can find essential oils, herbal extracts, or herbal teas being sold in pharmacies alongside conventional drugs. Essential oils, herbal extracts, and herbal teas are also quite popular in Europe, with Germany and France leading the continent in over-the-counter sales. Many studies have demonstrated that traditional plants enhance our immune system.

Syzygium aromaticum belongs to the Myrtaceae family. The most common name for *Syzygium aromaticum* is clove. The clove is native to the Maluku Islands (or Moluccas) in Indonesia. It is commonly used as a spice, flavoring or fragrance in consumer products, such as toothpaste, soaps, or cosmetics. Clove possesses medicinal and pharmacological activities such as anti-bacterial, anti-viral, anti-cancer, analgesic, anti-inflammatory, anti-thrombotic, antiprotozoal, hypoglycemic, gastroprotective, antiseptic in oral diseases, etc. (Johannah et al., 2015).

The identification of genes and pathways that may be linked to significant bioactive secondary metabolites from medicinal plants is facilitated using bioinformatics methods such as molecular modeling, dynamics simulation, docking, etc. Rath and her co-researchers (2021) have explored the interaction between phytoconstituents from the *Moringa oleifera* plant (ligands) and p53 (target protein) (PDB ID: 3DCY) using a molecular docking approach. They studied the role of p53 in the cell-death program of oral squamous cell carcinoma using bioinformatics tools. They evaluated the pharmacokinetics and toxicity of bioactive compounds according to Lipinski's Rule of Five. They predicted the active sites of p53. They found that quercetin had the lowest binding affinity (-6.72 kcal/mol) while kaempferol displays poor binding affinity (-5.29 kcal/mol) with p53 using Auto Dock Tools 1.5.6.

In this study, the molecular interaction between the SARS-CoV-2 main protease protein, M^{pro} (PDB ID: 6LU7), RNA-dependent RNA polymerase, RdRp (PDB ID: 7BW4) and spike receptor (PDB ID: 6M0J) and the fifteen phytocompounds from the clove plant was determined via docking, pharmacokinetics and toxicity prediction tools.

Methodology

Plant Selection

The clove (*Syzygium aromaticum*) plant was selected due to its high concentration of phytochemical compounds and related medicinal uses. The most identified phytochemicals in clove plant were phenols, terpenoids and alkaloids which were proven to have medicinal effects of reducing inflammation and inhibiting biochemical mechanisms. Fifteen phytochemical compounds (eugenol, acetyl eugenol, vanillin, eugenitin, caffeic acid, ferulic acid, ellagic acid, salicylic acid, β-pinene, limonene, farnesol, benzaldehyde, 2-heptanone, ethyl hexanoate and gallotannic acid) were identified.

Phytochemical Preparation

Bioactive compounds were used as ligands. The molecular structures of the 15 selected phytochemicals were downloaded in the three-dimensional (3-D) conformers from the PubChem database in SDF format for molecular docking studies (Kim et al., 2016).

Target Protein Preparation

M^{pro} is a cysteine protease. It plays an essential role in mediating the maturation of cleavage of polyproteins (Gorbalenya et al., 1989; Lee et al., 1991; Ziebuhr et al., 2000). SARS-CoV-2 requires RNA-dependent RNA polymerase (RdRp) for the replication of their genome and the transcription of the genes (Snijder et al. 2016; Posthuma et al. 2017). One of the main components of infection is the S protein found on the surface of the virus. During viral infection, the S protein of SARS-CoV-2 facilitates the recognition of receptor, cell attachment, and fusion. S protein consists of S1 and S2 subunits. The host receptor angiotensin-converting enzyme 2 (ACE2) is recognized and bound by the receptor-binding domain of the S1 subunit, while the two-heptad repeat domain of the S2 subunit facilitates the fusing of the viral cell membrane. The 3-D structure of the SARS-CoV-2 main protease protein, M^{pro} (PDB ID: 6LU7), RNA-dependent RNA polymerase (PDB ID: 7BW4) and spike receptor (PDB ID: 6M0J) were downloaded from the RCSB

Protein Data Bank (PDB) in PDB format (Berman et al., 2003). They were publicly available.

Preparation of Ligand

The retrieved 15 phytochemicals in SDF format; were prepared using the DS 4.0 'Prepare 'ligand' technique (Brooks et al., 2009). They are screened using the Lipinski rule of 5 (Lipinski, 2004; Veber et al., 2002; Jagtap et al., 2020). They were differentiated according to pharmacokinetics properties such as physicochemical properties, lipophilicity, water solubility, drug-likeness, medicinal chemistry and drug absorption. The pharmacokinetics properties of the phytochemical compounds were identified using a free web tool SwissADME from the Swiss institute of bioinformatics. The ligands which met the Lipinski rule of 5; prepare for molecular docking with the SARS-CoV-2 target proteins (PDB ID: 6LU, 7BW4 and 6M0J).

Molecular Docking

Autodock Vina in PyRx 0.8 application was used to perform the molecular docking between the target proteins (PDB ID: 6LU, 7BW4 and 6M0J) and phytochemical compounds which were used as the ligands. The calculation of binding energy was determined using PyRx Autodock Vina after selecting the entire crystal structure of the target protein (Dallakyan and Olson, 2015). Only 14 phytochemical molecules (eugenol, acetyl eugenol, vanillin, eugenitin, caffeic acid, ferulic acid, ellagic acid, salicylic acid, β-pinene, limonene, farnesol, benzaldehyde, 2-heptanone, ethyl hexanoate) were selected for performing molecular docking studies with the SARS-CoV-2 target proteins. The PyRx software gives an output with the lowest binding affinities for the phytochemical compounds that can be used for further toxicity analysis (Parmar et al., 2021).

Prediction of Toxicity

The toxicity of the chosen phytocomponents was assessed using the web-based admetSAR server (Tian et al., 2015).

Results and Discussion

The SARS-CoV-2 virus is an infectious disease that causes mild to life-threatening respiratory problems in individuals. The virus is an enclosed single-stranded RNA virus with structural and nonstructural protein gene components. The structural proteins encode the spike (S), envelope (E), membrane (M) and nucleocapsid (N) proteins which mediate viral attachment and entry while nonstructural proteins aid in viral replication and release (Mariano et al., 2020). SARS-CoV-2 spike protein has been identified as a first-line therapeutic target for medication and vaccine development. This is because the spike protein is involved in receptor identification and cell membrane fusion, both of which are required for viral entrance, and the virus must be linked to the host cell in order to multiply and infect. The two translated polyproteins are cleaved through the proteolysis process to generate components for packing new virions by M^{pro} whereas RNA-dependent RNA polymerase (RdRp) plays an important role in the replication of RNA genome. Thus, M^{pro}, RdRp and spike proteins are thought to make appealing therapeutic targets for the treatment of SARS-CoV-2 (Huang et al., 2020).

The clove plant (*Syzygium aromaticum*) includes a variety of phenolic compounds such as flavonoids, hydroxybenzoic acids, hydroxycinnamic acids and hydroxyphenyl propens, as well as other volatile compounds with antioxidant, anti-bacterial, antinociceptive, and anti-viral characteristics. It is one of the most precious spices, and it has been used for centuries for a variety of uses, including therapeutic purposes (Batiha et al., 2020).

Computational biology tools such as *in-silico* screening, docking, and toxicity evaluation analysis, can be regarded as a huge benefit that aids in the discovery of innovative and natural therapeutics against the SARS-CoV-2 virus, allowing for easy drug design. Molecular docking studies were conducted using free software such as PyRx to identify molecular interactions between target proteins and phytochemicals. The phytochemicals that are used as ligands potential to be the inhibitors of the SARS-CoV-2 target protein because they may inhibit the viral entry, attachment to the host receptor, proteolysis process and replication process.

Using molecular recognition principles to study and understand the drug likeliness, toxicity, binding affinity, and interactions of phytochemical compounds on various aspects allows for the development of effective structure-based drug designs that can also be used to remake or remodel pharmacological agents.

The phytochemical compounds were subjected to ADMET (absorption, distribution, metabolism, elimination and toxicity) analysis. Phytochemical compounds were analyzed on drug-likeness parameters of Lipinski's rule of five, number of hydrogen (H) donors and acceptors, LogP, LogS, Synthetic accessibility (SA) and gastrointestinal (GI) absorption. Plant compounds that fulfill Lipinski's rules are considered to be readily absorbable into the human body (good GI absorption); have easy binding; and have good permeability in developing orally active drugs (Sailah et al., 2021).

The partition coefficient or LogP expresses the lipophilicity and is referred to as the logarithm of the inorganic and aqueous phase partition coefficient of the target molecule. It influences the absorption and molecule pharmacokinetics. Pharmacokinetics can be named as the absorption, distribution, metabolism and excretion of a phytochemical compound. The logarithm of the solubility or LogS determines the solubility of the selected phytochemical compounds while the number of hydrogen bond donors and acceptors represents the capacity of a phytochemical molecule to cross the lipid bilayer of a cell membrane. The rotatable bonds are associated with the oral bioavailability of phytochemical compounds. An effective and safe drug can only be developed if these facts are taken into consideration (Ahammad et al., 2021). Based on the Lipinski's Rule of 5, all the plant compounds had passed the rule except gallotannic acid because it had molecular weight (1701.2 g/mol) higher than 500 g/mol (Table 1). It may have difficulty to cross the lipid bilayer of the cell membrane due to high value of LogP (6.2). Therefore, it was not a suitable drug to proceed with the molecular docking study.

Molecular docking studies were performed between SARS-CoV-2 target proteins and 14 screened phytochemicals. The binding affinity of natural compounds varied between each target protein, but all phytochemical compounds exhibited a good binding affinity with the target proteins (PDB ID: 6LU7; 7BW4 and 6M0J). Ellagic acid (CID:5281855) showed the most negative value and lowest binding affinity for all three target proteins. Ellagic acid (CID:5281855) had the lowest binding affinity for the SARS-CoV-2 main protease protein (PDB ID: 6LU7) at a value of -8.3 kcal/mol (Table 2). It is bound with 6LU7 through seven hydrogen bonds at the residues GLN110, THR111, ASP295 and SER158 along with Pi-Pi T-shaped bond (PHE294), Pi-Alkyl bond (VAL104) and Pi-Donor bond (ASN151) (Table 3).

Table 1. List of pharmacokinetics properties, molecular weight (MW), hydrogen bond donor (HBD), hydrogen bond acceptor (HBA), partitioning coefficient (LogP), number of rotatable bond (RB), number of heavy atoms (HA), number of aromatic heavy atoms (AHA), synthetic accessibility (SA), gastrointestinal (GI) absorption, and Lipinski's Rule of 5 of all plant compounds

Compound Name	PubChem CID	MW g/m (<500)	Num. heavy atoms	Num. arom. heavy atoms	RB (≤10)	HBA (≤10)	HBD (≤5)	LogP (<5)	LogS (EOSL)	SA	GI	LR
Eugenol	3314	164.20 g/mol	12	6	3	2	1	2.37	-2.46	1.58 Easy	High	YES
Acetyl eugenol	7136	206.24 g/mol	15	6	5	3	0	2.73	-2.53	1.94 Easy	High	YES
Vanillin	1183	152.15 g/mol	11	6	2	3	1	1.57	-1.82	1.15 Easy	High	YES
Eugenitin	3083581	220.22 g/mol	16	10	1	4	1	2.61	-3.15	2.72 Easy	High	YES
Caffeic acid	689043	180.16 g/mol	13	6	2	4	3	0.97	-1.89	1.81 Easy	High	YES
Ferulic acid	445858	194.18 g/mol	14	6	3	4	2	1.62	-2.11	1.93 Easy	High	YES
Ellagic acid	5281855	302.19 g/mol	22	16	0	8	4	0.79	-2.94	3.17 Easy	High	YES
Salicylic acid	338	138.12 g/mol	10	6	1	3	2	1.13	-2.50	1.00 Easy	High	YES
β-pinene	14896	136.23 g/mol	10	0	0	0	0	2.59	-3.31	3.73 Moderate	Low	YES
Limonene	22311	136.23 g/mol	10	0	1	0	0	2.72	-3.50	3.46 Easy	Low	YES
Farnesol	445070	222.37 g/mol	16	0	7	1	1	3.71	-4.17	3.17 Easy	High	YES
Benzaldehyde	240	106.12 g/mol	8	6	1	1	0	1.36	-1.92	1.00 Easy	High	YES
2-heptanone	8051	114.19 g/mol	8	0	4	1	0	2.09	-1.53	1.33 Easy	High	YES
Ethyl hexanoate	31265	144.21 g/mol	10	0	6	2	0	2.33	-1.89	1.67 Easy	High	YES
Beta-caryophyllene	5281515	204.35 g/mol	15	0	0	0	0	3.29	-3.87	4.51 Hard	Low	YES
Crategolic acid	3694932	472.70 g/mol	34	0	1	4	3	3.55	-6.81	6.22 Hard	High	YES
Gallotannic acid	16133892	1701.2 g/mol	122	-	31	46	25	6.2	-	-	-	NO

Table 2. Binding affinity score between phytocompound and target protein (6LU7)

No.	Compound name	Pubchem ID	Ligand	Binding Affinity (kcal/mol)
1.	Ellagic acid	5281855	6lu7_5281855_uff_E=227.58	-8.3
2.	Caffeic acid	689043	6lu7_689043_uff_E=98.60	-5.8
3.	Eugenitin	3083581	6lu7_3083581_uff_E=240.05	-5.8
4.	Ferulic acid	445858	6lu7_445858_uff_E=177.42	-5.7
5.	Salicylic acid	338	6lu7_338_uff_E=73.50	-5.7
6.	Eugenol	3314	6lu7_3314_uff_E=169.59	-5.5
7.	Acetyl eugenol	7136	6lu7_7136_uff_E=258.81	-5.1
8.	Farnesol	445070	6lu7_445070_uff_E=148.85	-5
9.	B-pinene	14896	6lu7_14896_ uff_E=713.31	-5
10.	Vanillin	1183	6lu7_1183_uff_E=79.78	-5
11.	Limonene	22311	6lu7_22311_uff_E=102.50	-4.8
12.	Benzaldehyde	240	6lu7_240_uff_E=52.88	-4.6
13.	2-heptanone	8051	6lu7_8051_uff_E=23.20	-3.9
14.	Ethyl hexanoate	31265	6lu7_31265_uff_E=33.85	-3.8

Table 3. List of hydrogen bond interactions between target protein (6LU7) and ligands (plant compounds)

No.	Compound name	PubChem ID	Residues	Distance	Bond category	Bond type
1.	Ellagic acid	5281855	GLN110	3.22	Hydrogen	Conventional
				2.85	Hydrogen	Conventional
			THR111	2.96	Hydrogen	Conventional
				2.72	Hydrogen	Conventional
				1.95	Hydrogen	Donor-donor
			ASP295	2.20	Hydrogen	Conventional
			SER158	2.88	Hydrogen	Conventional
			PHE294	5.74	Pi-Pi T shaped	Hydrophobic
			VAL104	5.19	Pi-Alkyl	Hydrophobic
			ASN151	3.76	Hydrogen	Pi-Donor
2.	Caffeic acid	689043	ASP153	2.19	Hydrogen	Conventional
			SER158	2.77	Hydrogen	Conventional
			THR292	3.01	Hydrogen	Conventional
			THR111	2.79	Hydrogen	Conventional
			GLN110	2.39	Hydrogen	Donor-donor
3.	Eugenitin	3083581	CYS145	4.79	Pi-Alkyl	Hydrophobic
			CYS145	3.07	Hydrogen	Conventional
			SER144	3.01	Hydrogen	Conventional
			GLY143	3.02	Hydrogen	Conventional
			HIS163	2.32	Hydrogen	Conventional
4.	Ferulic acid	445858	GLN110	2.80	Hydrogen	Conventional
			THR111	2.93	Hydrogen	Conventional
				3.03	Hydrogen	Conventional
				2.93	Hydrogen	Conventional

No.	Compound name	PubChem ID	Residues	Distance	Bond category	Bond type
			ASP295	2.97	Hydrogen	Conventional
			SER158	3.24	Hydrogen	Conventional
				2.29	Hydrogen	Conventional
			VAL104	5.22	Pi-Alkyl	Hydrophobic
			ASN151	4.13	Pi-Donor	Hydrophobic
5.	Salicylic acid	338	ASP295	3.01	Hydrogen	Conventional
			THR111	2.89	Hydrogen	Conventional
			THR111	2.34	Hydrogen	Conventional
			THR292	2.76	Hydrogen	Conventional
6.	Eugenol	3314	PHE294	4.40	Pi-Alkyl	Hydrophobic
			ILE106	5.03	Alkyl	Hydrophobic
			GLN110	3.14	Hydrogen	Conventional
			GLN110	2.59	Hydrogen	Donor-donor
			THR111	2.93	Hydrogen	Acceptor-Acceptor
7.	Acetyl eugenol	7136	GLN110	3.02	Hydrogen	Conventional
			THR111	3.02	Hydrogen	Conventional
				3.15	Hydrogen	Conventional
			PHE8	4.78	Pi-Alkyl	Hydrophobic
			PHE294	4.72	Pi-Alkyl	Hydrophobic
				4.46	Pi-Alkyl	Hydrophobic
				3.79	Pi-Pi Stacked	Hydrophobic
			ILE152	3.77	Carbon	Carbon-Hydrogen
8.	Farnesol	445070	PHE294	3.93	Pi-Sigma	Hydrophobic
				4.29	Pi-Alkyl	Hydrophobic
			VAL104	4.58	Alkyl	Hydrophobic
			ILE106	3.30	Carbon	Carbon-Hydrogen
9.	B-pinene	14896	LEU271	5.18	Alkyl	Hydrophobic
10.	Vanillin	1183	THR111	3.15	Hydrogen	Conventional
			GLN110	3.15	Hydrogen	Conventional
			ILE106	4.81	Alkyl	Hydrophobicc
11.	Limonene	22311	PHE294	4.26	Pi-Alkyl	Hydrophobic
				3.86	Pi-Alkyl	Hydrophobic
12.	Benzaldehyde	240	PHE294	3.81	Pi-Pi Packed	Hydrophobic
			ASN151	3.71	Carbon	Carbon-Hydrogen
			THR111	3.00	Hydrogen	Conventional
			GLN110	3.22	Hydrogen	Conventional
				3.07	Hydrogen	Conventional
13.	2-heptanone	8051	SER158	3.07	Hydrogen	Conventional
			PHE 294	3.86	Pi-Sigma	Hydrophobic
14.	Ethyl hexanoate	31265	HIS172	5.03	Pi-Alkyl	Hydrophobic
			HIS163	4.77	Pi-Alkyl	Hydrophobic
				5.34	Pi-Alkyl	Hydrophobic
			CIS145	5.50	Alkyl	Hydrophobic
				4.15	Alkyl	Hydrophobic
				3.10	Hydrogen	Conventional
			SER144	2.96	Hydrogen	Conventional
			HIS41	4.53	Pi-Alkyl	Hydrophobic

Table 4. Binding affinity score between phytocompound and target protein (7BW4)

No.	Compound name	Pubchem ID	Ligand	Binding Affinity (kcal/mol)
1.	Ellagic acid	5281855	7bw4_5281855_uff_E=227.58	-7.6
2.	Caffeic acid	689043	7bw4_689043_uff_E=98.60	-5.2
3.	Eugenitin	3083581	7bw4_445858_uff_E=177.42	-6.1
4.	Ferulic acid	445858	7bw4_3083581_uff_E=240.05	-6.0
5.	Salicylic acid	338	7bw4_338_uff_E=73.50	-5.8
6.	Eugenol	3314	7bw4_3314_uff_E=169.59	-5.6
7.	Acetyl eugenol	7136	7bw4_7136_uff_E=258.81	-5.5
8.	Farnesol	445070	7bw4_1183_uff_E=79.78	-5.4
9.	B-pinene	14896	7bw4_22311_uff_E=102.50	-5.4
10.	Vanillin	1183	7bw4_14896_uff_E=713.31	-5.2
11.	Limonene	22311	7bw4_445070_uff_E=148.85	-5.1
12.	Benzaldehyde	240	7bw4_240_uff_E=52.88	-4.7
13.	2-heptanone	8051	7bw4_31265_uff_E=33.85	-4.2
14.	Ethyl hexanoate	31265	7bw4_8051_uff_E=23.20	-4.0

The binding energy of other phytochemical compounds was within a range of -5.8 kcal/mol and -3.8 kcal/mol. The number of hydrogen bonds determines the stability of the target protein-ligand complex. Caffeic acid (CID:689043) and eugenitin (CID:3083581) had the second lowest binding affinity of -5.8 kcal/mol with a total of 5 hydrogen bonds at ASP153, SER158, THR292, THR111, GLN110 residues and 4 hydrogen bonds formed via CYS145, SER144, GLY143, HIS163 and a Pi-Alkyl bond with CYS145 residues of the target protein respectively.

Ellagic acid (CID:5281855) also had the lowest binding energy with 7BW4 at a value of -7.6 kcal/mol through 4 hydrogen bonds (Table 4). It interacted at residues of ARG555, THR556, ARG553 and SER682 with a distance of 2.23, 2.29, 1.91 and 2.06 respectively. All other phytochemical compounds' binding affinities were within a range of -6.2 kcal/mol and -4 kcal/mol (Table 5). Caffeic acid (CID:689043) had the second lowest binding affinity of -6.2 kcal/mol with a total of 5 hydrogen bonds. It was bound at residues of ASP623, ARG555, THR556, ARG553, SER682 and CYS622 with a distance of 2.55, 2.23, 2.64, 2.63 and 2.71 respectively.

Ellagic acid (CID:5281855) had the most negative value of -8.1 kcal/mol for binding energy (Table 6). It interacted with 6M0J via six hydrogen bonds. It had a hydrogen bond distance of 2.61, 2.56, 1.93, 2.50, 2.11 and 1.92 with GLN102, TYR202, TRP203, TRP203, ARG514 and SER511 residues respectively (Table 7). All other phytochemical compounds' binding affinities were within a range of -7.4 kcal/mol and -4.5 kcal/mol. Caffeic acid

(CID:689043) had the second lowest binding affinity of -7.4 kcal/mol with a total of 2 hydrogen bonds and 2 carbon-hydrogen bonds, 1 Pi-Alkyl bond and 1 Pi-Pi stacked bond at residues of ALA413, HIS540, ASN290, THR414, PRO415 and PHE438 residues respectively.

Table 5. List of hydrogen bond interactions between target protein (7BW4) and ligands (plant compounds)

No.	Compound name	PubChem ID	Residues	Distance	Bond category	Bond type
1.	Ellagic acid	5281855	ASP623	3.66	Pi-Anion	Hydrophobic
				348	Pi-Anion	Hydrophobic
				4.19	Pi-Anion	Hydrophobic
				3.59	Pi-Anion	Hydrophobic
			ARG555	4.04	Pi-Cation	Hydrophobic
				4.10	Pi-Cation	Hydrophobic
				2.23	Hydrogen	Conventional
			THR556	2.29	Hydrogen	Conventional
			ARG553	1.91	Hydrogen	Conventional
			SER682	2.06	Hydrogen	Conventional
			CYS622	4.61	Pi-Alkyl	Hydrophobic
2.	Caffeic acid	689043	ASP684	2.55	Hydrogen	Conventional
			THR565	2.23	Hydrogen	Conventional
			GLN573	2.64	Hydrogen	Conventional
			LYS577	2.63	Hydrogen	Conventional
			ARG569	2.71	Hydrogen	Conventional
				5.26	Pi-Alkyl	Hydrophobic
			ALA685	4.45	Pi-Alkyl	Hydrophobic
3.	Ferulic acid	445858	THR565	2.00	Hydrogen	Conventional
			ARG569	2.61	Hydrogen	Conventional
			LYS577	2.07	Hydrogen	Conventional
			GLN573	2.42	Hydrogen	Conventional
			ALA685	3.92	Alkyl	Hydrophobic
				4.24	Pi-Alkyl	Hydrophobic
			TYR689	5.19	Pi-Alkyl	Hydrophobic
			LEU576	4.56	Alkyl	Hydrophobic
4.	Eugenitin	3083581	TYR728	5.27	Pi-Alkyl	Hydrophobic
			TYR38	5.18	Pi-Pi T shaped	Hydrophobic
				3.92	Pi-Sigma	Hydrophobic
			VAL204	3.67	Pi-Sigma	Hydrophobic
				522	Pi-Alkyl	Hydrophobic
				459	Alkyl	Hydrophobic
			ASP221	2.34	Hydrogen	Conventional
			ILE37	3.54	Carbon	Carbon-Hydrogen
5.	Salicylic acid	338	ILE494	2.79	Hydrogen	Acceptor-Acceptor
			ASN496	2.17	Hydrogen	Donor-donor
				2.60	Hydrogen	Conventional
			GLN573	2.63	Hydrogen	Conventional
				2.06	Hydrogen	Conventional

Table 5. (Continued)

No.	Compound name	PubChem ID	Residues	Distance	Bond category	Bond type
5.	Salicylic acid (continued)		LYS577	2.37	Hydrogen	Conventional
			ALA685	4.51	Pi-Alkyl	Hydrophobic
			ARG569	4.98	Pi-Alkyl	Hydrophobic
6.	Eugenol	3314	VAL204	4.28	Pi-Alkyl	Hydrophobic
				3.82	Alkyl	Hydrophobic
			TYR38	5.24	Pi-Pi T shaped	Hydrophobic
				3.68	Pi-Sigma	Hydrophobic
				2.39	Hydrogen	Conventional
			ASP36	2.83	Hydrogen	Conventional
			TYR728	2.66	Hydrogen	Conventional
7.	Acetyl eugenol	7136	ALA685	4.17	Pi-Alkyl	Hydrophobic
				4.13	Alkyl	Hydrophobic
			LEU576	5.38	Pi-Alkyl	Hydrophobic
				5.25	Alkyl	Hydrophobic
			AS496	2.13	Hydrogen	Conventional
			THR689	2.82	Hydrogen	Conventional
8.	Vanillin	1183	ASP84	2.99	Hydrogen	Acceptor-Acceptor
				2.99	Hydrogen	Acceptor-Acceptor
			ARG569	4.75	Pi-Alkyl	Hydrophobic
				5.00	Alkyl	Hydrophobic
				3.08	Hydrogen	Conventional
			ASN568	2.54	Hydrogen	Conventional
			GLN573	2.71	Hydrogen	Conventional
9.	Limonene	22311	TYR515	4.17	Pi-Alkyl	Hydrophobic
				4.36	Pi-Alkyl	Hydrophobic
			LEU514	4.08	Alkyl	Hydrophobic
			TRP509	3.92	Pi-Sigma	Hydrophobic
10.	B-pinene	14896	TRP509	4.85	Pi-Alkyl	Hydrophobic
				4.54	Pi-Alkyl	Hydrophobic
				5.37	Pi-Alkyl	Hydrophobic
			PHE3.68	4.94	Pi-Alkyl	Hydrophobic
			LEU372	4.97	Alkyl	Hydrophobic
			TYR515	5.12	Pi-Alkyl	Hydrophobic
				3.81	Pi-Alkyl	Hydrophobic
11.	Farnesol	445070	TRP509	4.23	Pi-Alkyl	Hydrophobic
				3.92	Pi-Sigma	Hydrophobic
				3.54	Pi-Sigma	Hydrophobic
				5.01	Pi-Alkyl	Hydrophobic
				5.16	Pi-Alkyl	Hydrophobic
			TYR515	4.52	Pi-Alkyl	Hydrophobic
			ALA375	3.70	Alkyl	Hydrophobic
12.	Benzal-dehyde	240	VAL204	3.59	Pi-Sigma	Hydrophobic
			TYR38	5.02	Pi-Pi T shaped	Hydrophobic
13.	Ethyl hexanoate	31265	GLN573	2.26	Hydrogen	Conventional
			ARG569	2.33	Hydrogen	Conventional
				4.20	Alkyl	Hydrophobic

No.	Compound name	PubChem ID	Residues	Distance	Bond category	Bond type
14.	2-heptanone	8051	THR515	4.40	Pi-Alkyl	Hydrophobic
			ALA375	4.32	Alkyl	Hydrophobic
			PHE506	4.86	Pi-Alkyl	Hydrophobic
			TRP509	5.02	Pi-Alkyl	Hydrophobic
				4.86	Pi-Alkyl	Hydrophobic
			LEU372	5.41	Alkyl	Hydrophobic

Table 6. Binding affinity score between phytocompound and target protein (6M0J)

No.	Compound name	Pubchem ID	Ligand	Binding Affinity (kcal/mol)
1.	Ellagic acid	5281855	6m0j_5281855_uff_E=227.58	-8.1
2.	Caffeic acid	689043	6m0j_689043_uff_E=98.60	-7.4
3.	Eugenitin	3083581	6m0j_22311_uff_E=102.50	-7.0
4.	Ferulic acid	445858	6m0j_7136_uff_E=258.81	-6.9
5.	Salicylic acid	338	6m0j_3314_uff_E=169.59	-6.8
6.	Eugenol	3314	6m0j_3083581_uff_E=240.05	-6.4
7.	Acetyl eugenol	7136	6m0j_445858_uff_E=177.42	-6.3
8.	Farnesol	445070	6m0j_338_uff_E=73.50	-6.2
9.	B-pinene	14896	6m0j_1183_uff_E=79.78	-6.1
10.	Vanillin	1183	6m0j_445070_uff_E=148.85	-5.9
11.	Limonene	22311	6m0j_240_uff_E=52.88	-5.5
12.	Benzaldehyde	240	6m0j_14896_uff_E=713.31	-5.3
13.	2-heptanone	8051	6m0j_31265_uff_E=33.85	-5.0
14.	Ethyl hexanoate	31265	6m0j_8051_uff_E=23.20	-4.5

Apart from physiochemical properties, the toxicity prediction of plant compounds was analyzed. *In silico* toxicity analysis of phytochemical compounds requires less time, is inexpensive, and also cuts down the need for animal experimentation. Toxicity analysis of phytochemical compounds on hepatotoxicity, carcinogenicity and AMES toxicity and hERG inhibition mainly focuses on creating drugs with maximum human safety. ADMET prediction of the phytochemicals from clove showed 14 plant compounds (eugenol, acetyl eugenol, vanillin, caffeic acid, ferulic acid, ellagic acid, salicylic acid, β-pinene, limonene, farnesol, benzaldehyde, 2-heptanone, ethyl hexanoate) can be used as SARS-CoV-2 protein inhibitors. Most importantly the 14 bioactive compounds did not have any carcinogenicity and mutagenesis (Table 8).

Table 7. List of hydrogen bond interactions between target protein (6M0J) and ligands (plant compounds)

No.	Compound name	PubChem ID	Residues	Distance	Bond category	Bond type
1.	Ellagic acid	5281855	GLN102	2.61	Hydrogen	Conventional
			TYR202	2.56	Hydrogen	Conventional
			TRP203	1.93	Hydrogen	Conventional
				2.50	Hydrogen	Conventional
			ARG514	2.11	Hydrogen	Conventional
			SER511	1.92	Hydrogen	Conventional
				3.10	Carbon	Carbon-Hydrogen
2.	Caffeic acid	689043	ASN290	2.42	Carbon	Carbon-Hydrogen
			THR414	2.48	Carbon	Carbon-Hydrogen
			ALA413	2.37	Hydrogen	Conventional
			HIS540	2.38	Hydrogen	Conventional
			PHE438	3.78	Pi-Pi Stacked	Hydrophobic
			PRO415	4.20	Pi-Alkyl	Hydrophobic
3.	Limonene	22311	ILE291	4.21	Alkyl	Hydrophobic
				3.99	Alkyl	Hydrophobic
			PRO415	5.10	Alkyl	Hydrophobic
			PHE438	5.33	Pi-Alkyl	Hydrophobic
				3.80	Pi-Sigma	Hydrophobic
4.	Acetyl eugenol	7136	HIS540	4.55	Pi-Alkyl	Hydrophobic
			LYS541	4.37	Alkyl	Hydrophobic
			PRO415	5.14	Pi-Alkyl	Hydrophobic
				4.25	Alkyl	Hydrophobic
			PHE438	4.41	Pi-Pi Stacked	Hydrophobic
			ILE291	4.03	Pi-Alkyl	Hydrophobic
				2.11	Hydrogen	Conventional
			THR434	3.66	Carbon	Carbon-Hydrogen
5.	Eugenol	3314	PHE428	5.43	Pi-Alkyl	Hydrophobic
			PRO415	4.36	Alkyl	Hydrophobic
			PHE438	4.57	Pi-Pi Stacked	Hydrophobic
			ILE291	5.35	Alkyl	Hydrophobic
				3.86	Pi-Alkyl	Hydrophobic
				2.15	Hydrogen	Conventional
			ASN290	1.77	Hydrogen	Conventional
			THR434	3.55	Carbon	Carbon-Hydrogen
6.	Eugenitin	3083581	SER44	2.18	Hydrogen	Conventional
			PHE40	4.19	Pi-Alkyl	Hydrophobic
			SER43	5.29	Amide-Pi Stacked	Hydrophobic
			TRP349	5.11	Pi-Pi Stacked	Hydrophobic
				4.20	Pi-Pi Stacked	Hydrophobic
				4.90	Pi-Pi Stacked	Hydrophobic
				3.63	Pi-Sigma	Hydrophobic
				2.32	Carbon	Carbon-Hydrogen
7.	Ferulic acid	445858	LYS562	2.68	Hydrogen	Conventional
				2.76	Hydrogen	Conventional
			ASP206	2.31	Hydrogen	Conventional

No.	Compound name	PubChem ID	Residues	Distance	Bond category	Bond type
			ASN210	1.92	Hydrogen	Conventional
			LEU95	2.39	Pi-Sigma	Hydrophobic
			VAL209	3.99	Pi-Alkyl	Hydrophobic
			PRO565	4.67	Pi-Alkyl	Hydrophobic
8.	Salicylic acid	338	ASN290	2.85	Hydrogen	Conventional
				2.52	Hydrogen	Conventional
				2.52	Carbon	Carbon-Hydrogen
			ILE291	2.21	Hydrogen	Conventional
				2.12	Hydrogen	Conventional
				3.87	Pi-Alkyl	Hydrophobic
			PRO415	5.27	Pi-Alkyl	Hydrophobic
			PHE438	4.31	Pi-Pi Stacked	Hydrophobic
9.	Vanillin	1183	ASN290	2.66	Hydrogen	Conventional
			ILE291	2.76	Hydrogen	Conventional
				2.15	Hydrogen	Conventional
				4.02	Pi-Alkyl	Hydrophobic
			PRO289	3.56	Carbon	Carbon-Hydrogen
			THR434	3.61	Carbon	Carbon-Hydrogen
				2.86	Carbon	Carbon-Hydrogen
			PHE438	4.78	Pi-Pi Packed	Hydrophobic
			PRO415	5.34	Pi-Alkyl	Hydrophobic
10.	Farnesol	445070	LEU391	4.81	Alkyl	Hydrophobic
			ARG393	5.29	Alkyl	Hydrophobic
			PHE390	5.37	Pi-Alkyl	Hydrophobic
				4.18	Pi-Alkyl	Hydrophobic
			PHE40	4.87	Pi-Alkyl	Hydrophobic
			ASN394	2.71	Carbon	Carbon-Hydrogen
11.	Benzaldehyde	240	PHE438	3.65	Pi-Pi Packed	Hydrophobic
			THR434	2.90	Carbon	Carbon-Hydrogen
			PRO415	4.11	Pi-Alkyl	Hydrophobic
12.	B-pinene	14896	LYS31	5.15	Alkyl	Hydrophobic
			LEU79	4.67	Alkyl	Hydrophobic
13.	Ethyl hexanoate	31265	PHE 428	5.14	Pi-Alkyl	Hydrophobic
			PHE438	5.01	Pi-Alkyl	Hydrophobic
				5.21	Pi-Alkyl	Hydrophobic
			ILE291	5.08	Alkyl	Hydrophobic
				5.45	Alkyl	Hydrophobic
				4.24	Alkyl	Hydrophobic
			PRO415	4.37	Alkyl	Hydrophobic
				4.55	Alkyl	Hydrophobic
				2.42	Carbon	Carbon-Hydrogen
			THR414	2.73	Carbon	Carbon-Hydrogen
14.	2-heptanone	8051	PHE497	4.47	Pi-Alkyl	Hydrophobic
			TYR505	5.27	Pi-Alkyl	Hydrophobic
				5.42	Pi-Alkyl	Hydrophobic
			ARG403	5.43	Alkyl	Hydrophobic
				4.75	Alkyl	Hydrophobic
			TYR495	5.48	Pi-Alkyl	Hydrophobic

Table 8. List of the drug-induced hERG inhibition, AMES toxicity, carcinogens, *Tetrahymena pyriformis* (TP) toxicity, fish toxicity, honeybee (HB) toxicity, hepatotoxicity, rat acute toxicity (LD_{50} in mol/kg), plasma protein binding along with P-glycoprotein inhibitor (PGI) activity of all plant compounds

Compound Name	hERG inhibition	AMES	Carcinogenicity	Fish toxicity	Honeybee toxicity	Hepatotoxicity	PGI	Rat acute toxicity	TP Toxicity
Eugenol	No	No	No	Yes	Yes	No	No	1.645 kg/mol	1.124 (ug/L)
Acetyl eugenol	No	No	No	Yes	Yes	No	No	1.886 kg/mol	0.752 (ug/L)
Vanillin	No	No	No	Yes	Yes	No	No	1.537 kg/mol	0.596 (ug/L)
Eugenitin	No	No	No	Yes	Yes	Yes	No	2.412 kg/mol	1.077 (ug/L)
Caffeic acid	No	No	No	Yes	Yes	No	No	1.568 kg/mol	0.103 (ug/L)
Ferulic acid	No	No	No	Yes	Yes	No	No	1.407 kg/mol	0.368 (ug/L)
Ellagic acid	No	No	No	Yes	No	Yes	No	1.037 kg/mol	1.664 (ug/L)
Salicylic acid	No	No	No	Yes	Yes	No	No	2.092 kg/mol	-0.189 (ug/L)
β-pinene	No	No	No	Yes	Yes	No	No	1.405 kg/mol	0.733 (ug/L)
Limonene	No	No	No	Yes	Yes	No	No	1.856 kg/mol	0.113 (ug/L)
Farnesol	No	No	No	Yes	Yes	No	No	2.087 kg/mol	0.561 (ug/L)
Benzaldehyde	No	No	No	Yes	Yes	No	No	1.81 kg/mol	0.076 (ug/L)
2-heptanone	No	No	No	Yes	Yes	No	No	2.024 kg/mol	-0.267 (ug/L)
Ethyl hexanoate	No	No	No	Yes	Yes	No	No	1.59 kg/mol	0.563 (ug/L)

ADMET profiling of phytochemical compounds from plants can create a strong base for researchers and scientists in the pharmaceutical industry to create new plant-based drugs. (Roy et al., 2016).

Conclusion

In this study, the molecular docking study proved that ellagic acid (CID:5281855) had the lowest binding affinity at the value of -8.3, -7.6 and -8.1 kcal/mol with all three target proteins (PDB ID: 6LU7, 7BW4 and 6M0J) respectively. Ellagic acid shows no toxicity to humans. It could potentially be a suitable and safe drug for the treatment of SARS-CoV-2. It plays multiple roles in SARS-CoV-2 mechanisms such as viral entry; attachment to the hACE2 receptor; cleavage of translated polypeptides and viral replication process.

References

Ahammad F, Alam R, Mahmud R, Akhter S, Talukder EK, Tonmoy AM, Fahim S, Al-Ghamdi K, Samad A, Qadri I. Pharmacoinformatics and molecular dynamics simulation-based phytochemical screening of neem plant (*Azadiractha indica*) against human cancer by targeting MCM7 protein. *Briefings in Bioinformatics* (2021) 22(5):bbab098.

Allaq AA, Sidik NJ, Abdul-Aziz A, Alkamil AM, Elengoe A, Yahya EB, Abdulsamad MA. Epidemiological studies of the novel Coronavirus (COVID-19) in Libya. *Pakistan Journal of Biotechnology* (2021) 4:18(1-2) :7-16.

Batiha GE, Alkazmi LM, Wasef LG, Beshbishy AM, Nadwa EH, Rashwan EK. *Syzygium aromaticum L.* (Myrtaceae): Traditional uses, bioactive chemical constituents, pharmacological and toxicological activities. *Biomolecules* 2020 10(2):202.

Berman H, Henrick K, Nakamura H. Announcing the worldwide Protein Data Bank. *Nature Structural & Molecular Biology* (2003) 10(12):980.

Brooks BR, Brooks III CL, Mackerell Jr AD, Nilsson L, Petrella RJ, Roux B, Won Y, Archontis G, Bartels C, Boresch S, Caflisch A. CHARMM: The biomolecular simulation program. *Journal of Computational Chemistry* (2009) 30(10):1545-1614.

Dallakyan S, Olson AJ. Small-molecule library screening by docking with PyRx. *Methods in Molecular Biology* (2015).

De Smet P. Herbal medicine in Europe : Relaxing regulatory standards. *The New England Journal of Medicine* (2005) 352:1176-1178.

Elengoe A, Selvam S, Selvarajah M, Manikam H, Vijaya R. Sequence analysis and structure prediction of Malaysia SARS-CoV-2 strain's structural and accessory proteins. *Biointerface Research in Applied Chemistry* (2022) 12(1):3259-3304.

Elengoe A. Indian spices boost the immune system against COVID-19. *The Annals of the University Dunarea de Jos of Galati. Fascicle VI-Food Technology* (2020) 44(2):189-206.

Gorbalenya AE, Koonin EV, Donchenko AP, Blinov VM. Coronavirus genome: prediction of putative functional domains in the non-structural polyprotein by comparative amino acid sequence analysis. *Nucleic Acids Research* (1989) 17(12):4847-4861.

Harrison AG, Lin T, Wang P. Mechanisms of SARS-CoV-2 transmission and pathogenesis. *Trends in Immunology* (2020) 41(12):1100-1115.

Huang Y, Yang C, Xu XF, Xu W, Liu SW. Structural and functional properties of SARS-CoV-2 spike protein: potential antivirus drug development for COVID-19. *Acta Pharmacologica Sinica* (2020) 41(9):1141-1149.

Jagtap N, Yadav A, Mohite S. Synthesis, Molecular docking studies and anticancer activity of 1, 3, 4-oxadiazole-3 (2h)-thione derivatives. *Journal of University of Shanghai for Science and Technology* (2020) 22(11):535-550.

Johannah NM, RM R, G G, Maliakel B, D S, IM K. Beyond the flavour: a de-flavoured polyphenol rich extract of clove buds (*Syzygium aromaticum L*) as a novel dietary antioxidant ingredient. *Food Function Journal* (2015) 6(10) :3373-3382.

Khailany RA, Safdar M, Ozaslan M. Genomic characterization of a novel SARS-CoV-2. *Gene reports* (2020) 19:100682.

Kim S, Thiessen PA, Bolton EE, Chen J, Fu G, Gindulyte A, Han L, He J, He S, Shoemaker BA, Wang J. PubChem substance and compound databases. *Nucleic Acids Research (*2016) 44(D1):D1202-1213.

Lee HJ, Shieh CK, Gorbalenya AE, Koonin EV, La Monica N, Tuler J, Bagdzhadzhyan A, Lai MM. The complete sequence (22 kilobases) of murine coronavirus gene 1 encoding the putative proteases and RNA polymerase. *Virology* (1991) 180(2):567-582.

Li S, Li S, Disoma C, Zheng R, Zhou M, Razzaq A, Liu P, Zhou Y, Dong Z, Du A, Peng J. SARS-CoV-2: Mechanism of infection and emerging technologies for future prospects. *Reviews in Medical Virology* (2021) 31(2):e2168.

Lipinski CA. Lead-and Drug-Like Compounds : The rule-of-five revolution. *Drug Discovery Today: Technologies* (2004) 1(4):337-341.

Mariano G, Farthing RJ, Lale-Farjat SL, Bergeron JR. Structural characterization of SARS-CoV-2 : Where we are, and where we need to be. *Frontiers in Molecular Biosciences* (2020):344.

Michel CJ, Mayer C, Poch O, Thompson JD. Characterization of accessory genes in coronavirus genomes. *Virology Journal* (2020) 17(1):1-3.

Parmar DR, Soni JY, Guduru R, Rayani RH, Kusurkar RV, Vala AG, Talukdar SN, Eissa IH, Metwaly AM, Khalil A, Zunjar V. Discovery of new anticancer thiourea-azetidine hybrids: Design, synthesis, *in vitro* antiproliferative, SAR, *in silico* molecular docking against VEGFR-2, ADMET, toxicity, and DFT studies. *Bioorganic Chemistry* (2021) 115:105206.

Posthuma CC, Te Velthuis AJ, Snijder EJ. Nidovirus RNA polymerases: complex enzymes handling exceptional RNA genomes. *Virus Research* (2017) 234:58-73.

Rath S, Jagadeb M, Bhuyan R. Molecular docking of bioactive compounds derived from *Moringa oleifera* with p53 protein in the apoptosis pathway of oral squamous cell carcinoma. *Genomics & Informatics* (2021) 19(4):e46.

Roy N, Davis S, Narayanankutty A, Nazeem PA, Babu TD, Abida PS, Valsala PA, Raghavamenon A. Garlic phytocompounds possess anticancer activity by specifically targeting breast cancer biomarkers-an *in silico* study. *Asian Pacific Journal of Cancer Prevention* (2016) 17(6):2883-2888.

Sailah I, Tumilaar SG, Lombogia LT, Celik I, Tallei TE. Molecular docking and dynamics simulations study of selected phytoconstituents of "Pangi" (*Pangium edule Reinw*) leaf as anti-SARS-CoV-2. *Philippine Journal of Science* (2021) 150:925-937.

Shang J, Wan Y, Luo C, Ye G, Geng Q, Auerbach A, Li F. Cell entry mechanisms of SARS-CoV-2. *Proceedings of the National Academy of Sciences* (2020) 117(21):11727-11734.

Snijder EJ, Decroly E, Ziebuhr J. The nonstructural proteins directing coronavirus RNA synthesis and processing. *Advances in Virus Research* (2016) 96:59-126.

Tian S, Wang J, Li Y, Li D, Xu L, Hou T. The application of *in silico* drug-likeness predictions in pharmaceutical research. *Advanced Drug Delivery Reviews* (2015) 86:2-10.

Tilburt JC, Kaptchuk TJJ. Herbal medicine research and global health: An ethical analysis. *Bulletin of the World Health Organization* (2008) 86(8):594-599.

Veber DF, Johnson SR, Cheng HY, Smith BR, Ward KW, Kopple KD. Molecular properties that influence the oral bioavailability of drug candidates. *Journal of Medicinal Chemistry* (2002) 45(12):2615-2623.

Ziebuhr J, Snijder EJ, Gorbalenya AE. Virus-encoded proteinases and proteolytic processing in the Nidovirales. Journal *of General Virology* (2000) 81(4):853-879.

Chapter 16

Bioactive Phytoconstituents of *Andrographis paniculata* as Potential Inhibitors of SARS-CoV-2

Shanaya Amath[1], BSc
and Asita Elengoe[2,*], PhD

[1]School of Biotechnology, Manipal International University (MIU), Putra Nilai, Negeri Sembilan, Malaysia
[2]Department of Biotechnology, Faculty of Applied Science, Lincoln University College, Selangor, Malaysia

Abstract

Medicinal and herbal plants are often used as alternative medicines because they contain bioactive compounds that can be proven effective in the treatment of diseases with minimal side effects. They are easily obtained from the surrounding environment, as in the case of green chiretta (*Andrographis paniculata*) which is a medicinal plant that is indigenous to the nations of Sri Lanka and India. This plant is known to have antiviral activity. This study primarily explored the potential bioactive compounds from green chiretta as well as the pharmacokinetic properties of these bioactive compounds, which can be potentially used in the treatment of severe acute respiratory syndrome coronavirus 2 (SARS-CoV-2). The RNA-dependent RNA polymerase (RdRp) (PDB ID: 7BW4) and nine natural compounds of green chiretta were obtained from RCSB Protein Data Bank and PubChem database respectively. Molecular docking analysis was carried out to explore the molecular interaction between 7BW4 and neoandrographolide, andrographolide,

---

* Corresponding Author's Email: asitaelengoe@yahoo.com.

In: Medicinal Plants and COVID-19
Editor: Asita Elengoe
ISBN: 979-8-88697-958-9

paniculide-A, deoxyandrographolide, 5-hydroxy-7,8,2'-trimethoxy-flavone, paniculide-B and 14-deoxy-11-oxoandrographolide using SwissDock. The pharmacokinetics and toxicity analysis were also performed to find a safe and suitable drug against SARS-CoV-2 using SwissADME and admetSAR tools. Andrographolide had the highest number of hydrogen bonds (3) and a good binding affinity of -7.1 kcal/mol among the nine phytocompounds. It shows no carcinogenicity and mutagenesis to humans. There is prominent potential for andrographolide to inhibit virus replication. This could serve as a milestone in the initial development of a natural therapeutic medicine that can be used against COVID-19 disease.

Keywords: *Andrographis paniculata*, andrographolide, docking, SARS-CoV-2, toxicity

Introduction

Coronaviruses are a part of a huge category of single-stranded, enveloped, positive-sense RNA infections that possess the capability of contaminating a wide variety of creatures, inclusive of people, birds, rodents, carnivores, chiropters and other warm-blooded animals. Although they have been known for numerous years and have been categorized as one of the viral sources capable of causing respiratory illnesses, they caught the attention of the entire world in December 2019, when an epidemic scene of cases with respiratory tract contaminations was reported in Wuhan, which is the biggest metropolitan region within the province of Hubei in China (Muhammad Fayyaz ur Rehman, 2021; Elengoe, 2020). In late December of that year, an episode of puzzling pneumonia characterized by dry cough, fever, weakness, and incidental gastrointestinal side effects happened in a fish wholesale wet market, which was the Huanan Fish Wholesale Market, in Wuhan, Hubei, China. The introductory episode was detailed within the market in December 2019 and included approximately 66% of the staff who worked there. The market was closed down on January 1, 2020, after the declaration of an epidemiologic alarm by the neighborhood wellbeing specialist on December 31, 2019. Be that as it may, within the following month (January) many individuals in China, inclusive of numerous territories such as Hubei, Zhejiang, Guangdong, etc.; and cities such as Beijing and Shanghai were assaulted by the uncontrolled spreading of the infection (Wu et al., 2020).

Since the primary affirmed diagnosis of SARS-CoV-2 in China, more than 74.3 million individuals have been affected by the virus, from which more than 1.67 million lives have been claimed as of December 19, 2020. Even though more than 52 million individuals have vanquished COVID-19 and recuperated from the illness, the fight between SARS-CoV-2 and people is continuing today and no particular therapeutics are accessible. However, researchers hope that medicinal plants may be the potential to be anti-SARS-CoV-2 agents (Muhammad Fayyaz ur Rehman, 2021). Since ancient times, people have used plants as a source of medicine, and many widely used pharmaceuticals have a plant-based basis (Elengoe and Hamdan, 2018). Based on the Johnson et al., literature review study (2008), it has been demonstrated that a plant compound (prostratin) was used as an anti-HIV in Samoa. The prostratin activates the latent T-cell pools (Korin et al., 2002).

Andrographis paniculata belongs to the Acanthaceae family. Green chiretta is the common name for *Andrographis paniculata.* It has been used in ayurvedic treatment. It possesses many medicinal activities such as anti-microbial, antioxidant, anti-viral, anti-inflammatory, anti-cancer, anti-angiogenic, anti-diabetic, develops a stronger immune system, treats a variety of chronic diseases such as liver, kidney and heart related ailments, etc. The phytoconstituents of *Andrographis paniculata* are Diterpenes, flavonoids, xanthones, noriridoides, etc. (Okhuarobo et al., 2014).

Computational biology plays an essential role in plant-based drug design. The tools play a role in the identification of genes and mechanisms that may be linked to significant bioactive secondary metabolites. Shafi Mahmud and his colleagues (2022) used a deep literature mining method to generate a phytochemical dataset of 1145 phytoconstituents. They analyzed the molecular interaction between the plant compounds and the main protease of SARS-CoV-2 (M^{pro}). They determined the best inhibitory action of the phytocompound from the 1145 phytoconstituents against the main protease of SARS-CoV-2. Cosmosiine, pelargonidin-3-o-glucoside and cleomiscosin A show the best docking score through AutoDock version 1.2.0. They bound at the binding affinities of -8.4, -8.4 and -8.2 kcal/mol respectively. They found that these substances may bind to the GLN189, GLU166, CYS145, HIS41 and MET165 residues on the targeted protein's active site, causing a targeted inhibition. They also prove that these molecules do not contain carcinogenicity and mutagenicity activities.

In this study, the molecular interaction between SARS-CoV-2 main protease protein, M^{pro} (PDB ID: 7BW4) and the nine phytoconstituents from the *Andrographis paniculata* plant was explored using docking, pharmacokinetics and toxicity evaluation approaches.

Methodology

Plant Selection

The green chiretta (*Andrographis paniculata*) plant was selected due to its high concentration of bioactive compounds and related medicinal uses. It has been used for centuries to treat a variety of illnesses, including cancer, diabetes, high blood pressure, ulcer, leprosy, bronchitis, skin conditions, flatulence, colic, influenza, dysentery, dyspepsia, and malaria. Nine phytochemical compounds (neoandrographolide, andrographolide, paniculide-A, deoxyandrographolide, 5-hydroxy-7,8,2'-trimethoxyflavone, 5-hydroxy-7,8,2,'3'-tetramethoxyflavone, paniculide-B, 14-deoxy-11-oxoandrographolide and paniculide-C) were identified.

Phytochemical (Ligand) Preparation

The bioactive compounds of *Andrographis paniculata* were used to create a ligand library. The three-dimensional (3-D) models of the nine phytoconstituents were retrieved and downloaded in the SDF format from the PubChem database (Kim et al., 2016).

RNA-dependent RNA Polymerase (RdRp) (Target Protein) Preparation

RNA-dependent RNA polymerase (RdRp) controls viral replication and transcription machinery. It is also called non-structural protein 12 (nsp12). It appears to be remdesivir's main target when used as an antiviral medication. The 3-D structure of this protein (PDB ID: 7BW4) was downloaded in PDB format from the RCSB Protein Data Bank (Berman et al., 2003). Pre-preparation of the protein was carried out using Discovery Studio (DS) 4.0 (Brooks et al., 2009). This included the addition of polar bonds and hydrogen atoms, as well as minimizing protein energy. The protein model was publicly available.

Drug Likeliness and Toxicity Analysis

The phytoconstituents of green chiretta were screened using the Lipinski Rule of 5 (Lipinski, 2004; Veber et al., 2002; Jagtap et al., 2020). Physicochemical properties, lipophilicity, water solubility, drug-likeness, medicinal chemistry and drug absorption were also determined using the online server tool SwissADME. Besides that, the toxicological properties of the selected plant compounds were also predicted via the admetSAR server tool (Tian et al., 2015). The properties were carcinogenicity, AMES toxicity, hERG inhibition, hepatotoxicity, *Tetrahymena pyriformis* (TP), fish, honeybee toxicity, rat acute toxicity, plasma protein binding and P-glycoprotein inhibitor. This analysis was carried out to predict the potential bioactive compound to be a drug for COVID-19 disease. Only the plant compounds that noted 'yes' could be further analyzed for molecular docking because they passed the Lipinski Rule of 5. In addition, they should also pass the toxicity analysis for further investigation.

Molecular Docking

The docking of the target protein (PDB ID: 7BW4) with its relevant plant compound was performed using SwissDock (Grosdidier et al., 2011). The 3-D structure of the RdRp-phytoconstitute complex was viewed using DS 4.0. The binding affinity (kcal/mol), the number of hydrogen, hydrophobic, Pi Anion, Pi Cation, Pi-Pi T shaped and carbon bonds and the distance of the bonds between the RdRp and plant compound were determined (Parmar et al., 2021).

Results and Discussion

Green chiretta is a plant that can primarily be found in Sri Lanka and India. In the past, it has proven to be effective against bacterial infections. This plant has also been used in Ayurvedic medicine, which is an alternate medicinal system that emphasizes herbal remedies, originating from India. It was selected in this study because the leaves of this plant are proven to show antiviral activity, which can potentially be used in the fight against the COVID-19 pandemic. Green chiretta has been a part of Thailand's alternative treatment program. A pilot incentive could be an alternative medical treatment to fight COVID-19. The cabinet of Thailand approved this incentive to treat asymptomatic patients among prison inmates. However, there is still ample debate and dispute on whether or not this drug is sufficiently effective and safe to use against COVID-19. It is conclusive that substantial research must be carried out to potentially develop green chiretta into a form that is accepted worldwide as a drug candidate effective against SARS-CoV-2. The antiviral activity (H5N1) of green chiretta can serve as a stepping stone in the research that explores how the leaves of this plant can serve as an inhibitory component against SARS-CoV-2 using a molecular docking approach.

COVID-19 has persisted and spread worldwide for longer than predicted. Currently, there is prominent research based on therapeutic medicines that can be used to fight against COVID-19 being carried out. This particular study emphasizes blocking the viral life cycle. In this study, the molecular interaction between the target protein (PDB ID: 7BW4) and plant compounds from green chiretta was studied using computational biology tools such as molecular docking, pharmacokinetics and toxicity analysis.

To evaluate the possibility and potential of the phytoconstituents being developed as drugs, SwissADME and AdmetSAR were utilized. These are online server tools that can be used to assess the pharmacokinetics and drug-likeness of various compounds and molecules. Out of 9 plant compounds, 7 had passed the Lipinski Rule of 5 (Table 1).

An analysis of the toxicity of the seven selected phytoconstituents was carried out. This was performed using AdmetSAR, which can be used to identify and analyze the toxicity profile of various compounds. Table 2 shows that all plant compounds do not have carcinogenicity. Only 5-hydroxy-7, 8, 2,' 3'-tetramethoxyflavone and 5-hydroxy-7, 8, 2'-trimethoxyflavone were positive for P-glycoprotein inhibitor (PGI), AMES toxicity and hepatotoxicity. All phytoconstituents show fish and honeybee toxicity. Furthermore, neoandrographolide had hERG inhibition activity. Deoxyandrographolide, andrographolide and 14-deoxy-11-oxoandrographolide can be subjected to further experimentation and investigation, both *in vitro* and *in vivo* settings, to analyze its development as an inhibitory component against SARS-CoV-2.

Table 1. List of pharmacokinetics properties, molecular weight (MW), hydrogen bond donor (HBD), hydrogen bond acceptor (HBA), partitioning coefficient (LogP), number of rotatable bond (RB), number of heavy atoms (HA), number of aromatic heavy atoms (AHA), synthetic accessibility (SA), gastrointestinal (GI) absorption, and Lipinski's Rule of 5 of all plant compounds

Compound Name	PubChem CID	Mw/ g/mol	HBD	HBA	RB	Log P	Log S	SA	GI	LR	Num. HA	Num. arom. HA
Deoxyandrographolide	21679042	334.4	2	4	4	3.03	-3.48	4.95	High	Yes	24	0
Andrographolide	5318517	350.4	3	5	3	2.61	-3.18	5.06	High	Yes	25	0
14-Deoxy-11-oxoandrographolide	101593061	348.4	2	5	4	2.46	-2.79	4.87	High	Yes	25	0
Neoandrographolide	9848024	480.6	4	8	7	3.37	-4.01	6.36	High	Yes	34	0
5-Hydroxy-7,8,2,'3'-Tetramethoxyflavone	44258544	520.5	4	12	8	3.19	-3.43	5.85	Low	No	37	16
5-Hydroxy-7,8,2'-Trimethoxyflavone	44258542	490.5	4	11	7	2.95	-3.34	5.69	Low	No	35	16
Paniculide-A	11821485	264.3	1	4	3	2.88	-2.07	4.59	High	Yes	19	0
Paniculide-B	101289823	280.3	2	5	4	2.29	-1.45	4.64	High	Yes	20	0
Paniculide-C	101289824	278.3	1	5	4	2.31	-1.93	4.37	High	Yes	20	0

Table 2. List of the drug-induced hERG inhibition, AMES toxicity, carcinogens, *Tetrahymena pyriformis* (TP) toxicity, fish toxicity, honeybee (HB) toxicity, hepatotoxicity, rat acute toxicity (LD_{50} in mol/kg), plasma protein binding along with P-glycoprotein inhibitor (PGI) activity of all bioactive compounds

Compound Name	hERG inhibition	AMES	Carcinogenicity	Hepatotoxicity	Honeybee toxicity	Fish toxicity	Plasma protein binding (%)	Rat Acute Toxicity (kg/mol)	TP Toxicity (μg/L)	PG1
Deoxyandrographolide	No	No	No	Yes	Yes	No	0.68	2.179	0.635	No
Andrographolide	No	No	No	Yes	Yes	No	0.595	2.039	0.724	No
14-Deoxy-11-oxoandrographolide	No	No	No	Yes	Yes	No	0.562	2.479	0.69	No
Neoandrographolide	No	No	No	Yes	Yes	No	0.866	3.165	1.076	No
Paniculide-A	No	No	No	Yes	Yes	No	0.981	1.663	1.07	No
Paniculide-B	No	No	No	Yes	Yes	No	0.937	1.693	1.09	No
Paniculide-C	No	No	No	Yes	Yes	No	0.948	1.396	1.35	No

The selected seven plant compounds were docked with the target protein (PDB ID: 7BW4). The bond types and distances were analyzed. The observation of the binding interactions between the phytoconstituents of *Andrographis paniculata* and the target protein provided insight into the deeper understanding of how a therapeutic drug could be potentially developed through exploring the various types of interactions such as hydrophobic, hydrogen, aromatic interactions, etc.

The bioactive compound that showed the lowest binding affinity or the most negative value was neoandrographolide which had a binding affinity value of -7.8 kcal/mol (Table 3). This value may indicate the formation of various bonds including hydrogen and hydrophobic interactions. Neoandrographolide bound to 7BW4 at residue SER1 through a hydrogen bond with a distance of 2.32 A (Table 4). 14-deoxy-11-oxoandrographolide and 5-hydroxy-7, 8, 2'-trimethoxyflavone also had a relatively low binding affinity of -7.6 kcal/mol. 14-deoxy-11-oxoandrographolide had two hydrogen bonds with 7BW4 at the residus of ASN209 and ARG233 and ASP36 with bond lengths of 2.56 and 2.90 A respectively. 5-hydroxy-7, 8, 2'-trimethoxyflavone had the highest number of hydrogen bonds (4) among the seven plant compounds. It interacted with 7BW4 at the residues of ARG553, CYS622, ASP760 and SER759 through hydrogen bond lengths of 2.06, 1.72, 1.89 and 2.98 A respectively.

Table 3. Binding affinity score between phytocompound and target protein (7BW4)

No.	Compound name	Pubchem ID	Ligand	Binding Affinity (kcal/mol)
1.	14-Deoxy-11-Oxoandrographolide	101593061	7bw4_101593061_mmff94_E=89.74	-7.6
2.	Paniculide A	11821485	7bw4_11821485_mmff94_E=5.51	-6.1
3.	Deoxyandrographolide	21679042	7bw4_21679042_mmff94_E=61.06	-6.7
4.	Andrographolide	5318517	7bw4_5318517_mmff94_E=104.09	-7.1
5.	Neoandrographolide	9848024	7bw4_9848024_mmff94_E=168.04	-7.8
6.	Paniculide-B	101289823	7bw4_101289823_mmff94_E=4.61	-6.6
7.	Paniculide-C	101228924	7bw4_101289824_mmff94_E=1822.39	-5.9

Furthermore, andrographolide had a binding affinity value of -7.1 kcal/mol. It was bound with 7BW4 at residues of ASP761, TRP617 and ARG836 through three hydrogen bonds with distances of 3.06, 2.68d 2.44 A respectively. Deoxyandrographolide had a binding affinity of -6.7 kcal/mol. It was positioned by 2 residues, ASP618 and ARG836, with respective distances of 2.85 and 2.19 A, both of which are hydrogen bonds. Paniculide B had two hydrogen bonds with 7BW4 at residues of GLU23 and ASN447 through bond lengths of 2.55 and 2.12 A. Lastly, Paniculide A had a binding affinity of -6.1 kcal/mol. It had only one hydrogen bond which interacted at the residue of ILE266 with a distance of 2.37 A.

Table 4. List of hydrogen bond interactions between target protein (7BW4) and ligands (plant compounds)

No.	Compound name	PubChem ID	Residues	Distance	Bond category	Bond type
1.	Neoandrographolide	9848024	SER1	2.32	Hydrogen	Conventional
2.	14-Deoxy-11-Oxoandrographolide	101593061	ASN209	2.56	Hydrogen	Conventional
			ARG233	2.90	Hydrogen	Conventional
			ASP36	3.73	Carbon	Carbon-Hydrogen
3.	Paniculide A	11821485	ILE266	2.37	Hydrogen	Conventional Hydrogen
			PRO323	4.75	Hydrophobic	Alkyl
			TYR265	4.23	Hydrophobic	Alkyl
			TRP268	5.45	Hydrophobic	Alkyl
			PRO322	5.35	Hydrophobic	Alkyl
4.	Deoxyandrographolide	21679042	ASP618	2.85	Hydrogen	Conventional Hydrogen
			ARG836	2.19	Hydrogen	Conventional Hydrogen
5.	Andrographolide	5318517	ASP761	2.06	Hydrogen	Conventional Hydrogen
			TRP617	2.68	Hydrogen	Conventional Hydrogen
			ARG836	2.44	Hydrogen	Conventional Hydrogen
			ASP833	3.44	Carbon	Carbon Hydrogen bond
6.	Paniculide C	101289923	GLU23	2.55	Hydrogen	Conventional Hydrogen
			ASN447	2.12	Hydrogen	Conventional Hydrogen
			TRP29	3.79	Carbon	Carbon Hydrogen bond
			PRO183	3.64	Carbon	Carbon Hydrogen bond

Conclusion

An in-depth study of 9 phytoconstituents from the *Andrographis paniculata* plant has been successfully carried out to identify the potential natural compounds that can be used against SARS-CoV-2. Molecular docking results revealed that neoandrographolide shows the best potential to be an inhibitory component against the SARS-CoV-2 activity. However, this milestone is only the initial step in developing a potential drug that can inhibit SARS-CoV-2 replication. There has to be in depth *in vitro* and *in vivo* experiments should be carried out to validate the results from the computational biology approaches. Further analysis of the molecular mechanism and potential side effects of the natural compound should be performed.

References

Berman H, Henrick K, Nakamura H. Announcing the worldwide Protein Data Bank. *Nature Structural & Molecular Biology* (2003) 10(12):980.

Brooks BR, Brooks III CL, Mackerell Jr AD, Nilsson L, Petrella RJ, Roux B, Won Y, Archontis G, Bartels C, Boresch S, Caflisch A. CHARMM: The biomolecular simulation program. *Journal of Computational Chemistry* (2009) 30(10):1545-1614.

Elengoe A, Hamdan S. Computational approach towards exploring interaction of target protein-phytocompounds in drug development for breast cancer. In *Anticancer plants: mechanisms and molecular interactions* 2018 (pp. 131-141). Springer, Singapore.

Elengoe A. Indian spices boost the immune system against COVID-19. *The Annals of the University Dunarea de Jos of Galati. Fascicle VI-Food Technology* (2020) 44(2):189-206.

Grosdidier A, Zoete V, Michielin O. Swissdock, A protein-small molecule docking web service based on Eadock DSS. *Nucleic Acids Research* (2011) 39(suppl_2):W270-277.

Jagtap N, Yadav A, Mohite S. Synthesis, Molecular docking studies and anticancer activity of 1, 3, 4-oxadiazole-3 (2h)-thione derivatives. *Journal of University of Shanghai for Science and Technology* (2020) 22(11):535-550.

Johnson HE, Banack SA, Cox PA. Variability in content of the anti-AIDS drug candidate prostratin in Samoan populations of *Homalanthus nutans*. *Journal of Natural Products* (2008) 71(12):2041-2044.

Kim S, Thiessen PA, Bolton EE, Chen J, Fu G, Gindulyte A, Han L, He J, He S, Shoemaker BA, Wang J. PubChem substance and compound databases. *Nucleic Acids Research* (2016) 44(D1):D1202-1213.

Kasilingam T, Elengoe A. *In silico* molecular modeling and docking of apigenin against the lung cancer cell proteins. *Asian Journal of Pharmaceutical and Clinical Research* (2018) 11(9):246-252.

Korin YD, Brooks DG, Brown S, Korotzer A, Zack JA. Effects of prostratin on T-cell activation and human immunodeficiency virus latency. *Journal of Virology* (2002) 76(16):8118-8123.

Lipinski CA. Lead-and Drug-Like Compounds: The rule-of-five revolution. *Drug Discovery Today: Technologies* (2004) 1(4):337-341.

Mahmud S, Afrose S, Biswas S, Nagata A, Paul GK, Mita MA, Hasan MR, Shimu MS, Zaman S, Uddin MS, Islam MS. Plant-derived compounds effectively inhibit the main protease of SARS-CoV-2: An *in silico* approach. *Plos One* (2022) 17(8):e0273341.

Muhammad Fayyaz ur Rehman CF. Novel coronavirus disease (COVID-19) pandemic: A recent mini review. *Computational and Structural Biotechnology Journal* (2021) 19:612-623.

Okhuarobo A, Falodun JE, Erharuyi O, Imieje V, Falodun A, Langer P. Harnessing the medicinal properties of *Andrographis paniculata* for diseases and beyond: A review of its phytochemistry and pharmacology. *Asian Pacific Journal of Tropical Disease* (2014) 4(3):213-222.

Parmar DR, Soni JY, Guduru R, Rayani RH, Kusurkar RV, Vala AG, Talukdar SN, Eissa IH, Metwaly AM, Khalil A, Zunjar V. Discovery of new anticancer thiourea-azetidine hybrids: Design, synthesis, *in vitro* antiproliferative, SAR, *in silico* molecular docking against VEGFR-2, ADMET, toxicity, and DFT studies. *Bioorganic Chemistry* (2021) 115:105206.

Tian S, Wang J, Li Y, Li D, Xu L, Hou T. The application of *in silico* drug-likeness predictions in pharmaceutical research. *Advanced Drug Delivery Reviews* (2015) 86:2-10.

Veber DF, Johnson SR, Cheng HY, Smith BR, Ward KW, Kopple KD. Molecular properties that influence the oral bioavailability of drug candidates. *Journal of Medicinal Chemistry* (2002) 45(12):2615-2623.

Wu YC, Chen CS, Chan YJ. The outbreak of COVID-19: An overview. *Journal of the Chinese Medical Association* (2020):217-220.

Chapter 17

The Identification of Bioactive Phytoconstituents from Turmeric Plant as Potential Inhibitors of SARS-CoV-2: A Computational Biology Approach

Asita Elengoe[1,*], PhD
and Muskan Karmani[2], BSc

[1]Department of Biotechnology, Faculty of Applied Science, Lincoln University College, Petaling Jaya, Selangor, Malaysia
[2]Faculty of Biotechnology, INTI International University, Persiaran Perdana BBN, Putra Nilai, Malaysia

Abstract

Since ancient times, plants have been used as a source of medicine. Today, increasingly commercially significant medications have their roots in plants. Medicinal plants have been proven to treat a variety of diseases with low or no side effects. In this study, the physiochemical, pharmacokinetics and toxicity of plant compounds from the *Curcuma longa* (turmeric) plant was analyzed against severe acute respiratory syndrome coronavirus 2 (SARS-CoV-2) due to their pharmaceutical properties such as anti-inflammation, antiviral, etc. Three-dimensional (3-D) of twenty selected bioactive compounds from turmeric plant were retrieved from the PubChem database in SDF format. The pharmacokinetics and toxicity prediction analyses were carried out using SwissADME and admetSAR tools. The SwissADME results show that all the plant compounds passed Lipinski's Rule. Based on admetSAR

* Corresponding Author's Email: asitaelengoe@yahoo.com.

In: Medicinal Plants and COVID-19
Editor: Asita Elengoe
ISBN: 979-8-88697-958-9

analysis, all the phytocompounds show no carcinogenicity. However, only riboflavin (CID:493570) was identified to have AMES toxicity whereas borneol (CID:64685) and bisacurone (CID:14287397) had hERG inhibition activity. The information from the computational biology approach will aid in designing a plant-based drug for SARS-CoV-2.

Keywords: *Curcuma longa*, pharmacokinetics, SARS-CoV-2, toxicity

Introduction

The COVID-19 pandemic has contributed to a remarkable loss of human life globally and poses an immense threat to public health, food supplies and the working world. The social and economic damage caused by the pandemic is catastrophic: millions of people are at risk of descending into extreme poverty, whilst the amount of malnourished people in 2020 is reportedly projected to be nearly 690 million (ILO, 2020). The COVID-19 outbreak was declared a Public Health Emergency of International Concern (PHEIC). Severe acute respiratory syndrome coronavirus 2 (SARS-CoV-2) was the cause of COVID-19 disease (Elengoe et al., 2022). It has spread to many countries and territories (Bender, 2020). It is understood now that the virus is spread by direct contact with the respiratory droplets of an infected individual (generated through coughing and sneezing). Individuals can also be affected by handling objects contaminated with the virus and touching their faces (e.g., eyes, nose and mouth) (Allaq et al., 2021). The virus can persist on the surface for a few hours; however, basic disinfectants could kill it (Bender, 2020). Therefore, there is an urgency to develop a new drug to treat COVID-19 disease.

Phytoconstituents are becoming more popular in the treatment of a variety of diseases such as diabetes mellitus, hypertension, high-level cholesterol, kidney, heart and liver-related diseases, etc. (Elengoe and Hamdan, 2018). This is because the natural compounds are safe for humans; have high chemodiversity; have higher bioavailability; and are inexpensive. Bioactive compounds are developing into fresh candidates for the creation of brand-new drugs. A variety of biological activities are displayed by phytocompounds, which are employed as anti-inflammatory, antibacterial, anticancer, and immunomodulatory medications (Rallabandi et al., 2020). However, it is still difficult to identify them. Complete reliance on time-consuming *in vitro* and *in vivo* screening technologies is the main obstacle to the finding of an effective phytocompound.

Curcuma longa belongs to the Zingiberaceae family. Tumeric is the common name for *Curcuma longa*. It is the most common Indian spice and a major ingredient in chicken, mutton and fish curry powder. Numerous health benefits of turmeric have been scientifically demonstrated, including the ability to protect against Alzheimer's disease and cancer. Its most active component, curcumin, also has the potential to boost heart health. It works well as an antioxidant and anti-inflammatory. Additionally, it might lessen arthritis and depression symptoms (Hewlings and Kalman, 2017).

A computational drug development approach using bioinformatics and online databases would be less expensive and time-consuming. Currently, many naturally occurring drugs-like compounds and disease-related target macromolecules can be found in three-dimensional (3-D) structures due to the recent developments in computational and structural biology fields. These 3-D models have been deposited in online databases like RCSB-Protein Data Bank (RCSB-PDB), UniProt, PubChem, DNA Data Bank of Japan (DDBJ), DrugBank, ZINC, etc. With the help of bioinformatics tools, the field of drug development is quickly expanding, and the accessibility of digitalized molecular data is opening up new opportunities in the field of computer-aided drug discovery (CADD). Due to its efficiency, speed, and ease of use, virtual library screening (VLS) has gained popularity. It analyses drug candidates using computational tools like AutoDock Vina, SwissDock, HDock, CB-Dock, GOLD, Glide, etc (Rallabandi et al., 2020). Zígolo and co-researchers (2022) explored the molecular interaction between the target proteins (SARS-CoV-2 spike (S) glycoprotein and the main protease M^{pro}) and natural compounds from medicinal plants using the Autodock tool. They chose these target proteins because spike and M^{pro} play important in viral propagation and host cells' identification of the virus. They found that glycyrrhizin had the best binding energy with spike protein at the value of -19.22 kcal/mol (the most negative value). While M^{pro} had the best docking score (-9.36 kcal/mol) with spirostan. These findings imply that these natural substances may have applications as medications that can be experimentally tested against COVID-19.

In this study, the pharmacokinetics and toxicity evaluation analysis for the ten phytoconstituents from the *Curcuma longa* plant was determined using SwisADME and admetSAR tools.

Methodology

Plant Selection

The turmeric (*Curcuma longa*) plant was chosen due to its high concentration of secondary metabolites of bioactive compounds and related medicinal uses. It may delay aging and fight age-related chronic diseases. It also helps to prevent cardiovascular disease, cancer, arthritis, and Alzheimer's disease. It is known to possess biological activities such as antioxidant, anti-bacterial, anti-cancer, anti-fungal, anti-inflammatory and anti-viral. Twenty phytoconstituents (curcumin, 5'-methoxycurcumin, dihydrocurcumin, d-sabinene, borneol, zingiberene, germacrone, bisacurone, curcumenone, dehydrocurdione, procurcumadiol, curcumenol, isoprocurcumenol, epiprocurcumenol, procurcumenol, curlone, 2-hydroxymethyl, anthraquinone, riboflavin, niacin and ascorbic acid) were identified.

Phytochemical (Ligand) Preparation

The bioactive compounds of Curcuma longa were used to create a ligand library. The three-dimensional (3-D) models of the twenty phytochemical compounds were obtained and downloaded in the SDF format from the PubChem database (Kim et al., 2016).

Determination of Pharmacokinetics of the Plant Compounds

The phytoconstituents of turmeric were screened using the Lipinski Rule of 5 (Lipinski, 2004; Veber et al., 2002; Jagtap et al., 2020). Physicochemical properties, lipophilicity, water solubility, drug-likeness, medicinal chemistry and drug absorption were also determined using the SwissADME tool.

Toxicity Prediction Analysis

The toxicological properties of the selected plant compounds were also analyzed using admetSAR server (Tian et al., 2015). AMES mutagenesis, acute oral toxicity (c), androgen receptor binding, aromatase binding, avian toxicity, blood brain barrier, BRCP inhibitor, biodegradation, BSEP inhibitor, CACO-2, carcinogenicity (binary), carcinogenicity (trinary), crustacea aquatic toxicity, CYP1A2 inhibition, CYP2C19 inhibition, CYP2C9 inhibition, CYP2C9 substrate, CYP2D6 inhibition, CYP2D6 substrate, CYP3A4 inhibition, CYP3A4 substrate, CYP inhibitory promiscuity, eye corrosion, eye irritation, estrogen receptor binding, fish aquatic toxicity, glucocorticoid receptor binding, honey bee toxicity, hepatotoxicity, human either-a-go-go (hERG) inhibition, human intestinal absorption, human oral bioavailability, MATE1 inhibitor, micronuclear, acute oral toxicity, OATP1B1 inhibitor, OATP1B3 inhibitor, OATP2B1 inhibitor, OCT1 inhibitor, OCT2 inhibitor, P-glycoprotein inhibitor, P-glycoprotein substrate, PPAR gamma, plasma protein binding, subcellular localization, Tetrahymena pyriformis, thyroid receptor binding, UGT catalyzed and water solubility were determined to predict the potential of bioactive compound to be a safe and suitable drug for COVID-19 treatment.

Results and Discussion

In late 2019, a novel Coronavirus was found and given the designation 2019-nCoV; it was then dubbed severe acute respiratory syndrome coronavirus 2 (SARS-CoV-2) (Keni et al., 2020) which is a highly contagious disease caused by a new strain of coronavirus and is linked to the same family of viruses as a severe acute respiratory syndrome (SARS) and some types of the common cold (Hasöksüz et al., 2020; Mohan et al., 2021). Even though all coronaviruses can infect humans and cause severe respiratory illnesses, the latest SARS-CoV virus has been deemed superior to previous coronaviruses such as SARS, MERS, and influenza as the volume of confirmed SARS-CoV-2 infections surpasses the total number of probable SARS cases. In terms of public distribution and intensity, the novel coronavirus is anticipated to be more contagious than SARS (Acter et al., 2020).

Traditional medicine has long utilized turmeric (Curcuma longa). Apart from giving curry its vivid yellow color, turmeric is a flavorful spice that is generally grown from the rhizomes, or roots, of a flowering plant that grows in India and other regions of Southeast Asia. Turmeric is also well known for having strong anti-inflammatory and antioxidant effects. It possesses a variety of medicinal benefits such as reducing cardiovascular risk; may prevent different types of cancer such as prostate, gastric, breast, etc.; treating osteoarthritis; preventing diabetes mellitus; anti-Alzheimer's disease; antidepressant; treating rheumatoid arthritis; improves skin health; protects from free radicals; anti-aging, eye degeneration, etc (Hewlings and Kalman, 2017).

Table 1. List of bioactive compounds from turmeric (*Curcuma longa*)

No.	Name of bioactive compound	PubChem ID	Chemical Formula	3D structure
1.	curcumin	969516	$C_{21}H_{20}O_6$	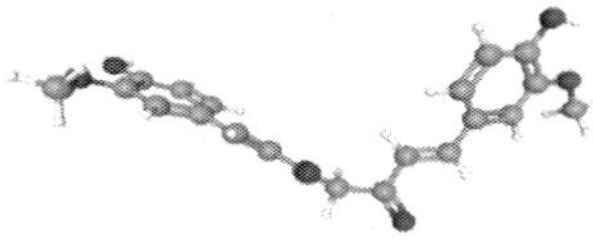
2.	5'-methoxycurcumin	10250249	$C_{22}H_{22}O_7$	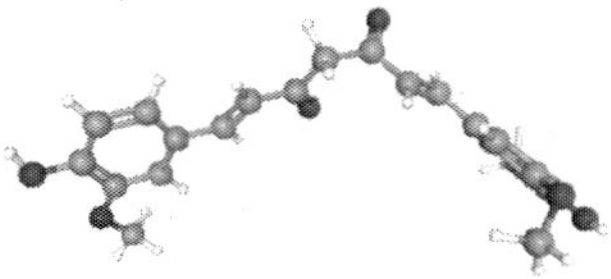
3.	dihydrocurcumin	10429233	$C_{21}H_{22}O_6$	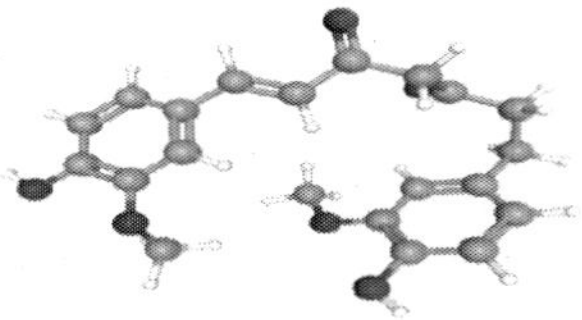
4.	d-sabinene	10887971	$C_{10}H_{16}$	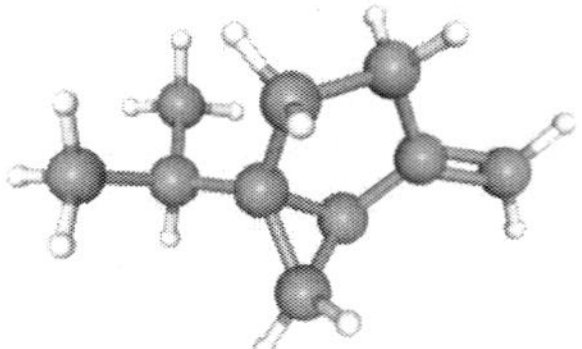
5.	borneol	64685	$C_{10}H_{18}O$	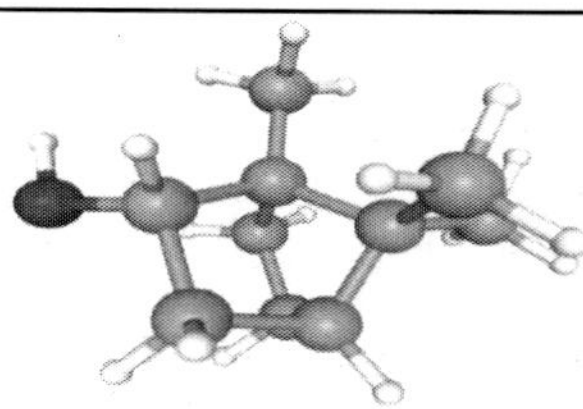
6.	zingiberene	92776	$C_{15}H_{24}$	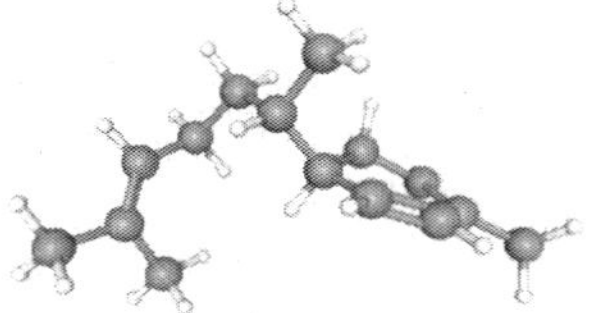

No.	Name of bioactive compound	PubChem ID	Chemical Formula	3D structure
7.	germacrone	6436348	$C_{15}H_{22}O$	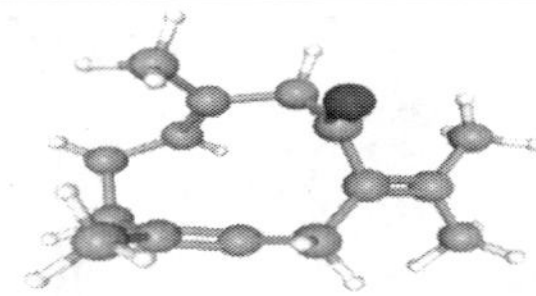
8.	bisacurone	14287397	$C_{15}H_{24}O_3$	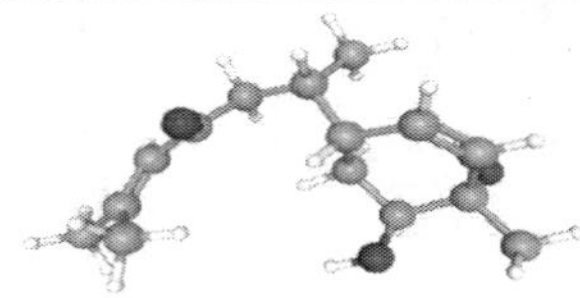
9.	curcumenone	153845	$C_{15}H_{22}O_2$	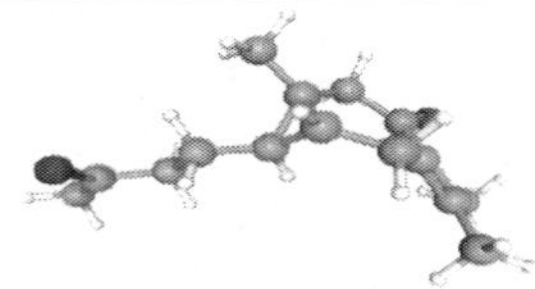
10.	dehydrocurdione	6442617	$C_{15}H_{22}O_2$	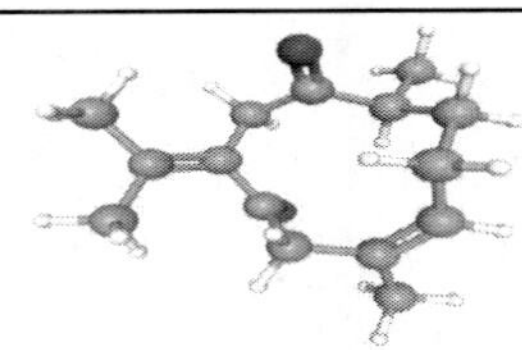
11.	procurcumadiol	14633011	$C_{15}H_{22}O_3$	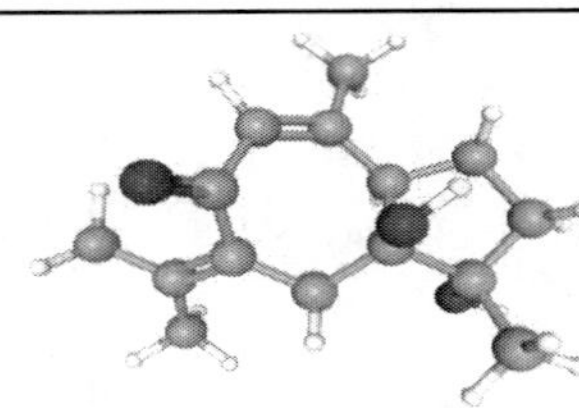
12.	curcumenol	167812	$C_{15}H_{22}O_2$	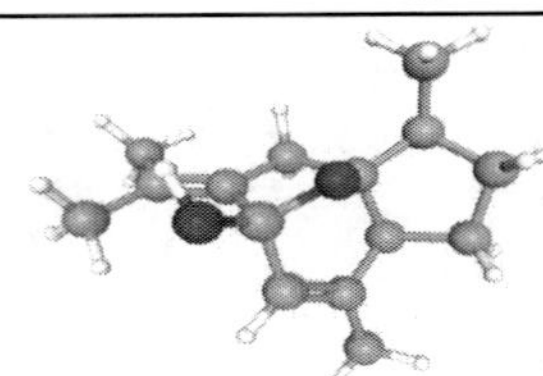
13.	Isoprocurcumenol	14543198	$C_{15}H_{22}O_2$	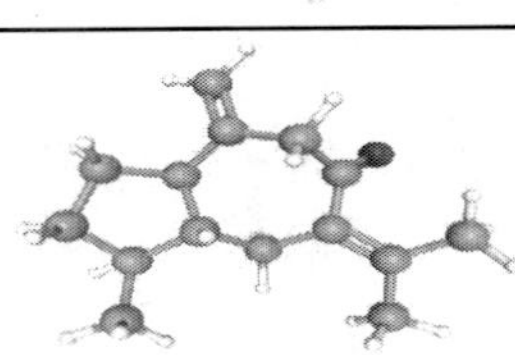

Table 1. (Continued)

No.	Name of bioactive compound	PubChem ID	Chemical Formula	3D structure
14.	epiprocurcumenol	10263440	$C_{15}H_{22}O_2$	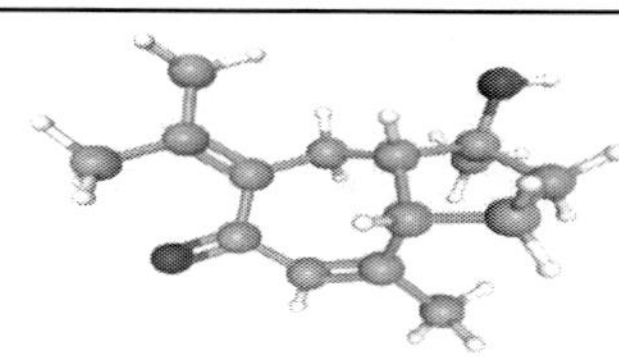
15.	procurcumenol	189061	$C_{15}H_{22}O_2$	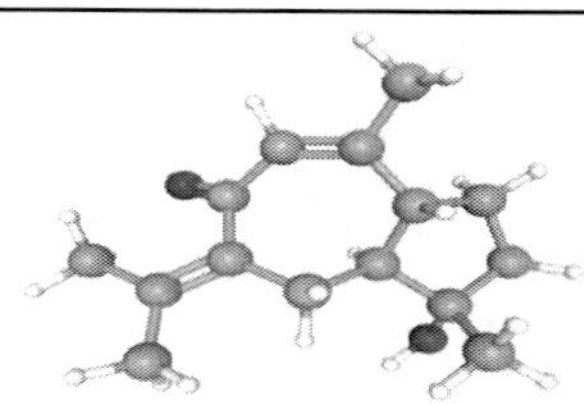
16.	curlone	196216	$C_{15}H_{22}O$	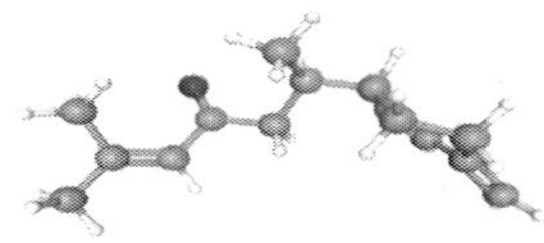
17.	2-hydroxymethyl anthraquinone	87014	$C_{15}H_{10}O_3$	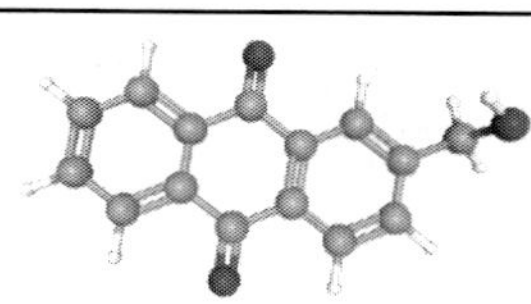
18.	riboflavin	493570	$C_{17}H_{20}N_4O_6$	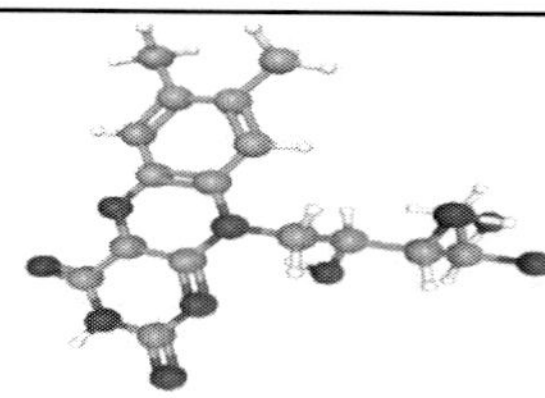
19.	niacin	938	$C_6H_5NO_2$	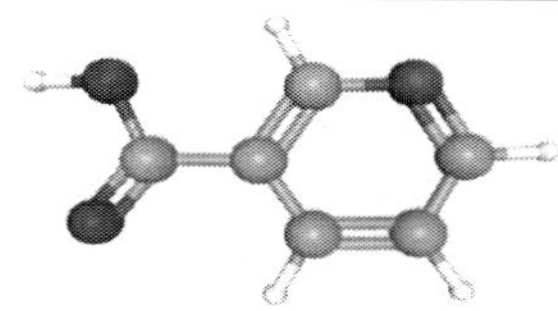
20.	ascorbic acid	54670067	$C_6H_8O_6$	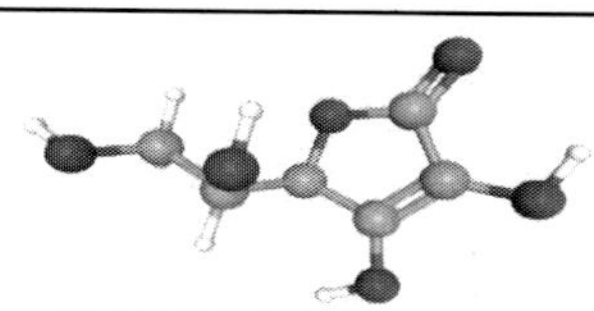

Table 2. List of pharmacokinetics properties, molecular weight (MW), hydrogen bond donor (HBD), hydrogen bond acceptor (HBA), partitioning coefficient (LogP), number of rotatable bond (RB), number of heavy atoms (HA), number of aromatic heavy atoms (AHA), synthetic accessibility (SA), gastrointestinal (GI) absorption, and Lipinski's Rule of 5 of all plant compounds

No.	Name of bioactive compound	PubChem ID	M/W g/m (<500)	No. Heavy Atoms	No. Aromatic Heavy Atoms	RB	HBA	HBD	Log P (<5)	Log S (EOSL)	SA	GI	LR
1.	curcumin	969516	368.38	27	12	8	6	2	3.27	-3.94	2.97	High	Yes, 0 violation
2.	5'-methoxycurcumin	10250249	398.41	29	12	9	7	2	3.33	-4.02	3.20	High	Yes, 0 violation
3.	dihydrocurcumin	10429233	370.40	27	12	9	6	2	2.87	-3.77	3.10	High	Yes, 0 violation
4.	d-sabinene	10887971	136.23	10	0	1	0	0	2.65	-2.57	2.87	Low	Yes, 1 violation
5.	borneol	64685	154.25	11	0	0	1	1	2.29	-2.51	3.43	High	Yes, 0 violation
6.	zingiberene	92776	204.35	15	0	4	0	0	3.65	-4.10	4.81	Low	Yes, 1 violation
7.	germacrone	6436348	218.33	16	0	0	1	0	2.96	-3.37	3.90	High	Yes, 0 violation
8.	bisacurone	14287397	252.35	18	0	4	3	2	2.74	-2.32	4.39	High	Yes, 0 violation
9.	curcumenone	153845	234.33	17	0	3	2	0	2.78	-2.61	3.66	High	Yes, 0 violation
10.	dehydrocurdione	6442617	234.33	17	0	0	2	0	2.53	-3.04	4.27	High	Yes, 0 violation
11.	procurcumadiol	14633011	250.33	18	0	0	3	2	2.30	-2.20	4.30	High	Yes, 0 violation
12.	curcumenol	167812	234.33	17	0	0	2	1	2.83	-2.70	5.73	High	Yes, 0 violation
13.	isoprocurcumenol	14543198	234.33	17	0	0	2	1	2.56	-2.72	3.89	High	Yes, 0 violation
14.	epiprocurcumenol	10263440	234.33	17	0	0	2	1	2.62	-2.77	4.04	High	Yes, 0 violation
15.	procurcumenol	189061	234.33	17	0	0	2	1	2.62	-2.77	4.04	High	Yes, 0 violation
16.	curlone	196216	218.33	16	0	4	1	0	3.14	-3.46	4.17	High	Yes, 0 violation
17.	2-hydroxymethyl anthraquinone	87014	238.24	18	12	1	3	1	1.95	-3.44	2.34	High	Yes, 0 violation
18.	riboflavin	493570	376.36	27	14	5	8	5	0.97	-1.31	3.84	Low	Yes, 0 violation
19.	niacin	938	123.11	9	6	1	3	1	0.86	-1.26	1.00	High	Yes, 0 violation
20.	ascorbic acid	54670067	176.12	12	0	2	6	4	0.39	0.23	3.47	High	Yes, 0 violation

Table 3. List of the drug-induced hERG inhibition, AMES toxicity, carcinogens, *Tetrahymena pyriformis* (TP) toxicity, fish toxicity, honeybee (HB) toxicity, hepatotoxicity, rat acute toxicity (LD_{50} in mol/kg), plasma protein binding along with P-glycoprotein inhibitor (PGI) activity of all natural compounds

	curcumin	5'-methoxy-curcumin	Dihydrocurcumin	d-sabinene	borneol	zingiberene	germacrone	Bisacurone	Curcumenone	Dehydro-curdione
Ames mutagenesis	-	-	-	-	-	-	-	-	-	-
Acute Oral Toxicity (c)	III	III	III	III	III	III	II	III	III	II
Androgen receptor binding	+	+	+	-	-	-	-	-	-	-
Aromatase binding	+	+	+	-	-	-	-	-	-	-
Avian toxicity	-	-	-	-	-	-	-	-	-	-
Blood Brain Barrier	-	-	+	+	+	+	+	+	+	+
BRCP inhibitior	+	-	-	-	-	-	-	-	-	-
Biodegradation	-	-	-	-	+	-	-	-	-	+
BSEP inhibitior	+	+	+	-	-	-	-	-	-	-
Caco-2	-	-	-	+	+	+	+	-	+	+
Carcinogenicity (binary)	-	-	-	-	-	-	-	-	-	-
Carcinogenicity (trinary)	Non-required	Non-required	Non-required	Non-required	Non-required	Warning	Non-required	Non-required	Non-required	Non-required
crustacea aquatic toxicity	-	-	-	+	+	-	-	-	-	-
CYP1A2 inhibition	+	+	+	-	-	-	-	-	-	-
CYP2C19 inhibition	+	+	+	-	-	-	-	-	-	-
CYP2C9 inhibition	+	-	+	-	-	-	-	-	-	-
CYP2C9 substrate	-	-	-	-	-	-	-	-	-	-
CYP2D6 inhibition	+	+	-	-	-	-	-	-	-	-
CYP2D6 substrate	-	-	-	-	-	-	-	-	-	-
CYP3A4 inhibition	-	+	+	-	-	-	-	-	-	-
CYP3A4 substrate	-	-	+	-	+	-	-	+	+	-
CYP inhibitory promiscuity	+	+	-	-	-	-	-	-	-	-

	curcumin	5'-methoxy-curcumin	Dihydrocurcumin	d-sabinene	borneol	zingiberene	germacrone	Bisacurone	Curcumenone	Dehydro-curdione
Eye corrosion	-	-	-	-	-	+	-	-	-	-
Eye irritation	+	+	+	+	+	-	+	-	+	+
Estrogen receptor binding	+	+	+	-	-	-	-	-	-	-
Fish aquatic toxicity	+	+	+	+	+	+	+	+	+	+
Glucocorticoid receptor binding	+	+	+	-	-	-	-	+	-	-
Honeybee toxicity	+	+	+	+	+	+	+	+	+	+
Hepatotoxicity	+	+	+	-	-	-	-	-	-	-
Human either-a-go-go inhibition	-	-	-	-	-	+	-	-	+	-
Human Intestinal Absorption	+	+	+	+	+	+	+	+	+	+
Human oral bioavailability	+	+	-	+	+	-	+	-	-	+
MATE1 inhibitior	-	-	-	-	-	-	-	-	-	-
micronuclear	+	+	-	-	-	-	-	-	-	-
Acute Oral Toxicity	1.9922305346	1.4809014797	1.90323174	2.3189282417	1.3422938585	1.827173233	2.3275892735	3.0167291164	1.4263353348	1.9158366919
OATP1B1 inhibitior	+	+	+	+	+	+	+	+	+	+
OATP1B3 inhibitior	+	+	+	+	+	+	+	+	+	+
OATP2B1 inhibitior	-	-	-	-	-	-	-	-	-	-
OCT1 inhibitior	-	-	-	-	-	-	-	-	-	-
OCT2 inhibitior	-	-	-	-	-	-	-	-	-	-
P-glycoprotein inhibitior	+	-	+	-	-	-	-	-	-	-
P-glycoprotein substrate	-	-	-	-	-	-	-	-	-	-
PPAR gamma	+	+	+	-	-	-	-	-	-	-
Plasma protein binding	0.8321629167	0.958293736	0.8974623084000001	0.5551440120000001	0.8529109359	1.0375547409	0.6413831115	0.7182907462	0.5342152715	0.7439326644

Table 3. (Continued)

	curcumin	5'-methoxy-curcumin	Dihydrocurcumin	d-sabinene	borneol	zingiberene	germacrone	Bisacurone	Curcumenone	Dehydro-curdione
Subcellular localzation	Mitochondria	Mitochondria	Mitochondria	Lysosomes	Lysosomes	Lysosomes	Mitochondria	Mitochondria	Mitochondria	Mitochondria
Tetrahymena pyriformis	0.8713880777	0.9034119248	0.6202898622	1.031681776	0.7262762189	0.6807615757000001	0.9544943571000001	-0.73116678	0.6792432070000001	0.9229826331000001
Thyroid receptor binding	+	+	+	-	-	-	-	-	-	-
UGT catelyzed	-	-	-	-	+	-	-	-	-	-
Water solubility	-3.3640974018	-3.7620831172	-3.5911969421	-3.9878399648	-2.3826087913	-4.9371671541	-3.0325952493	-2.643379953	-3.3045816704	-3.3074957276

In this study, physicochemical properties, lipophilicity, water solubility, drug-likeness, medicinal chemistry, drug absorption, pharmacokinetics and toxicity of twenty plant compounds from the turmeric plant (Table 1) were determined using computational biology tools such as SwissADME and admetSAR. Based on the Lipinski Rule of 5, all the twenty natural compounds passed the rule and proceed to toxicity prediction analysis (Table 2).

Toxicity evaluation analysis for all the twenty selected bioactive compounds was performed using admetSAR. The admetSAR is a free web-based online tool. Table 3 shows that all the phytoconstituents do not have carcinogenicity. Only riboflavin possesses AMES toxicity. Furthermore, borneol and bisacurone had hERG inhibition activity. Curcumin, 5'-methoxycurcumin, dihydrocurcumin, 2-hydroxymethyl anthraquinone and riboflavin had hepatotoxicity. The previous study has proved that some flavonoid molecules such as phloroglucinol, kaempferol, isobutyl isothiocyanate, taurine, and apigenin are hepatotoxic. Therefore, a low dosage of plant compounds could be an effective drug for the treatment of COVID-19 disease. It has a low harmful effect on the liver. Riboflavin, niacin and ascorbic acid have no fish toxicity while 2-hydroxymethyl anthraquinone and riboflavin have no honeybee toxicity. All the plant compounds except riboflavin, borneol and bisacurone can be subjected to molecular docking and dynamics simulation approaches. Further laboratory experiments, both *in vitro* and *in vivo,* should be carried out to validate the results from the computational biology tools. These results could help in developing a plant-based drug against SARS-CoV-2 viral replication in the host cell.

Conclusion

All the twenty selected plant compounds from the turmeric (Curcuma longa) plant show no carcinogenicity. Riboflavin, borneol and bisacurone show toxicity in humans. Therefore, they are not suitable drugs for humans. Computational biology approaches are the initial step to develop an effective drug against SARS-CoV-2. These methods could help in developing medicines and vaccines during an urgent period like the COVID-19 situation. However, the dry-laboratory results should be further analyzed in wet-laboratory to avoid any harmful effects on humans. Human safety is the most important matter.

References

Acter T, Uddin N, Das J, Akhter A, Choudhury T R, Kim S. Evolution of severe acute respiratory syndrome coronavirus 2 (SARS-CoV-2) as coronavirus disease 2019 (COVID-19) pandemic: A global health emergency. *Science of the Total Environment* (2020) 15:730:138996.

Allaq A A, Sidik N J, Abdul-Aziz A, Alkamil A M, Elengoe A, Yahya E B, Abdulsamad MA. Epidemiological studies of the novel Coronavirus (COVID-19) in Libya. *Pakistan Journal of Biotechnology* (2021) 4:18(1-2):7-16.

Bender L. Key messages and actions for COVID-19 prevention and control in schools. UNICEF: New York. (2021). Accessed 10 December 2020. https://www.who.int/docs/default-source/coronaviruse/key-messages-and-actions-for-covid-19-prevention-and-control-in-schools-march2020.pdf?sfvrsn=baf81d52_4#:~:text=COVID%2D19%20is%20a,2019%2DnCoV.

Elengoe A, Hamdan S. Computational approach towards exploring interaction of target protein-phytocompounds in drug development for breast cancer. In *Anticancer plants: mechanisms and molecular interactions* 2018 (pp. 131-141). Springer, Singapore.

Elengoe A, Selvam S, Selvarajah M, Manikam H, Vijaya R. Sequence analysis and structure prediction of Malaysia SARS-CoV-2 strain's structural and accessory proteins. *Biointerface Research in Applied Chemistry* (2022) 12(1):3259-3304.

Hasöksüz M, Kiliç S, Saraç F. Coronaviruses and SARS-CoV-2. *Turkish Journal of Medical Sciences* (2020) 50(SI-1):549-556.

Hewlings S J, Kalman D S. Curcumin: A review of its effects on human health. *Foods* (2017) 6(10):92.

ILO, FAO, IFAD, WHO. *Impact of COVID-19 on people's livelihoods, their health and our food systems.* Joint statement by ILO, FAO, IFAD and WHO. Food and Agriculture Organization of the United Nations. (2021). Accessed 15 December 2022. http://www.fao.org/news/story/en/item/1313598/icode/.

Jagtap N, Yadav A, Mohite S. Synthesis, Molecular docking studies and anticancer activity of 1, 3, 4-oxadiazole-3 (2h)-thione derivatives. *Journal of University of Shanghai for Science and Technology* (2020) 22(11):535-550.

Keni R, Alexander A, Nayak P G, Mudgal J, Nandakumar K. COVID-19: emergence, spread, possible treatments, and global burden. *Frontiers in Public Health* (2020) 8:216.

Kim S, Thiessen P A, Bolton E E, Chen J, Fu G, Gindulyte A, Han L, He J, He S, Shoemaker BA, Wang J. PubChem substance and compound databases. *Nucleic Acids Research (*2016) 44(D1):D1202-1213.

Lipinski C A. Lead-and Drug-Like Compounds: The rule-of-five revolution. *Drug Discovery Today*: *Technologies* (2004) 1(4):337-341.

Mohan S, Anjum M R, Kodidasu A, Prathyusha TVN, Mrunalini NV, Kishori B. SARS-CoV-2 infection: A global outbreak and its implication on public health. *Bulletin of the National Research Centre* (2021) 45(1):1-12.

Rallabandi H R, Mekapogu M, Natesan K, Saindane M, Dhupal M, Swamy M K, Vasamsetti BM. Computational Methods Used in Phytocompound-Based Drug Discovery. In *Plant-derived Bioactives* 2020 (pp. 549-573). Springer, Singapore.

Tian S, Wang J, Li Y, Li D, Xu L, Hou T. The application of *in silico* drug-likeness predictions in pharmaceutical research. *Advanced Drug Delivery Reviews* (2015) 86:2-10.

Veber D F, Johnson S R, Cheng H Y, Smith B R, Ward K W, Kopple K D. Molecular properties that influence the oral bioavailability of drug candidates. *Journal of Medicinal Chemistry* (2002) 45(12):2615-2623.

Zígolo M A, Goytia M R, Poma H R, Rajal V B, Irazusta V P. Virtual screening of plant-derived compounds against SARS-CoV-2 viral proteins using computational tools. *Science of The Total Environment* (2021) 781:146400.

Chapter 18

A Brief Overview of the COVID-19 Outbreak in Sri Lanka

Abdul Haleem Mohamad Wahhaj[1], BSc,
Asita Elengoe[2,*], PhD
and Shanaya Amath[1], BSc
[1]School of Biotechnology, Manipal International University (MIU), Putra Nilai, Negeri Sembilan, Malaysia
[2]Department of Biotechnology, Faculty of Applied Science, Lincoln University College, Petaling Jaya, Selangor, Malaysia

Abstract

Coronaviruses are a part of a category of single-stranded, enveloped, positive-sense RNA infections that possess the capability of contaminating a wide variety of creatures, inclusive of people, birds, rodents, carnivores, chiropters and other warm-blooded animals. Although they have been known for numerous years and have been categorized as one of the viral sources capable of causing respiratory illnesses, they caught the attention of the entire world in December 2019, when an epidemic scene of cases with respiratory tract contaminations was reported in Wuhan, which is the biggest metropolitan region within the province of Hubei in China. This article describes the outbreak of COVID-19 in Sri Lanka, taking into perspective the vaccination program as well as the state of mental health during the pandemic amongst Sri Lankans. To date, some vaccines have been developed to protect individuals from contracting COVID-19. However, preventative measures are still essential to prevent and curb the rampant spread of the

* Corresponding Author's Email: asitaelengoe@yahoo.com.

In: Medicinal Plants and COVID-19
Editor: Asita Elengoe
ISBN: 979-8-88697-958-9

virus in the community. Governments, as well as the local community, should be prepared in the event of future COVID-19 outbreaks and mutations of the SARS-CoV-2 virus.

Keywords: COVID-19, SARS-CoV-2, Sri Lanka

Introduction

The declaration of the World Health Organization (WHO) on 31st December 2019 on a worldwide pandemic situation after the spread of an unknown cluster of pneumonia cases that caused severe acute respiratory illness in China made the whole world panic (Guo et al., 2020). All countries were at risk due to the advanced travel methods used in border crossing and many Asian countries including Sri Lanka made pandemic response plans appropriately and proportionately to avoid an outbreak. But regardless of the active responses made many countries suffered and are still suffering due to the ill effects of the SARS-CoV-2 pandemic.

The ongoing global pandemic caused by the SARS-CoV-2 virus, or coronavirus disease, is now in its third phase in Sri Lanka. Since the very first SARS-CoV-2 case was identified on 27th January 2020, Sri Lanka has experienced negative effects on all fronts, including the economy (Sri Lanka: WHO Coronavirus Disease, 2021). During the first wave of the pandemic, the country was effective in establishing preventive measures, but following the second wave, the country suffered the most. People's daily lives were deteriorating, the economy was suffering, and the country's healthcare system was overburdened (Amaratunga et al., 2020). The number of SARS-CoV-2 cases increased uncontrollably during the second and third waves. All preventive measures were ineffective, and the only effective preventive step was to place the country on lockdown at any cost. Curfews were imposed across the nation from time to time, and Sri Lanka is currently under a CMCO (conditional movement control order) that began on August 20, 2021, and has been extended until October 1, 2021 (Liyanapathirana et al., 2020).

The pandemic prevention activity was executed by the Epidemiology Unit, Medical Research Institute (MRI), Health Promotion Bureau (HPB) and the Quarantine Unit in Sri Lanka (Liyanapathirana et al., 2020). These principal government facilities were playing a crucial role in disease prevention giving administration, technical guidance, advocacy, intra- and inter-sectoral coordination, hazard correspondence, monitoring and assessment along with the coordination and backing from Disaster Preparedness and Response Division. In addition to these, the authorities from the tri-forces of the country at all levels and different sectors were buckled down for the prevention of the coronavirus outbreak.

The principal targets of the outbreak reaction were to 1. Diminish importation of the pandemic infection into the nation; 2. fortify reconnaissance for ahead of schedule and composed reaction; 3. delay spread of infection or contain at the source; 4. lessen the effect of the infection on morbidity and mortality by early discovery and treatment; 5. ensure the weak gatherings and 6. screen and assess the advancing reaction to the episode (Hettiarachchi et al., 2021).

The Starting of an Outbreak

Sri Lanka is in the 16th position among countries with a high risk of contracting the virus, and the government's reaction to the pandemic has gone down since the second wave which began on October 4, 2020, since the country has been struck economically (Writer, 2020). Initially, identifying patients, isolating them, and establishing quarantine centers to isolate them was done immediately, but due to the rapid spread of the virus throughout the country and the risk involved, the Sri Lankan government decided to implement self-quarantine plans rather than quarantine centers because the number of patients had been outnumbered and hospitals were running out of space, and so was quarantine centers, which were unable to accommodate patients.

During the third wave, the country experienced an increase in SARS-CoV-2 deaths, and hospitals were running out of oxygen as well. This occurred shortly after health restrictions were relaxed during the Sinhala and Tamil New Year's Eve in April 2021, and the highly contagious delta variant was responsible for the country's elevated fatality rate (EconomyNext, 2021). The decision to enforce a total lockdown came after tremendous pressure from health experts and political parties, who pushed for it after Sri Lanka's SARS-CoV-2 outbreak resulted in 1000 deaths in only eight days. (France 24, 2021; Nadeera and Dilshan, 2021; EconomyNext, 2021).

Sri Lanka had the fourth-largest daily death rate in the world by population as of August 2021 and the government's failure to impose a lockdown, as well as the general public's negligence, have all led to the country's record number of SARS-CoV-2 cases and deaths. During the early part of the year, the government was determined not to impose a lockdown due mainly to the worsening economy, but the situation worsened, and in August 2021, Sri Lanka's Central Bank implemented a contractionary restrictive monetary policy approach, making it the first Asian country to do so during the epidemic (Reuters, 2021; The Hindu, 2021). When the daily death toll topped 200 for the first time since the epidemic began on August 27, 2021, officials extended the existing curfew (Sri Lanka Business News, 2021; The Times of India, 2021).

The First Wave (27th January 2020- June 2020)

The coronavirus had been identified in 11 countries by the 27th of January 2020, with 2798 confirmed cases and 80 deaths worldwide (Coronavirus Outbreak, 2020). The Sri Lankan Ministry of Health had instructed officials at Bandaranaike International Airport to examine travelers for symptoms before the 27th of January. Furthermore, the ministry recommended babies, children, pregnant women, the elderly, and those with chronic illnesses avoid crowded places if possible. The Ministry of Health organized a 22-member National Action Committee (Adaderana, 2021) to successfully halt the virus's spread. The Department of Immigration and Emigration urged all construction sites with Chinese resident visa holders to limit their Chinese workers to their specified workplaces and accommodations (Adaderana, 2020).

On January 27, 2020, the first verified case of SARS-CoV-2 was reported in Sri Lanka. The infected person was a 44-year-old Chinese national female who had arrived in Sri Lanka 2 weeks before the reported date as a tourist (Colombo Page, 2020). The Chinese national was admitted to the National Institute of Infectious Diseases (NIID) for treatment and five more individuals were suspected of having the coronavirus by the 28th January 2020, two of which were Sri Lankan nationals. At the NIID, these people were kept under observation. The first confirmed patient was fully recovered and discharged from the NIID on February 19, 2020. There were no further known cases in the nation until March 11, 2020, when a tour guide working with Italian tourists was diagnosed with the virus, becoming the country's second confirmed case [also the first local case] (Dailymirror, 2020). By March 31, 2020, the country's total number of confirmed cases had risen to 122, including two fatalities. By July 31, 2020, there were 2814 SARS-CoV-2 cases in Sri Lanka, including 951 imported cases reflecting returnees from other countries and 1833 indigenous patients.

Individuals from the Sri Lankan tri-forces and their close associates made up the majority of the local cases as the local forces worked in the first line against the pandemic. Following the first documented case of SARS-CoV-2, the country's demand for face masks spiked, resulting in a mask crisis. Some pharmacies around the country had sold out all face masks, with customers reporting that they were being offered for ten to twenty times the original price. Face masks were consequently subjected to price limits by the country's drug regulating agency (Hamza, 2020). On February 1, 33 Sri Lankan students and families were evacuated from Wuhan and sent to Sri Lanka, where they were held in quarantine at a military facility in Diyatalawa.

Since the first week of March, visitors from Italy, Iran, and South Korea have been obliged to quarantine for 2 weeks at one of two facilities. In Batticaloa, 186 persons were quarantined (164 Sri Lankan nationals, 20 Italian nationals, and 2 South Korean nationals) on March 10th (Colombopage, 2021). The first confirmed SARS-CoV-2 case in Jaffna was reported on March 22, and the patient was thought to have had close relations with a leading preacher who had tested positive for the virus. The first confirmed death from SARS-CoV-2 was reported in Sri Lanka on March 28th, a 60-year-old Marawila patient who had previously received a kidney transplant. Moreover, he suffered from diabetes and high blood pressure (EconomyNext, 2021). The body was incinerated within 15 hours.

A 60-year-old Negombo man was the second person to die from SARS-CoV-2 on March 30 (Hiru News, 2020). After a surprise check-up on April 20, 33 patients in Colombo tested positive for SARS-CoV-2, bringing the total case count to 303. This was the highest number of new SARS-CoV-2 cases in a single day until April 26 (India Today, 2020; News First, 2020). By the end of June, positive cases were reported all over the country and the total number of cases had risen to 1797, with 11 deaths as a result of the virus.

The Second Wave (4th October – December 2020)

After an outbreak in Minuwangoda and Divulapitiya, an island-wide curfew was imposed (Newsfirst, 2020). Governments were hesitant to impose another lockdown for fear of bankrupting firms and putting staff out of work. Since October 8, according to Sri Lanka's Bureau of Foreign Employment, 67 Sri Lankan migrant workers have died due to SARS-CoV-2, which has been documented predominantly in Middle Eastern countries. Many government workers on the front line were infected, and the country was placed in a crisis state. Except for hospitals, police stations, and armed forces, all government institutions were shut down (Newsfirst, 2020).

The WHO had informed the Health Ministry that different vaccinations for SARS-CoV-2 had been approved and that Sri Lanka could expect a shipment of them to be distributed to first-line staff.

A quarantine curfew was re-imposed on November 1, 2020, in the Western Province, Kurunegala, Ratapura, and Kuliyapitiya (Newsfirst, 2020). The overall number of cases had climbed to 3720 as of December 28, 2020, with 22 deaths.

Travel Restrictions and Movement Control

After an outbreak in Minuwangoda and Divulapitiya, an island-wide curfew was imposed (Newsfirst, 2020). Governments The emergence of the virus was sudden and the exact cause was suspense, but the virus effects and its impact were thoroughly studied to implement health care guidelines. The global spread of the virus was closely monitored and risk factors were assessed regularly and the Sri Lankan government implemented strategies based on evidence and public health and epidemiological principles to prevent and detect the SARS-CoV-2 virus. Control strategies were created with guidance provided by the WHO to prevent an outbreak in Sri Lanka. The Government promptly imposed many control measures from local and international travel restrictions to localized and countrywide lockdowns. This is whilst many other actions across several fields were taken covering (a) detection, isolation and contact tracing (b) medical surveillance, lab investigations and case management (c) right information to the public, confidence-building and consolidating solidarity and (d) supply of food and medicine and maintenance of essential service.

Travel from several countries in continental Europe, including Spain, Norway, Switzerland, Belgium, Sweden, Germany, France, Austria, and the Netherlands, was also prohibited, with the prohibition being extended to the United Kingdom on March 15 (BBC Sport, 2020; UK Foreign & Commonwealth Office, 2020). Foreign visitors from India, South Korea, and Iran were also subjected to the travel ban. Due to concerns about the coronavirus, Sri Lankan Airlines has canceled numerous scheduled flights from several countries, including India. The Sri Lankan government has imposed travel restrictions to Canada, Bahrain, and Qatar as of midnight on March 17, 2020 (EconomyNext, 2020). Except for the delivery of cargo products, aircraft landings at BIA airport were likewise banned for two weeks with immediate effect from March 19. The government with the support of the Airport and Aviation Services Limited, Civil Aviation Authority, Sri Lankan Airlines, Airport Health Office, Department of Immigration & Emigration, Sri Lanka Customs, Ports Authority and Sri Lanka Air Force imposed many new border control strategies and the administration also extended the validity of all foreigner visas till April 12 to avoid any inconvenience. The government again restricted the entrance of all inbound passenger ships and planes on March 22, 2020.

The travel restrictions have already been extended until December and Sri Lanka said that travel restrictions would be eased beginning from December 26, allowing foreigners and foreign passport holders to return without first registering with the Foreign Ministry if they have paid for quarantine. However, on December 21, the relaxation was suspended after a highly transmissible SARS-CoV-2 strain was discovered in the UK and other nations. The government barred the entrance of all inbound passenger ships and planes on March 2021 (EconomyNext, 2020).

The Third Wave (1 January 2021 – Present)

Sri Lanka's National Medical Regulatory Authority stated on January 22, 2021 that the Oxford–Astra-Zeneca vaccine had been certified, and front-line workers were already getting vaccinated since the end of February. The SARS-CoV-2 pandemic quickly turned from a health emergency to a social and economic disaster, resulting in many negative economic and social consequences in the country. For example, Sri Lanka's Gross Domestic Product (GDP) was anticipated to increase at a rate of just 2.2 percent when the pandemic struck, despite the fact that it was expected to rise by 4.5 percent to 5% following the 2019 Easter Sunday attack. (EconomyNext, 2020).

The lockdown also affected the country's most important economic sectors, especially manufacturing and services. The tourism sector is one of the country's most important service industries, which accounts for about 5% of the country's GDP. Visitors to the nation decreased by 71% in March 2020 as a result of worldwide travel prohibitions imposed during the epidemic, and visitors arrived at zero in April, May, and June. Similarly, the manufacturing sector's Purchasing Manager's Index (PMI) declined from 54 in January 2020 to 24.2 in April 2020 due to lower new orders, production, and employment. Furthermore, industrial exports in the country declined by 74% year over year, while agricultural exports fell by 32%, resulting in a negative impact on export revenues in the country. It's also worth mentioning that migrant workers' private remittances to Sri Lanka contribute considerably to the country's foreign exchange earnings. Migrant workers' private remittances, for example, account for 63 percent of the country's total export profits. Remittances from workers, on the other hand, plummeted by 32% in April 2020, thus impacting the country's foreign exchange earnings.

Similarly, a labor market assessment of 2764 private sector establishments found that throughout the pandemic period, 1465 establishments were entirely closed, 1025 companies were operating at maximum capacity, and only 94 establishments were operating at full capacity. During the epidemic in the country, 1084 employers out of the total number of businesses investigated were unable to pay their employees' salaries. High levels of unemployment, job insecurity, and salary cuts have caused anxiety and economic stress among the country's citizens.

The pandemic had a significant impact on Small and Medium Scale Enterprises (SMEs) and the informal sector, particularly daily wage earners. Most SMEs in Sri Lanka have faced challenges such as a lack of materials to continue manufacturing or provide services, a drop in local and global demand for their products, and issues repaying loans (Robinson and Kengatharan, 2020).

The disruption of educational activities is one of the most serious issues to be noted in terms of the pandemic's social impacts. The closure of primary, secondary, and tertiary educational institutions (Ex: schools, universities and support classes) came from the lockdown limitations that accompanied the demands for social distancing. As a result, the government advocated for the ongoing provision of educational services via the internet. While online education would have been the most convenient option, it has worsened educational inequality and fueled social instability since some people, particularly those in rural regions, lacked access to the facilities and infrastructure required for online learning (Jayasena and Chinthaka, 2020).

The disgrace or disapproval of population groups exposed to the public is another social implication that cannot be ignored. Certain categories of people, such as healthcare professionals, tri-force officials and grocery cashiers, were condemned in their communities. This is because they were present in important services that continued to be offered despite the country's pandemic condition, were unable to self-isolate, and were thus seen as potential disease carriers (Palihawadana, 2020; Warnapura, 2020; Godman, 2020; Hewage, 2020).

As a result of the pandemic, the country is experiencing unavoidable negative consequences and the overall number of SARS-CoV-2 cases documented as of July 8, 2021 had risen to over 268,000, with over 3,300 deaths. In Sri Lanka, from 3rd January 2020 to 5:41 pm CEST, 21st September 2021, there have been 506,009 confirmed cases of SARS-CoV-2 with 12,218 deaths, reported to WHO [15]. As of 12th September 2021, a total of 24,226,161 vaccine doses have been administered.

Treatment of COVID-19 in Sri Lanka

Sri Lanka's In the initial stages of clinical management and treatment, ninety-four patients, which is 68% were given HCQ, whereas 39, which is 28% were given antibiotics and six patients, which is 4% were given antiviral drugs such as oseltamivir and lopinavir-ritinovir. 15% of patients received mellow symptomatic treatment, such as steam inhalation and gargling with salt water. Concurring to therapeutic records, all patients with deteriorating clinical conditions were instantly admitted to the intensive care unit (ICU). They were eight in number, of whom five advanced to the critical illness category and unfortunately surrendered to the illness. Their middle (IQR) term of ICU stay was approximately 8 days. The other three patients who were admitted to the ICU were not ventilated and were released alive. One patient in the critical infection category died on confirmation to the hospital. Concurring to medical record data, no patient was denied treatment, inclusive of ICU care (Arambepola et al., 2021).

Preventative Measures in Sri Lanka

Adequate consideration has been given to preventative measures such as spraying disinfectants, aimed at destroying the infection. However, the major control procedures executed in Sri Lanka were directed at minimizing the movement of people. This was not a procedure utilized in the nation amid the different plagues of dengue. Be that as it may, with the rise of the COVID-19 episode, strict measures were taken to minimize human movement. On the 20th of March, an islandwide curfew was imposed until the 24th of March and it was extended in conjunction with a strict ban on travel between districts. The curfew was a lawful component accessible to the recently elected government since parliament had been broken down pending an election. Extremely crowded areas of Colombo, Kalutara, and Gampaha were distinguished as zones at high risk because of their high population. Other preventative measures involved making wearing a mask obligatory in indoor and open-air public spaces, as well as in public transport. The police and armed force staff have been encouraged to carefully observe whether people are complying with the rules. In addition, sanitary measures such as having wash taps and sanitizers near grocery stores and markets for clients have been made obligatory. Moreover, general stores were instructed to use thermometers to verify the temperatures of customers before they entered. A distance of a minimum of 1m is kept in the counter lines between customers to minimize close contact (Erandi et al., 2021).

An island-wide curfew was imposed throughout the nation to try to stem the spread of COVID-19. This includes a two-month lockup with severe movement limitations. During this period, medical workers used tight contact tracking and separation of suspicious patients with the help of the armed and military forces. In places with high populations, community medical teams worked diligently to undertake random polymerase chain response testing, or PCR tests, to screen for suspected community spread. By building and managing isolation units, the Sri Lankan military performed a substantial and laudable role in alleviating the load placed on overworked and exhausted medical professionals. Sri Lanka was also one of the first countries to establish repatriation missions to securely return its nationals who were trapped abroad (Jayasena and Chinthaka, 2020).

Vaccination Status of Sri Lanka

The National Medicines Regulatory Authority (NMRA) ensures the standards, safety, quality and efficacy of vaccines and medicinal products in the country. Due to the outbreak, the State Pharmaceuticals Corporation of Sri Lanka (SPC) joined hands with NMRA to validate vaccine candidates for emergency use in Sri Lanka. The review of the quality, safety, immunogenicity and efficacy of SARS-CoV-2 vaccines by the Advisory Committee on Communicable Diseases in Sri Lanka was sought and thoroughly inspected when selecting and approving different vaccine candidates.

In response to the current pandemic, the Sri Lankan government launched an immunization campaign against SARS-CoV-2 in late February 2021. The immunization campaign was initiated under the COVAX facility (SARS-

CoV-2 Vaccines Global Access) and Sri Lanka received its first batch of Oxford–AstraZeneca vaccines on January 28, 2021 from the Serum Institute of India. The health ministry undertook studies on vaccine candidates as early as December 2020 and the first phase of SARS-CoV-2 immunization began in February 2021 and was addressed by health, community workers, security forces personnel and frontline professionals who played a critical role engaged with pandemic control activities in safeguarding the public from the virus. In March 2021, after vaccinating all medical professionals and front-line workers within one month, phase two of SARS-CoV-2 immunization began with people over the age of 30 with priority to people with underlying health conditions, pregnant women, employees of entry and exit points, harbors, airports, economic sector employees, factory workers, employees in essential services, school and university teachers and lectures respectively.

In phase two, senior citizens over the age of 60 with more attention on those who have underlying health conditions were given a higher priority, and by April 2021, every person over the age of 30 in the Western Province had received at least one dose of vaccine, and nearly half of those over the age of 30 in the Central and Southern Provinces had received the first dose of vaccine. The Sinopharm vaccine was approved in Sri Lanka in March 2021, after the Chinese government donated 1.1 million units as a gesture of goodwill to the Sri Lankan government. As the third wave of the SARS-CoV-2 virus had a major impact on all parts of Sri Lanka, the government of Sri Lanka faced a vaccination shortage.

The vaccine scarcity was due to the unavailability of the Astra-Zeneca vaccine as India imposed an export ban and the Russian government stopped the export of the Gamaleya-Sputnik V vaccine after a rapid increase in cases in Russia. As a result, the majority of the population who received the first dose did not receive the second dose, and the number of cases in Sri Lanka continued to rise. The country ordered 14 million doses of the vaccine in May 2021, on top of 1.1 million doses previously provided by the Chinese government, and received 10 million doses by June 2021.

The Government of Sri Lanka provided immunization opportunities for certain group of individuals in July, after a two-month registration period. Pfizer-BioNTech and Moderna vaccinations were available to those aged 18 to 30 who were students planning to travel overseas for studies or migrant workers planning to travel overseas for work.

The Government of Sri Lanka approved 7 Vaccines for use in Sri Lanka

1. Oxford-AstraZeneca
2. Pfizer-BioNTech
3. Sinopharm
4. Gamaleya-Sputnik V
5. Moderna
6. Covishield (Oxford/AstraZeneca formulation)
7. Sinovac-CoronaVac

On the 13th August 2021, the Army Commander and Head of the National Operations Centre for SARS-CoV-2 Prevention stated that authorities have made it mandatory for citizens over the age of 30 to carry their vaccination cards when visiting public places beginning from September 15 to minimize the rapid spread of SARS-CoV-2 pandemic.

As of September 21, 2021, Sri Lanka is in phase three of its vaccination program, with vaccination of individuals aged 18 to 30 years having begun recently. According to the Ministry of Health, 63.3 percent of the Sri Lankan population has received at least one dose of any type of approved vaccine; with 51 percent covering 11.6 million individuals receiving both doses (The Hindu, 2021).

Conclusion

This brief review of the SARS-CoV-2 outbreak in Sri Lanka is presented to create awareness, knowledge and preparedness for a pandemic situation. As a country, Sri Lanka has undergone many bad side effects due to the SARS-CoV-2 outbreak and this can be considered as a lesson to take from a developing country. Being prepared and rapid yet strict actions are needed to reduce the fear, anxiety and losses that accompany a pandemic. As responsible citizens, we should know our role in the community and act accordingly to avoid an outbreak like SARS-CoV-2.

References

Amaratunga, D., Fernando, N., Haigh, R., and Jayasinghe, N. (2020). The SARS-CoV-2 outbreak in Sri Lanka: A synoptic analysis focusing on trends, impacts, risks and science-policy interaction processes. *Progress in Disaster Science*, 8, 100133.

Arambepola, C., Wickramasinghe, N. D., Jayakody, S., Hewage, S. A., Wijewickrema, A., Gunawardena, N., Dhanapala, S., and Prathapan, S. (2021). Sri Lanka's early success in the containment of COVID-19 through its rapid response: Clinical & epidemiological evidence from the initial case series. *Plos One*, 16(7), e0255394.

Chinese construction workers in Sri Lanka to be restricted to workplace & lodgings. adaderana.lk. (2020). Accessed 31 January 2020. http://www.adaderana.lk/news/60545/chinese-workers-in-sri-lanka-to-be-restricted-to-workplace-lodgings.

Coronavirus outbreak. (2020). https://www.worldometers. info/coronavirus/.

Coronavirus: Sri Lanka records highest single-day jump in positive cases. *India Today.* (2020). Accessed 2 December 2020. https://www.indiatoday.in/world/story/coronavirus-sri-lanka-records-highest-single-day-jump-in-positive-cases-1662138-2020-04-01.

Coronavirus: Sri Lanka border closed to UK visitors. BBC Sport. (2020). Accessed 15 March 2020. https://www.bbc.com/sport/cricket/51901086.

COVID- 19 patients in SL rises up to 146. News First. (2020). Accessed 3 April 2020. https://www.newsfirst.lk/2020/04/01/covid-19-patients-in-sl-rises-upto-146/.

Erandi, K. K., Mahasinghe, A. C., Perera, S. S., and Jayasinghe, S. (2020). Effectiveness of the strategies implemented in Sri Lanka for controlling the COVID-19 outbreak. *Journal of Applied Mathematics*, 2020, 1-0.

First coronavirus patient dies in Sri Lanka, total COVID-19 count 113. EconomyNext. (2020). Accessed 28 March 2020. https://economynext.com/first-coronavirus-patient-dies-in-sri-lanka-total-covid-19-count-113-62718/.

First patient with coronavirus reported in Sri Lanka. Colombo Page. (2020). Accessed 27 January 2020. http://www.colombopage.com/archive_20A/Jan27_1 580144354CH.php.

First ever COVID-19 patient identified in Sri Lanka. Dailymirror. (2020). Accessed 15 December 2022. www.dailymirror.lk.

Foreign travel advice (Sri Lanka). UK Foreign & Commonwealth Office. (2020). Accessed 29 February 2020. https://www.gov.uk/foreign-travel-advice/sri-lanka.

Godman, B. (2020). Combating COVID-19: Lessons learnt particularly among developing countries and the implications. *Bangladesh Journal of Medical Science*, 103-S.

Guo, Y. R., Cao, Q. D., Hong, Z. S., Tan, Y. Y., Chen, S. D., Jin, H. J., Tan, K. S., Wang, D. Y., and Yan, Y. (2020). The origin, transmission and clinical therapies on coronavirus disease 2019 (COVID-19) outbreak-an update on the status. *Military Medical Research*, 7, 1-0.

Hamza, M. *Sri Lanka plans price controls on face masks as virus fears drive demand.* EconomyNext. (2020). Accessed 30 January 2020. https://economynext.com/sri-lanka-plans-price-controls-on-face-masks-as-virus-fears-drive-demand-44393/.

Hettiarachchi, D., Noordeen, N., Gamakaranage, C., Somarathne, E. R., and Jayasinghe, S. (2021). Ethical responses to the COVID-19 pandemic-lessons from Sri Lanka. *Asian Bioethics Review*, 13(2), 225-233.

Hewage, S. A. (2020). COVID-19: the world held at gun point. *Journal of the College of Community Physicians of Sri Lanka*, 26(1).

James, R., and Kengatharan, N. (2020). Exploring the effect of COVID-19 on small and medium enterprises: Early evidence from Sri Lanka. *Journal of Applied Economics and Business Research*, 10(2), 115-124.

Jayasena, H., and Chinthaka, W. (2020). COVID-19 and developing countries: lessons learnt from the Sri Lankan experience. *Journal of the Royal Society of Medicine*, 113(11), 464-465.

Kelaniya University and several institutes in Gampaha closed: UGC. *Sri Lanka News - Newsfirst*. (2020). Accessed 4 October 2020. https://www.newsfirst.lk/2020/10/04/kelaniya-university-and-several-institutes-in-gampaha-closed-ugc/.

Liyanapathirana, A., Ferdinando, R., and Palihawadana, P. (2020). COVID-19 outbreak in Sri Lanka: an epidemiological response. *Nutrition*, 1, 2.

Nadeera, Dilshan. *Sri Lankan lives matter-Sajith*. (2021). Accessed 4 December 2022. https://island.lk/sri-lankan-lives-matter-sajith/.

National Action Committee to prevent spread of coronavirus in Sri Lanka. adaderana.lk. (2021). Accessed 30 November 2022. http://www.adaderana.lk/news_ intensedebate.php?nid=60442.

Palihawadana, P. (2020). COVID-19 pandemic 2019-20. *Journal of the College of Community Physicians of Sri Lanka*, 26(1).

Police curfew imposed in Divulapitiya and Minuwangoda Police jurisdictions. *Sri Lanka News - Newsfirst*. (2020). Accessed 4 October 2020. https://www.newsfirst.lk/2020/10/04/police-curfew-imposed-in-divulapitiya-and-minuwangoda-police-jurisdictions/.

Quarantine Curfew re-imposed for the Western Province. *Sri Lanka News - Newsfirst*. (2020). Accessed 1 November 2020. Available from: https://www.newsfirst.lk/2020/11/01/quarantine-curfew-re-imposed-for-the-western-province/.

Sri Lanka: WHO Coronavirus Disease (COVID-19). (2021). Accessed 2 December 2022. https://SARS-CoV-219.who.int/region/searo/country/lk.

Sri Lanka records world's fourth-highest daily deaths by population amid lockdown calls. EconomyNext. (2021). Accessed 15 August 2021. https://economynext.com/sri-lanka-records-worlds-fourth-highest-daily-deaths-by-population-amid-lockdown-calls-84902/.

Sri Lanka rejects lockdown to beat COVID 'bomb'. France 24. (2021). Accessed 10 August 2021. https://www.france24.com/en/live-news/20210810-sri-lanka-rejects-lockdown-to-beat-covid-bomb.

Sri Lanka confirms highest daily COVID-19 deaths yet; lockdown calls grow louder. EconomyNext. (2021). Accessed 19 August 2021. https://economynext.com/sri-lanka-confirms-highest-daily-covid-19-deaths-yet-lockdown-calls-grow-louder-852 32/.

Sri Lanka becomes first in Asia to tighten policy in pandemic-era. Reuters. (2021). 19 August 2021. https://www.reuters.com/world/asia-pacific/sri-lanka-becomes-first-asia-raise-interest-rates-since-pandemic-2021-08-19/.

Sri Lanka extends COVID-19 lockdown till September 6 as deaths rise. *The Hindu*. PTI. (2021). Accessed 27 August 2021. ISSN 0971-751X. https://www.thehindu.com/news/international/sri-lanka-extends-covid-19-lockdown-till-september-6-as-deaths-rise/article36136243.ece.

Sri Lanka becomes first in Asia to tighten policy in pandemic-era. Adaderana Biz English | *Sri Lanka Business News*. (2021). Accessed 19 August 2021. http://bizenglish.adaderana.lk/sri-lanka-becomes-first-in-asia-to-tighten-policy-in-pandemic-era/.

Sri Lanka Lockdown News: Sri Lanka extends COVID-19 lockdown till September 6 as deaths rise. *The Times of India*. PTI. (2021). Accessed 27 August 2021. https://timesofindia.indiatimes.com/world/south-asia/sri-lanka-extends-covid-19-lockdown-till-september-6-as-deaths-rise/articleshow/85685266.cms.

Sri Lanka: *Coronavirus: Sri Lanka quarantines over 160 passengers arrived from South Korea in new facility*. www.colombopage.com. (2021). Accessed 18 December 2022. http://www.colombopage.com/archive_20A/Mar10_1583816169CH.php.

Sri Lanka bars travelers from Qatar, Bahrain and Canada to control Coronavirus. EconomyNext. (2020). Accessed 16 March 2020. https://economynext.com/sri-lanka-bars-travelers-from-qatar-bahrain-and-canada-to-control-coronavirus-59229/.

Sri Lanka tightens repatriation after new COVID-19 strain Foreign Ministry says. EconomyNext. (2020). https://economynext.com/sri-lanka-reverses-travel-relaxation-after-new-covid-19-strain-foreign-ministry-says-77148/.

Sri Lanka GDP to shrink 6.7-pct in 2020, Maldives 19.5-pct: World Bank. EconomyNext. (2020). Accessed 8 October 2020. https://economynext.com/sri-lanka-gdp-to-shrink-6-7-pct-in-2020-maldives-19-5-pct-world-bank-74593,%20https:/economynext.com/sri-lanka-gdp-to-shrink-6-7-pct-in-2020-maldives-19-5-pct-world-bank-74593/.

Sri Lanka receives 500,000 doses of COVID-19 vaccines from India. *The Hindu*. PTI. (2021). Accessed 25 February 2021. https://www.thehindu.com/news/international/sri-lanka-receives-500000-doses-of-covid-19-vaccines-from-india/article33933291.ece.

Warnapura, C. (2020). COVID 19; challenges to critical care and lessons learnt. *Sri Lankan Journal of Anaesthesiology*, 28(2).

Chapter 19

Conclusion

Asita Elengoe[*]**, PhD**
Department of Biotechnology, Faculty of Applied Science,
Lincoln University College, Selangor, Malaysia

Abstract

In conclusion, this book gives a quick review of SARS-CoV-2 and COVID-19, including their morphology, pathogenic mechanisms, causes, transmission, symptoms, diagnoses, therapeutic possibilities, clinical research on COVID-19 infection, and preventative measures and mental health care. SARS-CoV-2 is a deadly and highly contagious virus. COVID-19 disease is transmitted by air easily. A variety of ORFs, including ORF1ab, ORF8, ORF6, ORF7a, ORF3a, and ORF10, are present in SARS-CoV-2. Four important structural proteins-S, E, M, and N proteins-as well as a number of accessory proteins are encoded by the remaining portion of the genome. The Pfizer/BioNTech SARS-CoV-2 vaccine was first developed and approved for use in the globe by the UK on December 2, 2020. It was administered successfully all over the world. The COVID-19 patients, however, may also be treated with a number of additional complementary drugs, including Remdesivir, Hydroxychloroquine, Lopinavir/Ritonavir, Ribavarin, and Interferon Beta. The results advise developing a better vaccine and taking preventative steps to stop the spread of SARS-CoV-2. In this book, medicinal plants and their mechanism to suppress viral entry into the host, viral attachment, viral replication, etc., were discussed. This book supports early planning for potential pandemic breakouts in the future.

Keywords: COVID-19, medicinal plants, SARS-CoV-2

---

[*] Corresponding Author's Email: asitaelengoe@yahoo.com.

In: Medicinal Plants and COVID-19
Editor: Asita Elengoe
ISBN: 979-8-88697-958-9

Worldwide, COVID-19 has resulted in more than 68 million confirmed illnesses and more than 6.8 million fatalities since the discovery of SARS-CoV-2 in China in late 2019 (Zhou et al., 2020). SARS-CoV-2 can cause severe, potentially fatal pneumonia, even though the majority of infections are mild. This is especially true for older adults and people with pre-existing illnesses. Although the precise pathophysiology of severe COVID-19 is still unknown, it frequently includes an uncontrolled, excessively inflammatory response after viral infection (Domingo et al., 2020). Variations in the viral strain, however, may also influence the severity of the illness and the speed at which it spreads, in addition to the host reaction.

Since the COVID-19 pandemic occurred, different virus strains were detected in different countries. As an RNA virus, coronavirus evolves and gradually mutates over time as the mutation produces new variants (Roy et al., 2021). One variant of concern is D614G which was detected in the United States of America (USA) 10 times more highly infectious than other strains although there are several new variants detected worldwide (Abidemi et al., 2021). There are four different types of variants such as alpha, beta, gamma and delta. From a worldwide view, the delta variant is more contagious and lethal than alpha, beta and gamma due to its ability to cause vulnerable individuals to succumb to death who contracted the virus (Roy et al., 2021).

As an infectious virus, several measures have been taken to contain the spread of the COVID-19 virus. These measures include the introduction of a Movement Control Order, MCO. There were three phases of MCOs namely Movement Control Order, Conditional Movement Control Order and Recovery Movement Control Order (Waiho et al., 2020). In each of these phases, the restriction varies for example in MCO, complete closure of economy and social activities with exception given to essential services whereas in CMCO, partial reopening of the economic sector to revive the economy whereas in RMCO, it involves the full reopening of the economy and social sector. Even though various MCOs have been introduced, there are several preventive measures that are to be followed by the public such as avoiding contact with a person infected with COVID-19, using public transport when it is necessary, working from home, avoiding large crowds and confined space, avoid gatherings with friends and family and use of online services. Despite preventive measures introduced to contain the spread of COVID-19, several SOPs are currently practiced worldwide such as physical distancing, personal hygiene, mandatory use of face masks, immediate reporting of COVID-19 cases to the Ministry of Health (MOH), isolation of symptomatic and asymptomatic person and health screening (Hashim et al., 2021).

Herbs are traditionally employed in a variety of therapeutic procedures, either as the primary treatment or as a supplemental therapy used in conjunction with pharmaceuticals to increase immunity for prevention. Numerous studies have demonstrated the efficacy of phytotherapy, including its capacity to treat infectious illnesses (Zhang et al., 2020). There are medicinal herbs whose extracts have an inhibitory impact on viral strains resistant to traditional antiviral medications, as well as on viruses like herpes simplex virus type 2, HIV, hepatitis B virus (HBV), smallpox virus, and severe acute respiratory syndrome (Mukhtarad et al., 2020; Lubbe and Verpoorte, 2011). Therefore, phytochemicals have been shown to suppress coronaviruses via mechanisms such as viral entry inhibition, inhibition of replication enzymes and virus release blockage; which were discussed in this book. The effects of plant compounds on the angiotensin converting enzyme-2 (ACE-2) receptor, $3CL^{pro}$ and other SARS-CoV-2 viral protein targets also were described in this book.

References

Abidemi A, Zainuddin Z M, Aziz N A. Impact of control interventions on COVID-19 population dynamics in Malaysia : a mathematical study. *The European Physical Journal Plus* (2021) 136 :1-35.

Domingo P, Mur I, Pomar V, Corominas H, Casademont J, de Benito N. The four horsemen of a viral Apocalypse : The pathogenesis of SARS-CoV-2 infection (COVID-19). *EBioMedicine* (2020) 58 :102887.

Hashim J H, Adman M A, Hashim Z, Mohd Radi M F, Kwan S C. COVID-19 epidemic in Malaysia : epidemic progression, challenges, and response. *Frontiers in Public Health* (2021) 9 :560592.

Lubbe A, Verpoorte R. Cultivation of medicinal and aromatic plants for specialty industrial materials. *Industrial Crops and Products* (2011) 34(1) :785-801.

Mukhtar M, Arshad M, Ahmad M, Pomerantz R J, Wigdahl B, Parveen Z. Antiviral potentials of medicinal plants. *Virus Research* (2008) 131(2) :111-120.

Roy B, Dhillon J K, Habib N, Pugazhandhi B. Global variants of COVID-19 : Current understanding. *Journal of Biomedical Sciences* (2021) 8(1) :8-11.

Waiho K, Fazhan H, Ishak S D, Kasan N A, Liew H J, Norainy M H, Ikhwanuddin M. Potential impacts of COVID-19 on the aquaculture sector of Malaysia and its coping strategies. *Aquaculture Reports* (2020) 18 :100450.

Zhang D H, Wu K L, Zhang X, Deng S Q, Peng B. *In silico* screening of Chinese herbal medicines with the potential to directly inhibit 2019 novel coronavirus. *Journal of Integrative Medicine* (2020) 18(2) :152-158.

Zhou P, Yang X L, Wang X G, Hu B, Zhang L, Zhang W, Si H R, Zhu Y, Li B, Huang C L, Chen HD. A pneumonia outbreak associated with a new coronavirus of probable bat origin. *Nature* (2020) 579(7798) :270-273.

Index

A

B

C

D

E

F

G

H

I

L

M

N

O

P

About the Editor

Dr. Asita Elengoe is currently the head of Biotechnology Department at the Faculty of Applied Science, Lincoln University College Malaysia. She obtained her PhD in Bioscience (2015) and Master of Science in Biotechnology (2010) from Universiti Teknologi Malaysia. She received her Bachelor of Science in Biotechnology (Hons) (2008) from Universiti Tunku Abdul Rahman. Her research interest includes cancer, gene therapy, computational biology, natural products and medicinal plants. She is a member of the Royal Society of Biology (MRSB) United Kingdom. She is the journal editor for Current Issues in Molecular Biology and assistant executive editor for the International Journal of Advancement in Life Sciences Research (IJALSR). She is the reviewer for indexed and non-indexed journals. She is actively involved in research and publications. She published articles in indexed and non-indexed journals. She is also book editor for Obesity and its impact on health (Springer Nature, Singapore), Rabies: Simple handbook of rabies management (Tamil), Laboratory Manual (Instrumentation) Bachelor of Biotechnology (Hons.), Laboratory Manual (Basic Cytology) Bachelor of Biotechnology (Hons.), Laboratory Manual (Human Genetics) Bachelor of Biotechnology (Hons.) and Laboratory Manual (Biochemistry) Bachelor of Biotechnology (Hons.). She received an award of the 'Young Researcher in Biotechnology' at International Healthcare Awards 2018, Chennai, India. Currently, she carries out research on drug designing using computational biology approach.